Introduction to Numerical Analysis

Numerical analysis is an increasingly important link between pure mathematics and its application in science and technology. This textbook provides an introduction to the justification and development of constructive methods that provide sufficiently accurate approximations to the solution of numerical problems, and the analysis of the influence that errors in data, finite-precision calculations, and approximation formulas have on results, problem formulation, and the choice of method. It also serves as an introduction to scientific programming in MATLAB, including many simple and difficult, theoretical and computational exercises.

A unique feature of this book is the consequent development of interval analysis as a tool for rigorous computation and computer-assisted proofs, along with the traditional material.

Arnold Neumaier is a Professor of Computational Mathematics at the University of Wien. He has written two books and published numerous articles on the subjects of combinatorics, interval analysis, optimization, and statistics. He has held teaching positions at the University of Freiburg, Germany, and the University of Wisconsin, Madison, and he has worked on the technical staff at AT&T Bell Laboratories. In addition, Professor Neumaier maintains extensive Web pages on public domain software for numerical analysis, optimization, and statistics.

Introduction to Numerical Analysis

ARNOLD NEUMAIER

University of Wien

CAMBRIDGE
UNIVERSITY PRESS

PUBLISHED BY THE PRESS SYNDICATE OF THE UNIVERSITY OF CAMBRIDGE
The Pitt Building, Trumpington Street, Cambridge, United Kingdom

CAMBRIDGE UNIVERSITY PRESS
The Edinburgh Building, Cambridge CB2 2RU, UK
40 West 20th Street, New York, NY 10011-4211, USA
10 Stamford Road, Oakleigh, VIC 3166, Australia
Ruiz de Alarcón 13, 28014 Madrid, Spain
Dock House, The Waterfront, Cape Town 8001, South Africa

http://www.cambridge.org

First published 2001

Printed in the United Kingdom at the University Press, Cambridge

Typeface Times Roman 10/13 pt.   *System* LATEX 2_ε [TB]

A catalog record for this book is available from the British Library.

Library of Congress Cataloging in Publication Data
Neumaier, A.
Introduction to numerical analysis / Arnold Neumaier.
p.   cm.
Includes bibliographical references and index.
ISBN 0-521-33323-7 – ISBN 0-521-33610-4 (pb.)
1. Numerical analysis. I. Title.
QA297 .N48   2001
519.4 – dc21   00-066713

ISBN 0 521 33323 7 hardback
ISBN 0 521 33610 4 paperback

Contents

Preface   *page* vii

1. The Numerical Evaluation of Expressions   1
  1.1 Arithmetic Expressions and Automatic Differentiation   2
  1.2 Numbers, Operations, and Elementary Functions   14
  1.3 Numerical Stability   23
  1.4 Error Propagation and Condition   33
  1.5 Interval Arithmetic   38
  1.6 Exercises   53

2. Linear Systems of Equations   61
  2.1 Gaussian Elimination   63
  2.2 Variations on a Theme   73
  2.3 Rounding Errors, Equilibration, and Pivot Search   82
  2.4 Vector and Matrix Norms   94
  2.5 Condition Numbers and Data Perturbations   99
  2.6 Iterative Refinement   106
  2.7 Error Bounds for Solutions of Linear Systems   112
  2.8 Exercises   119

3. Interpolation and Numerical Differentiation   130
  3.1 Interpolation by Polynomials   131
  3.2 Extrapolation and Numerical Differentiation   145
  3.3 Cubic Splines   153
  3.4 Approximation by Splines   165
  3.5 Radial Basis Functions   170
  3.6 Exercises   182

4. Numerical Integration   179
  4.1 The Accuracy of Quadrature Formulas   179

4.2 Gaussian Quadrature Formulas 187
4.3 The Trapezoidal Rule 196
4.4 Adaptive Integration 203
4.5 Solving Ordinary Differential Equations 210
4.6 Step Size and Order Control 219
4.7 Exercises 225

5. Univariate Nonlinear Equations 233
5.1 The Secant Method 234
5.2 Bisection Methods 241
5.3 Spectral Bisection Methods for Eigenvalues 250
5.4 Convergence Order 256
5.5 Error Analysis 265
5.6 Complex Zeros 273
5.7 Methods Using Derivative Information 281
5.8 Exercises 293

6. Systems of Nonlinear Equations 301
6.1 Preliminaries 302
6.2 Newton's Method and Its Variants 311
6.3 Error Analysis 322
6.4 Further Techniques for Nonlinear Systems 329
6.5 Exercises 340

References 345
Index 351

Preface

They...explained it so that the people could understand it.
Good News Bible, Nehemiah 8:8

Since the introduction of the computer, numerical analysis has developed into an increasingly important connecting link between pure mathematics and its application in science and technology. Its independence as a mathematical discipline depends, above all, on two things: the justification and development of constructive methods that provide sufficiently accurate approximations to the solution of problems, and the analysis of the influence that errors in data, finite-precision calculations, and approximation formulas have on results, problem formulation, and the choice of method. This book provides an introduction to these themes.

A novel feature of this book is the consequent development of interval analysis as a tool for rigorous computation and computer-assisted proofs. Apart from this, most of the material treated can be found in typical textbooks on numerical analysis; but even then, proofs may be shorter than and the perspective may be different from those elsewhere. Some of the material on nonlinear equations presented here previously appeared only in specialized books or in journal articles.

Readers are expected to have a background knowledge of matrix algebra and calculus of several real variables, and to know just enough about topological concepts to understand that sequences in a compact subset in $\mathbb{R}^n$ have a convergent subsequence. In a few places, elements of complex analysis are used.

The book is based on course lectures in numerical analysis that the author gave repeatedly at the University of Freiburg (Germany) and the University of Vienna (Austria). Many simple and difficult theoretical and computational exercises help the reader gain experience and deepen the understanding of the techniques presented. The material is a little more than can be covered in a European winter term, but it should be easy to make suitable selections.

The presentation is in a rigorous mathematical style. However, the theoretical results are usually motivated and discussed in a more leisurely manner so that many proofs can be omitted without impairing the understanding of the algorithms. Notation is almost standard, with a bias towards MATLAB (see also the index). The abbreviation "iff" frequently stands for "if and only if."

The first chapter introduces elementary features of numerical computation: floating point numbers, rounding errors, stability and condition, elements of programming (in MATLAB), automatic differentiation, and interval arithmetic. Chapter 2 is a thorough treatment of Gaussian elimination, including its variants such as the Cholesky factorization. Chapters 3 through 5 provide the tools for studying univariate functions – interpolation (with polynomials, cubic splines, and radial basis functions), integration (Gaussian formulas, Romberg and adaptive integration, and an introduction to multistep formulas for ordinary differential equations), and zero finding (traditional and less traditional methods ensuring global and fast local convergence, complex zeros, and spectral bisection for definite eigenvalue problems). Finally, Chapter 6 discusses Newton's method and its many variants for systems of nonlinear equations, concentrating on methods for which global convergence can be proved.

In a second course, I usually cover numerical data analysis (least squares and orthogonal factorization, the singular value decomposition and regularization, and the fast Fourier transform), unconstrained optimization, the eigenvalue problem, and differential equations. Therefore, this book contains no (or only a rudimentary) treatment of these topics; it is planned that they will be covered in a companion volume.

I want to thank Doris Norbert, Andreas Schäfer, and Wolfgang Enger for their help in preparing the German lecture notes; Michael Wolfe and Baker Kearfott for their English translation, out of which the present text grew; Carl de Boor, Baker Kearfott, Weldon Lodwick, Günter Mayer, Jiří Rohn, and Siegfried Rump for their comments on various versions of the manuscript; and Stefan Dallwig and Waltraud Huyer for computing the tables and figures and proofreading. I also want to thank God for giving me love and power to understand and use the mathematical concepts on which His creation is based, and for the discipline and perseverance to write this book.

I hope that working through the material presented here gives my readers insight and understanding, encourages them to become more deeply interested in the joys and frustrations of numerical computations, and helps them apply mathematics more efficiently and more reliably to practical problems.

Vienna, January 2001
Arnold Neumaier

1

The Numerical Evaluation of Expressions

In this chapter, we introduce the reader on a very elementary level to some basic considerations in numerical analysis. We look at the evaluation of arithmetic expressions and their derivatives, and show some simple ways to save on the number of operations and storage locations needed to do computations.

We demonstrate how finite precision arithmetic differs from computation with ideal real numbers, and give some ideas about how to recognize pitfalls in numerical computations and how to avoid the associated numerical instability. We look at the influence of data errors on the results of a computation, and how to quantify this influence using the concept of a condition number. Finally we show how, using interval arithmetic, it is possible to obtain mathematically correct results and error estimates although computations are done with limited precision only.

We present some simple algorithms in a pseudo-MATLAB® formulation very close to the numerical MATrix LABoratory language MATLAB (and some – those printed in `typewriter font` – in true MATLAB) to facilitate getting used to this excellent platform for experimenting with numerical algorithms. MATLAB is very easy to learn once you get the idea of it, and the online help is usually sufficient to expand your knowledge. We explain many MATLAB conventions on their first use (see also the index); for unexplained MATLAB features you may try the online help facility. Type `help` at the MATLAB prompt to find a list of available directories with MATLAB functions; add the directory name after `help` to get information about the available functions, or type the function name (without the ending .m) after `help` to get information about the use of a particular function. This should give enough information, in most cases. Apart from that, you may refer to the MATLAB manuals (for example, [58,59]). If you need more help, see, for example, Hanselman and Littlefield [38] for a comprehensive introduction.

For those who cannot afford MATLAB, there are public domain variants that can be downloaded from the internet, SCILAB [87] and OCTAVE [74]. (The syntax and the capabilities are slightly different, so the MATLAB explanations given here may not be fully compatible.)

1.1 Arithmetic Expressions and Automatic Differentiation

Mathematical formulas occur in many fields of interest in mathematics and its applications (e.g., statistics, banking, astronomy).

1.1.1 Examples.

(i) The absolute value $|x + iy|$ of the complex number $x + iy$ is given by

$$|x + iy| = \sqrt{x^2 + y^2}.$$

(ii) A capital sum K_0 invested at $p\%$ per annum for n years accumulates to a sum K given by

$$K = K_0(1 + p/100)^n.$$

(iii) The altitude h of a star with spherical coordinates (ψ, δ, t) is given by

$$h = \arcsin(\sin \psi \sin \delta + \cos \psi \cos \delta \cos t).$$

(iv) The solutions x_1, x_2 of the quadratic equation

$$ax^2 + bx + c = 0$$

are given by

$$x_{1,2} = \frac{-b \pm \sqrt{b^2 - 4ac}}{2a}.$$

(v) The standard deviation σ of a sequence of real numbers $x_1, \ldots, x_n$ is given by

$$\sigma = \sqrt{\frac{1}{n} \sum_{i=1:n} \left(x_i - \frac{1}{n} \sum_{i=1:n} x_i \right)^2}$$

$$= \sqrt{\frac{1}{n} \left(\sum_{i=1:n} x_i^2 - \frac{1}{n} \left(\sum_{i=1:n} x_i \right)^2 \right)}.$$

Here we use the MATLAB notation `1:n` for a list of consecutive integers from 1 to n.

(vi) The area θ under the normal distribution curve $e^{-t^2/2}/\sqrt{2\pi}$ between $t = -x$ and $t = x$ is given by

$$\theta = \frac{1}{\sqrt{2\pi}} \int_{-x}^{x} e^{-t^2/2} \, dt.$$

With the exception of the last formula, which contains an integral, only elementary arithmetic operations and elementary functions occur in the examples. Such formulas are called *arithmetic expressions*. Arithmetic expressions may be defined recursively with the aid of variables $x_1, \ldots, x_n$, the binary operations $+$ (addition), $-$ (subtraction), $*$ (multiplication), $/$ (division), $\hat{}$ (exponentiation), forming the set

$$O = \{+, -, *, /, \hat{}\,\},$$

and with certain unary elementary functions in the set

$$J = \{+, -, \sin, \cos, \exp, \log, \text{sqrt}, \text{abs}, \ldots\}.$$

The set of elementary functions are not specified precisely here, and should be regarded as consisting of the set of unary continuous functions that are defined on the computer used.

1.1.2 Definition. The set $\mathcal{A} = \mathcal{A}(x_1, \ldots, x_n)$ of arithmetic expressions in $x_1, \ldots, x_n$ is defined by the rules

(E1) $\mathbb{R} \subseteq \mathcal{A}$,
(E2) $x_i \in \mathcal{A}$ $(i = 1, \ldots, n)$,
(E3) $g, h \in \mathcal{A}$, $\circ \in O \Rightarrow (g \circ h) \in \mathcal{A}$,
(E4) $g \in \mathcal{A}$, $\varphi \in J \Rightarrow \varphi(g) \in \mathcal{A}$,
(E5) $\mathcal{A}(x_1, \ldots, x_n)$ is the smallest set that satisfies (E1)–(E4). (This rule excludes objects not created by the rules (E1)–(E4).)

Unnecessary parentheses may be deleted in accordance with standard rules.

1.1.3 Example. The solution

$$\frac{-b + \sqrt{b^2 - 4ac}}{2a}$$

of a quadratic equation is an arithmetic expression in a, b, and c because we can write it as follows:

$$(-b + \text{sqrt}(b\hat{}2 - 4 * a * c))/(2 * a) \in \mathcal{A}(a, b, c).$$

Evaluation of an arithmetic expression means replacing the variables with numbers. This is possible on any machine for which the operations in O and the elementary functions in J are realized.

Differentiation of Expressions

For many numerical problems it is useful and sometimes absolutely essential to know how to calculate the *value* of the derivative of a function f at given points. If no routine for the calculation of $f'(x)$ is available then one can determine approximations for $f'(x)$ by means of numerical differentiation (see Section 3.2). However, because of the higher accuracy attainable, it is usually better to derive a routine for calculating $f'(x)$ from the routine for calculating $f(x)$. If, in particular, f is given by an arithmetic expression, then an arithmetic expression for $f'(x)$ can be obtained by analytic (symbolic) differentiation using the rules of calculus. We describe several useful variants of this process, confining ourselves here to the case of a single variable x; the case of several variables is treated in Section 6.1. We shall see that, for all applications in which a closed formula for the derivative of f is not needed but one must be able to evaluate $f'(x)$ at arbitrary points, a recursive form of generating this value simultaneously with the value of $f(x)$ is the most useful way to proceed.

(a) The Construction of a Closed Expression for f'

This is the traditional way in which the formula for the derivative is calculated by hand and the expression that is obtained is then programmed as a function subroutine. However, several disadvantages outweigh the advantage of having a closed expression.

(i) Algebraic errors can occur in the calculation and simplification of formulas for derivatives by hand; this is particularly likely when long and complicated formulas for the derivative result. Therefore, a correctness test is necessary; this can be implemented, for example, as a comparison with a sequence of values obtained by numerical differentiation (cf. Exercise 8).

(ii) Often, especially when both $f(x)$ and $f'(x)$ must be calculated, certain subexpressions appear several times, and their recalculation adds needlessly to the running time of the program. Thus, in the example

$$f(x) = e^x/(1 + \sin x),$$
$$f'(x) = \frac{e^x(1 + \sin x) - e^x \cos x}{(1 + \sin x)^2}$$
$$= e^x(1 + \sin x - \cos x)/(1 + \sin x)^2$$

the expression $1 + \sin x$ occurs three times.

The susceptibility to error is considerably reduced if one automates the calculation and simplification of closed formulas for derivatives using symbolic algebra (i.e., the processing of syntactically organized strings of symbols). Examples of high-quality packages that accomplish this are MAPLE and MATHEMATICA.

(b) The Construction of a Recursive Program for f and f'

In order to avoid the repeated calculation of subexpressions, it is advantageous to give up the closed form and calculate repeatedly occurring subexpressions using auxiliary variables. In the preceding example, one obtains the program segment

$$f_1 = \exp x;$$
$$f_2 = 1 + \sin x;$$
$$f = f_1/f_2;$$
$$f' = f_1 * (f_2 - \cos x)/f_2\hat{\,}2;$$

in which one can represent the last expression for f' more briefly in the form $f' = f * (f_2 - \cos x)/f_2$ or $f' = f * (1 - \cos x/f_2)$. One can even arrange to have this transformation done automatically by programs for symbolic manipulation and dispense with closed intermediate expressions. Thus, in decomposing the expression for f recursively into its constituent parts, a sequence of assignments of the form

$$f = -g, \quad f = g \circ h, \quad f = \varphi(g)$$

is obtained. One can differentiate this sequence of equations, taking into account the fact that the subexpressions f, g, and h are themselves functions of x and bearing in mind that a constant has the derivative 0 and that the variable x has the derivative 1. In addition, the well-known rules for differentiation (Table 1.1)

Table 1.1. *Differentiation rules for expressions f*

f	f'
$g \pm h$	$g' \pm h'$
$g * h$	$g' * h + g * h'$
g/h	$(g' - f * h')/h$ (see main text)
$g\hat{\ }2$	$2 * g * g'$
$g\hat{\ }h$	$f * (h' * \log(g) + h * g'/g)$
$\pm g$	$\pm g'$
$\mathrm{sqrt}(g)$	$g'/(2 * f)$ if $g > 0$
$\exp(g)$	$f * g'$
$\log(g)$	g'/g
$\mathrm{abs}(g)$	$\mathrm{sign}(g) * g'$
$\varphi(g)$	$\varphi'(g) * g'$

are used. The unusual form of the quotient rule results from

$$
\begin{aligned}
f = g/h \Rightarrow f' &= (g'h - gh')/h^2 \\
&= (g' - (g/h)h')/h \\
&= (g' - fh')/h
\end{aligned}
$$

and is advantageous for machine calculation, saving two multiplications. In the preceding example, one obtains

$$
\begin{aligned}
f_1 &= \exp x; \\
f_2 &= \sin x; \\
f_3 &= 1 + f_2; \\
f &= f_1/f_3; \\
f_1' &= f_1; \\
f_2' &= \cos x; \\
f_3' &= f_2'; \\
f' &= (f_1' - f * f_3')/f_3;
\end{aligned}
$$

and from this, by means of a little simplification and substitution,

$$
\begin{aligned}
f_1 &= \exp x; \\
f_3 &= 1 + \sin x; \\
f &= f_1/f_3; \\
f' &= (f_1 - f * \cos x)/f_3;
\end{aligned}
$$

with the same computational cost as the old, intuitively derived recursion.

(c) Computing with Differential Numbers

The compilers of all programming languages can do a syntax analysis of arbitrary expressions, but usually the result of the analysis is not accessible to the programs. However, in programming languages that provide user-definable data types, operators, and functions for objects of these types, the compiler's syntax analysis can be utilized. This possibility exists, for example, in the programming languages MATLAB 5, FORTRAN 90, ADA, C++, and in the PASCAL extension PASCAL-XSC, and leads to automatic differentiation methods.

In order to understand how this can be done, we observe that Table 1.1 constructs, for each operation $\circ \in O$, from two pairs of numerical values (g, g') and (h, h'), a third value, namely (f, f'). We may regard this pair simply as the result of the operation $(g, g') \circ (h, h')$. Similarly, one finds in Table 1.1 for each elementary function $\varphi \in J$ a new pair (f, f') from the pair (g, g'), and we regard this as a definition of the value $\varphi((g, g'))$. The analogy with complex numbers is obvious and motivates our definition of differential numbers as pairs of real numbers.

Formally, a differential number is a pair (f, f') of real numbers. We use the generic form df to denote such a differential number, and regard similarly the variables h and h' as the components of the differential number dh, so that $dh = (h, h')$, and so on. We now define, in correspondence with Table 1.1, operations and elementary functions for differential numbers.

$$dg \pm dh := (g \pm h, g' \pm h');$$

$$dg * dh := (g * h, g' * h + g * h');$$

$$dg/dh := (f, (g' - f * h')/h) \quad \text{with } f = g/h \quad (\text{if } h \neq 0);$$

$$dg\hat{\ }n := \begin{cases} (k * g, n * k * g') & \text{with } k = g\hat{\ }(n-1) \quad (\text{if } 1 \leq n \in \mathbb{R}); \\ (f, n * f * g'/g) & \text{with } f = g\hat{\ }n \quad (\text{if } 1 > n \in \mathbb{R}, g > 0); \end{cases}$$

$$dg\hat{\ }dh := (f, f * (h' * k + h * g'/g))$$
$$\text{with } k = \log(g), \quad f = \exp(h * k) \quad (\text{if } g > 0);$$

$$\pm dg := (\pm g, \pm g');$$

$$\text{sqrt}(dg) := (f, g'/(2 * f)) \quad \text{with } f = \text{sqrt}(g) \quad (\text{if } g > 0);$$

$$\exp(dg) := (f, f * g') \quad \text{with } f = \exp(g);$$

$$\log(dg) := (\log(g), g'/g) \quad (\text{if } g > 0);$$

$$\text{abs}(dg) := (\text{abs}(g), \text{sign}(g) * g') \quad (\text{if } g \neq 0);$$

$$\varphi(dg) := (\varphi(g), \varphi'(g) * g')$$
$$\text{for other } \varphi \in J \text{ for which } \varphi(g), \varphi'(g) \text{ exists.}$$

The next result follows directly from the definitions.

1.1.4 Proposition. *Let q, r be real functions of one variable, differentiable at $x_0 \in \mathbb{R}$, and let*

$$dq = (q(x_0), q'(x_0)), \quad dr = (r(x_0), r'(x_0)).$$

(i) If $dq \circ dr$ is defined (for $\circ \in O$), then the function p given by

$$p(x) := q(x) \circ r(x)$$

is defined and differentiable at x_0, and

$$(p(x_0), p'(x_0)) = dq \circ dr.$$

(ii) If $\varphi(dq)$ is defined (for $\varphi \in J$), then the function p given by

$$p(x) := \varphi(q(x))$$

is defined and differentiable at x_0, and

$$(p(x_0), p'(x_0)) = \varphi(dq).$$

The reader interested in algebra easily verifies that the set of differential numbers with the operations $+, -, *$ forms a commutative and associative ring with null element $0 = (0, 0)$ and identity element $1 = (1, 0)$. The ring has zero divisors; for example, $(0, 1) * (0, 1) = (0, 0) = 0$. The differential numbers of the form $(a, 0)$ form a subring that is isomorphic to the ring of real numbers.

We may call differential numbers of the form $(a, 0)$ *constants* and identify them with the real numbers a. For operations with constants, the formulas simplify

$$dg \pm a = (g \pm a, g'), \quad a \pm dh = (a \pm h, h'),$$
$$dg * a = (g * a, g' * a), \quad a * dh = (a * h, a * h'),$$
$$dg/a = (g/a, g'/a), \quad a/dh = (f, -f * h'/h), \quad \text{with } f = a/h.$$

The arithmetic for differential numbers can be programmed without any difficulty in any of the programming languages mentioned previously. With their help, we can compute the values $f(x_0)$ and $f'(x_0)$ for any arithmetic expression f and at any point x_0. Indeed, we need only initialize the independent variable as the differential number $dx = (x_0, 1)$ and substitute this into f.

1.1.5 Theorem. *Let f be an arithmetic expression in the variable x, and let the differential number $f(dx)$ that results from inserting $dx = (x_0, 1)$ into f be defined. Then f is defined and is differentiable at x_0, and*

$$(f(x_0), f'(x_0)) = f(dx).$$

Proof. Recursive application of Proposition 1.1.4. □

The conclusion is that, using differential numbers, one can calculate the derivative of any arithmetic expression, at any admissible point, *without knowing an expression for* f'!

1.1.6 Example. Suppose we want to find the value and the derivative of

$$f(x) = \frac{(x-1)(x+3)}{x+2}$$

at the point $x_0 = 3$. The classical method (a) starts from an explicit formula for the derivative; for example,

$$f'(x) = \frac{x^2 + 4x + 7}{(x+2)^2}$$

and finds by substitution that

$$f(3) = \frac{12}{5} = 2.4, \quad f'(3) = \frac{28}{25} = 1.12.$$

Substituting the differential number $dx = (3, 1)$ into the expression for f gives

$$(f(3), f'(3)) = f(dx) = \frac{((3, 1) - 1) * ((3, 1) + 3)}{(3, 1) + 2}$$

$$= \frac{(2, 1) * (6, 1)}{(5, 1)} = \frac{(12, 8)}{(5, 1)}$$

$$= (2.4, (8 - 2.4 * 1)/5) = (2.4, 1.12),$$

without knowing an expression for f'. In the same way, substituting $dx = (3, 1)$ into the equivalent expression

$$f(x) = x - \frac{3}{x+2},$$

we obtain, in spite of completely different intermediate results, the same final result

$$(f(3), f'(3)) = f(dx) = (3, 1) - \frac{3}{(3, 1) + 2}$$

$$= (3, 1) - \frac{(3, 0)}{(5, 1)} = (3, 1) - (0.6, (0 - 0.6 * 1)/5)$$

$$= (3, 1) - (0.6, -0.12) = (2.4, 1.12).$$

Finally, we note that differential numbers can be generalized without difficulty to compute derivatives of higher order. For the computation of $f''(x), \ldots,$ $f^{(n)}(x)$ from an expression f containing N operations or functions, the number of operations and function calls needed grows like a small multiple of $n^2 N$.

The formalism handles differentiation with respect to any parameter. Therefore, it is also possible to compute the partial derivatives $\partial f(x)/\partial x_k$ of a function that is given by an expression $f(x)$ in several variables. Of course, one need not redo the function value part of the calculation when calculating the partial derivatives. In addition to this "forward" mode of automatic differentiation, there is also a "reverse" or "backward" mode that may further increase the efficiency of calculating partial derivatives (see Section 6.1).

A MATLAB implementation of automatic differentiation is available in the INTLAB toolbox by Rump [85]. An in-depth discussion of all aspects of automatic differentiation and its applications is given in Griewank and Corliss [33] and Griewank [32].

The Horner Scheme

A polynomial of degree at most n is a function of the form

$$f(x) = a_0 x^n + a_1 x^{n-1} + \cdots + a_{n-1} x + a_n \tag{1.1}$$

with given coefficients $a_0, \ldots, a_n$. In order to evaluate (1.1) for a given value of x, one does not proceed naively by forming each power $x^2, x^3, \ldots, x^n$, but one uses the following scheme. We define

$$f_i := a_0 x^i + a_1 x^{i-1} + \cdots + a_i \quad (i = 0, 1, \ldots, n).$$

Obviously, $f_n = f(x)$. The advantage of f_i lies in its recursive definition

$$\begin{aligned} f_0 &= a_0, \\ f_i &= f_{i-1} x + a_i \quad (i = 1, \ldots, n), \\ f(x) &= f_n. \end{aligned}$$

This is the Horner scheme for calculating the value of a polynomial. A simple count shows that only $2n$ operations are needed, whereas evaluating (1.1) directly requires at least $3n - 1$ operations.

The derivative of a polynomial can be formed recursively by differentiation of the recursion:

$$\begin{aligned} f_0' &= 0, \\ f_i' &= f_{i-1}' x + f_{i-1} \quad (i = 1, \ldots, n), \\ f'(x) &= f_n'. \end{aligned}$$

This is the Horner scheme for calculating the first derivative of a polynomial. Analogous results hold for the higher derivatives (see Exercise 4).

In MATLAB notation, we get $f = f(x)$ and $g = f'(x)$ as follows:

1.1.7 Algorithm: Horner Scheme

```
% computes f=f(x) and g=f'(x) for
% f(x)=a(1)*x^n+a(2)*x^(n-1)+...+a(n)*x+a(n+1)
f=a(1); g=0;
for i=1:n,
   g=g*x+f;
   f=f*x+a(i+1);
end;
```

Note the shift in the index numbering, needed because in MATLAB the vector indices start with index 1, in contrast to C, in which vector indices start with 0, and to FORTRAN, which allows the user to specify the starting index.

The reader unfamiliar with programming should also note that a statement such as $g = g * x + f$ is not an equation in the mathematical sense, but an *assignment*. In the right side, g refers to the contents of the associated storage location before evaluation of the right side; after its evaluation, the contents of g are replaced by the result of the calculation. Thus, the same variable takes different values at different times, and this allows to make most efficient use of the capacity of the computer, saving storage and index calculations. Note also that the update of f must come *after* the update of g because the formula for updating g requires the *old* value of f, which would be no longer available after the update of f.

1.1.8 Example. We evaluate the polynomial $f(x) = (1 - x)^6$ at 101 equidistant points in the range $0.995 \leq x \leq 1.005$ by calculating $(1 - x)^6$ directly and by using the Horner scheme for the equivalent expanded polynomial expression $1 - 6x + 15x^2 - 20x^3 + 15x^4 - 6x^5 + x^6$. The results are shown in Figure 1.1. The effect of (simulated) machine precision on the evaluation of the polynomial $p(x) = 1 - 6x + 15x^2 - 20x^3 + 15x^4 - 6x^5 + x^6$ without using the Horner scheme is demonstrated in Figure 1.2.

In MATLAB, the figure can be drawn easily with the following commands. (Text is handled as an array of characters, and concatenated by placing the pieces, separated by blanks or commas, within square brackets. num2str transforms a number into a string denoting that number, in some standard format. The number of rows or columns of a matrix can be found with size. The period

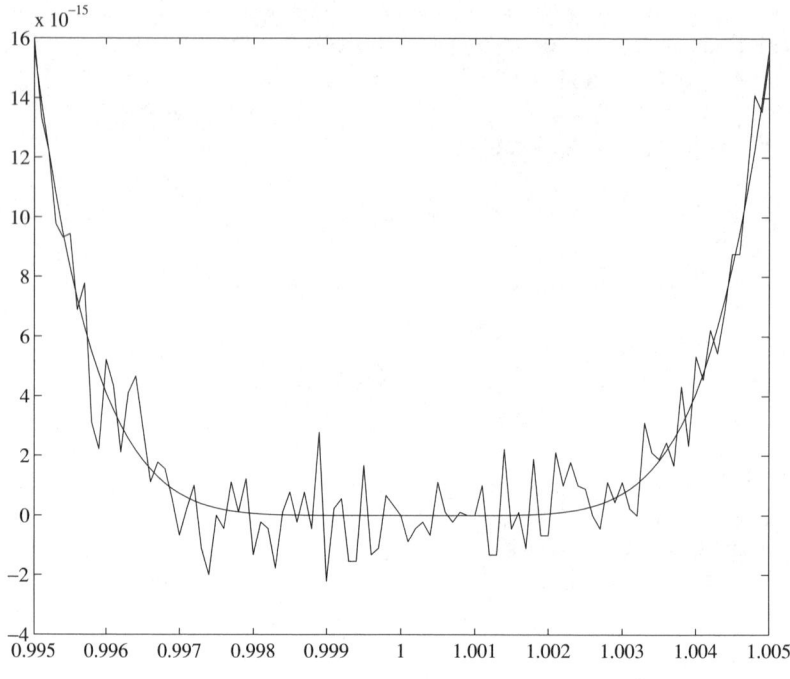

Figure 1.1. Two ways of evaluating $f(x) = (1-x)^6$.

before an operation denotes componentwise operations. The % sign indicates that the remainder of the line is a comment and does not affect the computations. The actual plot is produced by plot, and saved as a postscript file using print.)

1.1.9 Algorithm: Draw Figure 1.1

```
x=(9950:10050)/10000;
disp(['number of evaluation points: ',num2str(size(x,2))]);
y=(1-x).^6;
% a compact way of writing the Horner scheme:
z=((((((x-6).*x+15).*x-20).*x+15).*x-6).*x+1);
plot(x,[y;z]); % display graph on screen
print -deps horner.ps % save figure in file horner.ps
```

The figures illustrate a typical problem in numerical analysis. The simple expression $(1-x)^6$ produces the expected curve. However, for the expanded expression, monotonicity of f is destroyed through effects of finite precision arithmetic, and instead of a single minimum of zero at $x = 1$, we obtain more

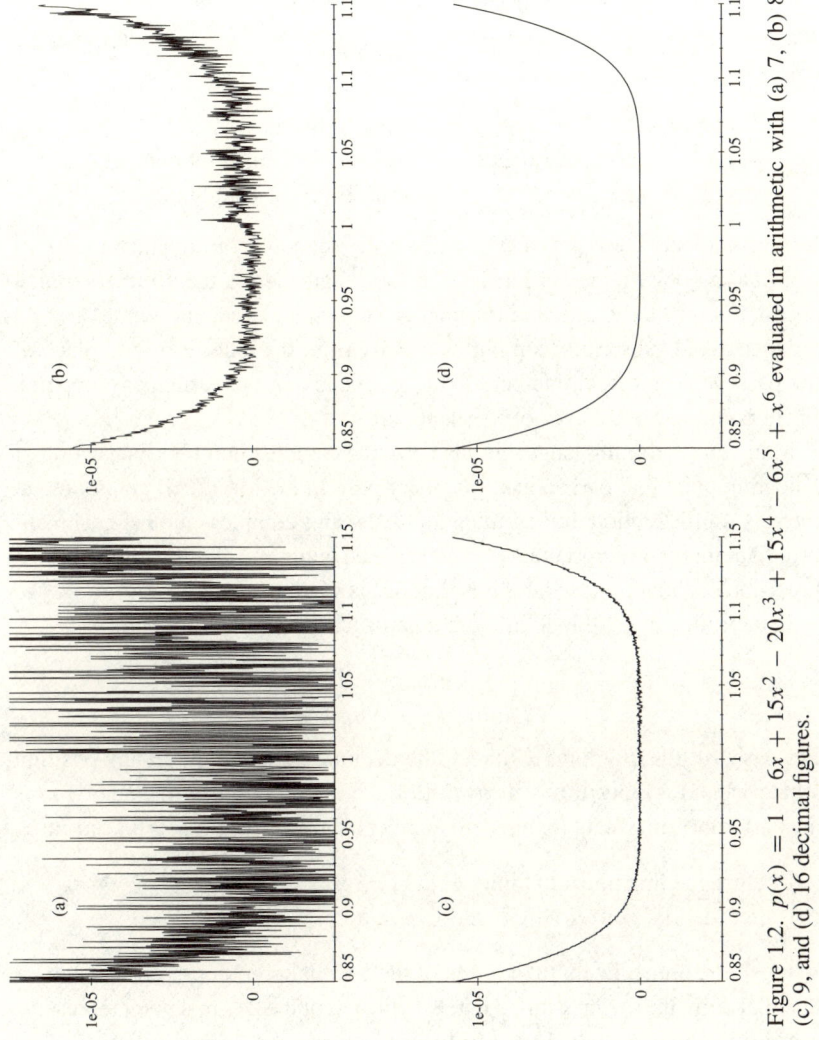

Figure 1.2. $p(x) = 1 - 6x + 15x^2 - 20x^3 + 15x^4 - 6x^5 + x^6$ evaluated in arithmetic with (a) 7, (b) 8, (c) 9, and (d) 16 decimal figures.

13

than 30 changes of sign in its neighborhood. This shows that we must look more closely at the evaluations of arithmetic expressions on a computer and with the rounding errors that create the observed inaccuracies.

1.2 Numbers, Operations, and Elementary Functions

Machine numbers of the following types (where each x denotes a digit) can be read and printed by any computer:

integer	$\pm xxxx$	(base 10 integer)
real	$\pm xx.xxx_{10} \pm xxx$	(base 10 floating point number)
real	$\pm xxxx.xx$	(base 10 fixed point number)

Integer numbers consist of a sign and a finite sequence of digits, and base 10 real floating point numbers consist of a sign, digits before the decimal point, a decimal point, digits after the decimal point, and an exponent with a sign. In order to avoid subscripts, computers usually print the letter d or e in place of the basis $_{10}$. The actual number of digits depends on the particular computer. Fixed point numbers have no exponent part.

Many programming languages also provide complex numbers, consisting of a floating point real part and an imaginary part. In the MATLAB environment, there is no distinction among integers, reals, and complex numbers. The appropriate internal representation is determined automatically; however, double precision calculation (see later this section) is used throughout.

In the following, we look in more detail at real floating point numbers.

Floating Point Numbers

For base 10 floating point numbers, the decimal point can be in any position whatever, and as input, this is always allowed. For the output, two standard forms are customary, in which, for nonzero numbers, the decimal point is placed either

(i) before the first nonzero digit: $\pm.xxx_{10} \pm xxx$, or
(ii) after the first nonzero digit: $\pm x.xx_{10} \pm xxx$.

In (i), the x, immediately to the right of the decimal point represents a nonzero decimal digit. In (ii), the x immediately to the left of the decimal point represents a nonzero decimal digit. In the following, we use the normalization (i). The sequence of digits before the exponent part is referred to as the *mantissa of the number.*

Internally, floating point numbers often have a different base B in which, typically, $B \in \{2, 8, 16, \ldots, 2^{31}, \ldots\}$. A huge base such as $B = 2^{31}$ or any other large number is used, for example, in the multi-precision arithmetic package of BRENT available through the electronic repository of mathematical software NETLIB [67]. (This valuable repository contains much other useful numerical software, too, and should be explored by the serious student.)

1.2.1 Example. The general (normalized) internal floating point number format has the form

$$\pm.\underbrace{xxxxxxxxxx}_{\substack{\text{Mantissa of} \\ \text{length } L}} \quad \text{B} \quad \overbrace{\pm xxxxxxxxxxxxxxx}^{\text{Exponent} \in [-E, F]}$$

where L is the mantissa length, B the base, and $[-E, F]$ the exponent range of the number format used. The arithmetic of personal computers (and of many workstations) conforms to the so-called IEEE floating point standard [45] for binary arithmetic (base $B = 2$).

Here the *single precision format* (also referred to as *real* or *real**4) uses 32 bits per number: 24 for the mantissa and 8 for the exponent. Because the sign of the mantissa takes 1 bit, the mantissa length is $L = 23$, and the exponents lie between $E = -126$ and $F = 127$. (The exponents -127 and 128 are reserved for floating point exceptions such as improper numbers like $\pm\infty$ and NaN; the latter ("**not a number**") signifies results of nonsensical calculations such as $0/0$.) The single precision format allows one to represent numbers of absolute value between approximately 10^{-38} and 10^{38}, with an accuracy of about seven decimals. (There are also some even tinier numbers with less accuracy that cannot be normalized with the given exponent range. Such numbers are called *denormalized*. In particular, zero is a denormalized number.)

The *double precision format* (also referred to as *double* or *real**8) uses 64 bits per number: 53 for the mantissa and 11 for the exponent. Because the sign of the mantissa takes 1 bit, the mantissa length is $L = 52$, and the exponents lie between $E = -1022$ and $F = 1023$. The double precision format allows one to represent numbers of absolute value between approximately 10^{-308} and 10^{308}, with an accuracy of about 16 decimals. (Again, there are also denormalized numbers.)

MATLAB works generally with double precision numbers. In particular, the selective use of double precision in predominantly single precision calculations cannot be illustrated with MATLAB, and we comment on such problems using FORTRAN language.

Ignoring the sign for a moment, we consider machine numbers of the form $.x_1 \ldots x_L {}_B e$ with mantissa length L, base B, digits $x_1, \ldots, x_L$ and exponent $e \in \mathbb{Z}$. By definition, such a number is assigned the value

$$\sum_{i=1:L} x_i B^{e-i} = x_1 B^{e-1} + x_2 B^{e-2} + \cdots + x_L B^{e-L}$$

$$= B^{e-L}(x_1 B^{L-1} + x_2 B^{L-2} + \cdots + x_L)$$

that is, the number is essentially the value of a polynomial at the point B. Therefore, the Horner scheme is a suitable method for converting from decimal numbers to base B numbers and conversely.

Rounding

It is obvious that between any two distinct machine numbers there are many real numbers that cannot be represented. In these cases, rounding must be used; that is, a "nearby" machine number must be found. Two rounding modes are most prevailing: optimal rounding and rounding by chopping.

 (i) *Optimal Rounding.* In the case of optimal rounding, the closest machine number is chosen. Ties can be broken arbitrarily, but in case of ties, the closest number with x_L even is most suitable on statistical grounds. (Indeed, in a long summation, one expects the last digit to have random parity; hence the errors tend to balance out.)
(ii) *Rounding by Chopping.* In the case of rounding by chopping, the digits after the Lth place are omitted. This kind of rounding is easier to realize than optimal rounding, but it is slightly less accurate.

We call a rounding *correct* if no machine number lies between a number x and its rounded value $\tilde{x}$. Both optimal rounding and rounding by chopping are correct roundings. For example, rounding with $B = 10$, $L = 3$ gives

	Optimal	Chopping
$x_1 = .123456_{10}5$	$\tilde{x}_1 = .123_{10}5$	$\tilde{x}_1 = .123_{10}5$
$x_2 = .567890_{10}5$	$\tilde{x}_2 = .568_{10}5$	$\tilde{x}_2 = .567_{10}5$
$x_3 = .123500_{10}5$	$\tilde{x}_3 = .124_{10}5$	$\tilde{x}_3 = .123_{10}5$
$x_4 = .234500_{10}5$	$\tilde{x}_4 = .234_{10}5$	$\tilde{x}_4 = .234_{10}5$

As measures of accuracy we use the *absolute error* $|x - \tilde{x}|$ and the *relative error* $|x - \tilde{x}|/|x|$ of an approximation $\tilde{x}$ for x. For general floating point numbers of highly varying magnitude, the relative error is the more important measure.

1.2.2 Proposition. *Under the hypothesis of an unrestricted exponent set, a correctly rounded value $\tilde{x}$ of x in $\mathbb{R}\backslash\{0\}$ satisfies*

$$\frac{|x - \tilde{x}|}{|x|} \leq \varepsilon := B^{1-L},$$

and an optimally rounded value $\tilde{x}$ of x in $\mathbb{R}\backslash\{0\}$ satisfies

$$\frac{|x - \tilde{x}|}{|x|} \leq \varepsilon := \frac{1}{2} B^{1-L}.$$

Thus, the relative error of the rounded value is bounded by a small machine dependent number ε.

The number ε is referred to as the *machine precision* of the computer. In MATLAB, the machine precision is available in the variable eps (unless eps is overwritten by something else).

Proof. We give the proof for the case of correct rounding; the case of optimal rounding is similar and left as Exercise 6. Suppose, without loss of generality, that $x > 0$ (the case $x < 0$ is analogous), where x is a real number that has, in general, an infinite base B representation. Suppose, without loss of generality, that

$$B^{e-1} \leq x < B^e$$

and

$$x = .x_1 x_2 \cdots x_L x_{L+1} \cdots {}_B e,$$

with $x_1 \neq 0$. Set

$$x' := .x_1 x_2 \cdots x_L {}_B e.$$

Then $\tilde{x} \in [x', x' + B^{e-L}]$ because the rounding is correct, and $\tilde{x} \leq x < \tilde{x} + B^{e-L}$. So

$$|x - \tilde{x}| \leq B^{e-L} = B^{e-1} B^{1-L} \leq |x| B^{1-L},$$

and the assertion follows. $\square$

This proposition is a basis for all rigorous error analysis of numerical methods; see in particular Higham [44] and Stummel and Hainer [92]. In this book, we usually keep such analyses on a more heuristic level (which is enough to expose the pitfalls in most numerical calculations); a fully quantitative treatment is only given for Gaussian elimination (see Section 2.1).

Overflow and Underflow

The preceding result is valid only for an unrestricted set of exponents. In practice, however, the exponent range is finite, and there are problems when $|x|$ is too large or too small to be representable by a normalized number within the exponent range. In this case, we speak of *overflow* or *underflow*, respectively.

The behavior resulting from overflow depends on the programming language and its implementation. Often, it results in an appropriate error message; alternatively, it may result in an improper number $\pm\infty$. (Some old compilers even set the result to zero, without warning!)

In case of underflow, numbers that are too small are either rounded to denormalized numbers (gradual underflow) or replaced by zero; this is, in general, reasonable, but leads to much larger relative errors, and occasionally can therefore be dangerous.

Overflow and underflow can usually be avoided by suitable scaling of the problem; hence we assume in later discussions that the exponent range is unrestricted and Proposition 1.2.2 is generally valid.

Operations and Elementary Functions

After the consideration of the representation of numbers on a computer, we look at the machine results of binary operations and elementary functions.

For correctly rounded results of the binary operations $\circ \in O$, we have

$$\frac{|x \circ y - x \widetilde{\circ} y|}{|x \circ y|} \leq \varepsilon,$$

where $x \widetilde{\circ} y$ represents the computed value of $x \circ y$ and ε is the machine precision. In the following, the validity of this property are assumed for $\circ \in \{+, -, *, /\}$ (but not for the power); it is satisfied by most modern computers as long as neither overflow nor underflow occur. For example, the IEEE floating point standard requires the results of these operations (and the square root) to be even optimally rounded.

The power $x\hat{\ }y = x^y$ (except for the square, $y = 2$) is usually slightly less accurate because it is computed as a composite expression. For small integers y, x^y is computed by repeated multiplication, and otherwise from $\exp(y \log x)$. For theoretical analysis, we suppose that the relative error of the power is bounded by $C\varepsilon$ for a constant not much larger than 1; this holds in practice except for extreme values of the operands.

For the calculation of the elementary functions $\varphi \in J$, one must use approximative methods. For a thorough treatment of the approximations of standard functions, see Hart [41]. We treat only two typical examples here, namely the square root and the exponential function.

The following three steps are usually employed for the realization of the elementary function $\varphi(x)$ on a computer: (i) *Argument reduction* to a number x_0 in a standard range, followed by (ii) *approximation* of $\varphi(x_0)$, and (iii) a subsequent *result adaptation*.

The argument reduction (i) serves mainly to reduce the amount of work needed in the approximation step (ii), and the result adaptation (iii) corrects the result of (ii) to account for the argument reduction. To demonstrate the ideas, we look in more detail at the computation of the square root and the exponential function.

Suppose we want to compute $\sqrt{x} = \text{sqrt}(x)$ with $x = m_B e$, where $m \in [1/B, 1[$ is the mantissa, B is the base, and e is the exponent of x.

Argument Reduction and Result Adaptation for $\sqrt{x}$

We may express $\sqrt{x}$ in the form

$$\sqrt{x} = (\sqrt{x_0})B^s$$

with $x_0 = m$, $s = e/2$ if e is even, and $x_0 = m/B$, $s = (e+1)/2$ if e is odd. This is the argument reduction. Because $x_0 \in [B^{-2}, 1]$, it is sufficient to determine the value of the square root in the interval $[B^{-2}, 1]$ to obtain a result which may then be adapted to give the required value $\sqrt{x}$ by multiplying the result $\sqrt{x_0} \in [B^{-1}, 1]$ (the mantissa) by B^s.

Approximation of $\sqrt{x}$

(a) The simplest but inefficient possibility is the expansion in a power series about 1. The radius of convergence of the series

$$\sqrt{1-z} = 1 - \frac{1}{2}z - \frac{1}{8}z^2 - \frac{1}{16}z^3 - \frac{5}{128}z^4 - \cdots$$

is 1. The series converges sufficiently rapidly only for $x \approx 1$ ($z \approx 0$), but converges very slowly for $x_0 = 0.01(z = 0.99)$, say (an essential value when $B = 10$), and is therefore useless for practical purposes. The cause of the slow convergence is the vertical tangent to the graph of $x^{1/2}$ at $x = 0$.

(b) Another typical way to approximate function values is offered by an iterative method. Here, iteration of the same process over and over again successively improves an approximation until a desired accuracy is reached.

In order to find $w = \sqrt{x}$ for a given value of $x > 0$, let w_0 be an initial estimate of w; for example, we may take $w_0 = 1$ when $x \in [B^{-2}, 1]$). Note that $w^2 = x$ and so $w = x/w$. To find a suitable iteration formula, we define $\bar{w}_0 := x/w_0$. If $\bar{w}_0 = w_0$, then we are finished ($w = w_0$); otherwise, we have $\bar{w}_0 = w^2/w_0 = w(w/w_0) < w$ (or $> w$) if $w_0 > w$ or $w_0 < w$, respectively. This suggests that the arithmetic mean $w_1 = (w_0 + \bar{w}_0)/2$ might be a suitable new estimate for w. Repeating the process leads to the following iterative method, already known to Babylonian mathematicians.

1.2.3 Proposition. *Let $x > 0$. For arbitrary $w_0 > 0$, define*

$$\bar{w}_i := x/w_i \quad (i \geq 0), \tag{2.1}$$

$$w_{i+1} := (w_i + \bar{w}_i)/2 \quad (i \geq 0). \tag{2.2}$$

If $w_0 = \sqrt{x}$, then

$$w_i = \bar{w}_i = \sqrt{x} \quad (i \geq 0);$$

otherwise,

$$\bar{w}_1 < \bar{w}_2 < \cdots < \bar{w}_i < \sqrt{x} < w_i < \cdots < w_2 < w_1.$$

The absolute error $\delta_i := w_i - \sqrt{x}$ satisfies

$$\delta_{i+1} = \delta_i^2/2w_i \quad (i \geq 0),$$

and the relative error $\varepsilon_i := (w_i - \sqrt{x})/\sqrt{x}$ satisfies

$$\varepsilon_{i+1} = \varepsilon_i^2/2(1 + \varepsilon_i) \quad (i \geq 0). \tag{2.3}$$

Here we use the convention that a product has higher priority than division if it is written by juxtaposition, and lower priority if it is written with explicit multiplication sign. Thus,

$$x^2/2y = \frac{x^2}{2y}, \quad \text{but } x^2/2 \cdot y = \frac{x^2}{2}y.$$

Proof. With $w := \sqrt{x}$, we have (for all $i > 0$)

$$w_i = \delta_i + w = (1 + \varepsilon_i)w,$$

hence

$$\delta_{i+1} = w_{i+1} - w = (w_i + x/w_i)/2 - w$$
$$= \left(w_i^2 + x - 2ww_i\right)/2w_i$$
$$= \{(\delta_i + w)^2 + w^2 - 2w(\delta_i + w)\}/2w_i$$
$$= \delta_i^2/2w_i \geq 0$$

and

$$\varepsilon_{i+1} = \delta_{i+1}/w = \delta_i^2/2ww_i = \varepsilon_i^2/2(1 + \varepsilon_i).$$

If now $w_0 \neq \sqrt{x}$, then $\delta_0 = w_0 - w \neq 0$ and $\delta_{i+1} > 0$. So,

$$0 < \delta_{i+1} = \delta_i^2/2w_i \leq \delta_i/2 < \delta_i \quad (i > 0)$$

whence $\sqrt{x} < w_{i+1} < w_i$ and $\bar{w}_i < \bar{w}_{i+1} < \sqrt{x}$ $(i > 0)$. This proves the proposition. $\square$

1.2.4 Example. The iteration of Proposition 1.2.3 was executed on a pocket calculator with $B = 10$ and $L = 12$ for $x = 0.01$, starting with $w_0 := 1$. The results displayed in Table 1.2 were obtained. One observes a rapid increase in the number of correct digits as predicted by (2.3). This is so-called quadratic convergence, and the iteration is in fact a special case of Newton's method (cf. Chapter 5).

For $B = 10$, $x_0 \in [0.01, 1]$ is a consequence of argument reduction. Clearly, $x_0 = 0.01$ is the most unfavorable case; hence seven iterations (i.e., 21 operations) always suffice to calculate $\sqrt{x}$ accurate to twelve decimal

Table 1.2.   *Calculation of $\sqrt{0.01}$*

i	w_i
0	1
1	0.505...
2	0.2624...
3	0.1502...
4	0.1084...
5	0.1003...
6	0.1000005...
7	0.1

numbers or calculations; however, this says nothing at all about the quality
of the numbers involved in relation to their intended meaning. Estimating the
accuracy of the latter is the subject of an error analysis that should accompany
any extended calculation, at least in a qualitative way.

Because in what follows inexact numbers occur frequently, we introduce
the notation $\approx$ for *approximately equal to*. We shall also write $x \gg y$ ($x \ll y$)
when x, y are positive and x/y is *much greater than* 1 (and, respectively, *much
smaller than* 1). These are qualitative notions only; it must be decided from the
context or application what "approximately" or "much" means quantitatively.

1.3.1 Example. All of the following examples were computed on a pocket calculator with $B = 10$ and $L = 12$. (Calculators with different rounding charac-
teristics probably give similar but not identical results.)

(i) $f(x) := (x + 1/3) - (x - 1/3)$.

x	$\widetilde{f(x)}$
1	0.666 666 666 . . .
10^3	0.666 666 663 . . .
10^6	0.666 663 . . .
10^9	0.663 . . .
10^{10}	0.33 . . .
10^{11}	0

Because $f(x) = 2/3$ for all x, there is an increasing loss of accuracy for
increasing x.

(ii) $f(x) := ((3 + x^2/3) - (3 - x^2/3))/x^2$.

x	$\widetilde{f(x)}$
10^{-1}	0.666 666 667 . . .
10^{-2}	0.666 666 7 . . .
10^{-3}	0.666 67 . . .
10^{-4}	0.667 . . .
10^{-5}	0.675 . . .
10^{-6}	0

Because $f(x) = 2/3$ for all $x \neq 0$, there is an increasing loss of accuracy
for decreasing x.

There are two reasons for the increasing loss of accuracy in (i) and (ii):

1. There is a large difference in order of magnitude between the numbers to be added or subtracted, respectively (e.g., $10^{10}+1/3$). If $x = m_{10}e$, $\bar{x} = \bar{m}_{10}\,\bar{e}$, with $|x| > |\bar{x}|$, and $k := e - \bar{e} - 1$, then $\bar{x}$ is too small to have any influence on the first k places of the sum or difference of x and $\bar{x}$, respectively. Thus, information on $\bar{x}$ is in the less significant digits of the intermediate results only, and the last k places of $\bar{x}$ are completely lost in the rounding process.

2. The subtraction of numbers of almost equal size leads to the cancellation of leading digits and promotes the wrong low-order digits in the intermediate results to a much more significant position. This leads to a drastic magnification of the relative error.

   In case (i), the result of the subtraction already has order unity, so that relative errors and absolute errors have the same magnitude. In case (ii), the result of the subtraction has small absolute (but large relative) error, and the division by the small number x^2 leads to a result with large absolute error.

(iii) $f(x) := \sin^2 x/(1 - \cos^2 x)$.

x	$\widetilde{f(x)}$
$0.01°$	$1.0007\ldots$
$0.001°$	$1.0879\ldots$
$0.0005°$	$1.2692\ldots$
$0.0004°$	$2.3166\ldots$
$0.0003°$	ERROR

Because $f(x) = 1$ for all x not a multiple of π, there is a drastic decrease in accuracy for decreasing x due to division by the small number $1 - \cos^2 x$, and ultimately an error occurs due to division by a computed zero.

(iv) $f(x) := \sin x/\sqrt{1 - \sin^2 x}$.

x	$\widetilde{f(x)}$	$\tan x$
$89.9°$	$572.9588\ldots$	$572.9572\ldots$
$89.95°$	$1145.87\ldots$	$1145.91\ldots$
$89.99°$	5735.39	$5729.57\ldots$

The correct digits are underlined. We have $f(x) = \tan x$ for all x, and here the loss of accuracy is due to the fact that $f(x)$ has a pole at $x = 90°$, together with the cancellation in the denominator.

(v) $f(x) := e^{x^2/3} - 1$. For comparison, we give together with $f(x)$ also the result of the first two terms of the power series expansion of $f(x)$.

x	$\widetilde{f(x)}$	$x^2/3 + x^4/18$	$f(x)$, correctly rounded
0.1	$3.338\,895_{10} - 3$	$3.338\,888\,889_{10} - 3$	$3.338\,895\,066\,88_{10} - 3$
0.01	$3.333\,388_{10} - 5$	$3.333\,388\,889_{10} - 5$	$3.333\,388\,889\,51_{10} - 5$
0.001	$3.333\,300_{10} - 7$	$3.333\,333\,889_{10} - 7$	$3.333\,333\,888\,89_{10} - 7$
0.0001	$3.330\,000_{10} - 9$	$3.333\,333\,339_{10} - 9$	$3.333\,333\,338\,89_{10} - 9$

Although the absolute errors of $\widetilde{f(x)}$ appear acceptable, the relative errors grow as x approaches zero (cancellation!). However, the truncated power series gives results with increasing relative accuracy.

(vi) The formula

$$\sigma_n = \sqrt{\frac{1}{n}\left(\sum_{i=1:n} x_i^2 - \frac{1}{n}\left(\sum_{i=1:n} x_i\right)^2\right)}$$

gives the standard deviation of a sequence of data $x_1, \ldots, x_n$ in a way as implemented in many pocket calculators. (The alternative formula mentioned in Example 1.1.1 is much more stable but does not allow the input of many data with little storage!) For $x_i := x$ where x is a constant, $\sigma_n = 0$, but on the pocket calculator, the corresponding keys yield the results shown in Table 1.3. The zero in the table arises because the calculator checks whether the computed argument of the square root is negative (which cannot happen in exact arithmetic) and, if so, sets σ_n equal to zero, but if the argument is positive, the positive result is purely due to round off.

Table 1.3.   *Standard deviation of $x_i := x$ where x is a constant*

x	$\tilde{\sigma}_{10}$	$\tilde{\sigma}_{20}$
100/3	$1.204 \cdots_{10} - 3$	$1.131 \cdots_{10} - 3$
1000/29	$8.062 \cdots_{10} - 4$	0

We conclude from the examples that the quality of a computed function value depends strongly on the expression used. To prevent, if possible, the magnification of old errors, one should avoid the following:

- cancellation; that is, subtraction of numbers of almost equal size that leads to a loss of leading digits and thereby to magnification of the relative error;
- division by a very small number leading to magnification of the absolute error;
- multiplication by a number of very large magnitude, leading to magnification of the absolute error.

In the formulation and analysis of algorithms, one must look out for these problems to see whether one can avoid them by a suitable reformulation. In general, one can say that *caution is appropriate in any computation in which some intermediate result has a much larger or smaller magnitude than a final result.* Then a more detailed investigation must reveal whether this really leads to numerical instability, or whether the further course of the computation reduces the effect to a harmless level.

Note that, in the case of cancellation, the difference is usually calculated *without error*; the instability comes from the *magnification of old errors*. This is shown in the following example.

1.3.2 Example. Subtraction of two numbers with $B = 10$ and with $L = 7$ and $L = 6$, respectively:

(i) $L = 7$

1st number:	$0.2789014_{10}3$	
2nd number:	$0.2788876_{10}3$	
Difference:	$0.0000138_{10}3$	No rounding error
Normalized:	$0.1380000_{10} - 1$	

(ii) $L = 6$

	Optimally rounded	Relative error
1st number:	$0.278901_{10}3$	$<2_{10} - 6$
2nd number:	$0.278888_{10}3$	$<2_{10} - 6$
Difference:	$0.000013_{10}3$	No rounding error
Normalized:	$0.130000_{10} - 1$	$>5_{10} - 2$

Although there is no rounding error in the subtraction, the relative error in the final result is about 25,000 times bigger than both the initial errors.

1.3.3 Example. In spite of division by a small number, the following formulas are stable:

(i) $f(x) := \begin{cases} 1 & \text{if } x = 0 \\ \sin x / x & \text{if } x \neq 0 \end{cases}$ is stable for $x \approx 0$:

x	$\widetilde{f}(x)$
0.1	0.998 3
0.01	0.999 983 ...
0.001	0.999 999 83 ...
0.0001	0.999 999 998 ...
0.00001	1

The reason for the stable behavior is that in both numerator and denominator only one operation (in our usage of the term) is performed; stability follows from the form of the relative error that is more or less independent of the way in which the elementary functions are implemented. The same behavior is visible in the next case.

(ii) $f(x) := \begin{cases} 1 & \text{if } x = 1 \\ (x-1)/\ln x & \text{if } x \neq 1 \end{cases}$ is stable for $x \approx 1$:

x	$\widetilde{f}(x)$
1.001	1.000 49 ...
1.000 1	1.000 049 ...
1.000 01	1.000 005 ...
1.000 000 001	1.000 000 001

Stabilizing Unstable Expressions

There is no general theory for stabilizing unstable expressions, but some useful recipes can be gleaned from the following examples.

(i) In the equality

$$\sqrt{x+1} - \sqrt{x} = \frac{1}{\sqrt{x+1} + \sqrt{x}},$$

the left side is unstable and the right side is stable for $x \gg 1$.

(ii) In the equalities

$$1 - \cos x = \frac{\sin^2 x}{1 + \cos x} = 2\sin^2 \frac{x}{2},$$

the left side is unstable and both of the right sides are stable for $x \approx 0$.

In both cases, the difference of two functions is rearranged so that there is a difference in the numerator that can be simplified analytically. This difference is thus evaluated without rounding error. For example,

$$\sqrt{x+1} - \sqrt{x} = \frac{(x+1) - x}{\sqrt{x+1} + \sqrt{x}} = \frac{1}{\sqrt{x+1} + \sqrt{x}}.$$

(iii) The expression $e^x - 1$ is unstable for $x \approx 0$. We substitute $y := e^x$, whence $e^x - 1 = y - 1$, which is still unstable.

For $x \neq 0$,

$$e^x - 1 = (y - 1)x/x = \frac{y - 1}{\ln y} x;$$

see Example 1.3.3(ii). Therefore we can obtain a stable expression as follows:

$$e^x - 1 = \begin{cases} x & \text{if } y = 1, \\ \frac{y-1}{\ln y} x & \text{if } y \neq 1 \end{cases}$$

where $y = e^x$.

(iv) The expression $e^x - 1 - x$ is unstable for $x \approx 0$. Using the power series expansion

$$e^x = \sum_{k=0}^{\infty} x^k/k!$$

we obtain

$$e^x - 1 - x = \begin{cases} e^x - 1 - x & \text{if } |x| > c, \\ \frac{x^2}{2} + \frac{x^3}{6} + \cdots & \text{otherwise} \end{cases}$$

in which a suitable threshold c is determined by means of an error estimate: We expect the first formula to have an error of $\varepsilon = B^{1-L}$, and the second, truncated after the terms shown, to have an error of $\approx x^4/24 \leq c^4/24$; hence, the natural threshold in this case is $c = \sqrt[4]{24\varepsilon}$, which makes the worst case absolute error minimal ($\approx \varepsilon$), and gives a worst case relative error of $\approx \varepsilon/(c^2/2) \approx \sqrt{\varepsilon/6}$.

Finally, we consider the stabilization of two widely used unstable formulas.

The Solution of a Quadratic Equation

The quadratic equation

$$ax^2 + bx + c = 0 \quad (a \neq 0)$$

has the two solutions

$$x_{1,2} = \frac{-b \pm \sqrt{b^2 - 4ac}}{2a}.$$

If $b^2 \gg 4ac$, then this expression is unstable when the sign before the square root is the same as the sign of b (the expression obtained by choosing the other sign before the square root is stable).

On multiplying both the numerator and the denominator by the algebraic conjugate $-b \mp \sqrt{b^2 - 4ac}$, the numerator simplifies to $b^2 - (b^2 - 4ac) = 4ac$, leading to the alternative formula

$$x_{1,2} = \frac{2c}{-b \mp \sqrt{b^2 - 4ac}}.$$

Now the hitherto unstable solution has become stable, but the hitherto stable solution has become unstable. Therefore, it is sensible to use a combination of both. For real b, the rule

$$q := -(b + \text{sign}(b)\sqrt{b^2 - 4ac})/2,$$

$$x_1 := q/a, \quad x_2 := c/q$$

calculates both solutions by evaluating a stable expression. The computational cost remains the same. (For complex b this formula does not work, because $\text{sign}(b) = b/|b| \neq \pm 1$; cf. Exercise 12.)

Mean Value and Standard Deviation

The mean value s_n and the standard deviation σ_n of $x_1, \ldots, x_n$ are usually defined by

$$s_n = \frac{1}{n} \sum_{i=1:n} x_i,$$

$$\sigma_n = \sqrt{t_n/n}$$

or, sometimes,

$$\sigma_n = \sqrt{t_n/(n-1)}$$

with

$$t_n = \sum_{i=1:n} (x_i - s_n)^2,$$

cf. Example 1.1.1. This formula for the standard deviation is impractical because all the x_i must be stored. This can be avoided by a transformation to a recursive form in which each x_i is needed only until x_{i+1} is read.

$$t_n = \sum_{i=1:n} x_i^2 - 2s_n \sum_{i=1:n} x_i + s_n^2 \sum_{i=1:n} 1$$
$$= \sum x_i^2 - 2s_n \cdot ns_n + s_n^2 \cdot n = \sum x_i^2 - ns_n^2,$$

cf. Example 1.3.1(vi). These are the formulas for calculating the standard deviation stated in most statistics books. Their advantage is that t_n can be computed recursively without storing the x_i, and their greater disadvantage lies in their instability due to cancellation, because (for data with small variance) $\sum x_i^2$ and ns_n^2 have several common leading digits.

The transformation into a stable expression is inspired by the observation that the numbers s_n and t_n are expected to change very little due to the addition of new data values x_i; therefore, we consider the differences $s_n - s_{n-1}$ and $t_n - t_{n-1}$. We have

$$s_n - s_{n-1} = \frac{(n-1)s_{n-1} + x_n}{n} - s_{n-1}$$
$$= \frac{x_n - s_{n-1}}{n} = \frac{\delta_n}{n},$$

where $\delta_n := x_n - s_{n-1}$, and

$$t_n - t_{n-1} = \left(\sum_{i=1:n} x_i^2 - ns_n^2 \right) - \left(\sum_{i=1:n-1} x_i^2 - (n-1)s_{n-1}^2 \right)$$
$$= x_n^2 - ns_n^2 + (n-1)s_{n-1}^2$$
$$= (\delta_n + s_{n-1})^2 - n \left(s_{n-1} + \frac{\delta_n}{n} \right)^2 + (n-1)s_{n-1}^2$$
$$= \delta_n^2 + 2\delta_n s_{n-1} - 2s_{n-1}\delta_n - \frac{\delta_n^2}{n}$$
$$= \delta_n \left(\delta_n - \frac{\delta_n}{n} \right) = \delta_n((x_n - s_{n-1}) - (s_n - s_{n-1}))$$
$$= \delta_n(x_n - s_n).$$

From these results, we obtain a new recursion for the calculation of s_n and t_n:

$$s_1 = x_1, \quad t_1 = 0,$$

and for $i \geq 2$

$$\delta_i = x_i - s_{i-1},$$
$$s_i = s_{i-1} + \delta_i / i,$$
$$t_i = t_{i-1} + \delta_i (x_i - s_i).$$

The differences $x_i - s_{i-1}$ and $x_i - s_i$, which have still to be calculated, are now harmless: the cancellation can bring about no great magnification of relative errors because the difference $(x_i - s_i)$ is multiplied by a small number δ_i and is then added to the (as a sum of many small positive terms generally larger) number t_{i-1}.

The new recursion can be programmed with very little auxiliary storage, and in an interactive mode there is no need for storage of old values of x_i, as the following MATLAB program shows.

1.3.4 Algorithm: Stable Mean and Standard Deviation

```
i=1; s=input('first number?>');
t=0;
x=input('next number? (press return for end of list)>');
while ~isempty(x),
   i=i+1;
   delta=x-s;
   s=s+delta/i;
   t=t+delta*(x-s);
   x=input('next number? (press return for end of list)>');
end;
disp(['mean: ',num2str(s)]);
sigma1=sqrt(t/i);sigma2=sqrt(t/(i-1));
disp('standard deviation');
disp(['   of sample: ',num2str(sigma1)]);
disp(['   estimate of distribution: ',num2str(sigma2)]);
```

(~, & and | code the logical "not," "and," and "or." The code also gives an example of how MATLAB allows one to read and print numbers and text.)

1.4 Error Propagation and Condition

In virtually all numerical calculations many small errors are made. Hence, one is interested in how these errors influence further calculations. As already seen, the final error depends strongly on the formulas that are used. Among other things, especially for unstable methods, this is because of the finite number of digits used by the computer.

However, in many calculations, already the input data are known only approximately; for example, when they are determined by measurement. Then even an exact calculation gives an inaccurate answer, and in some cases, as in the following example, the relative error grows strongly even when exact arithmetic is used, and hence independently of the method that is used to calculate the result.

1.4.1 Example. Let

$$f(x) := \frac{1}{1-x}.$$

If $x := 0.999$, then $f(x) = 1000$. An analytical error analysis shows that for $\tilde{x} = 0.999 + \varepsilon$, with ε small,

$$f(\tilde{x}) = \frac{1000}{1 - 1000\varepsilon}$$
$$= 1000(1 + 10^3\varepsilon + 10^6\varepsilon^2 + \cdots).$$

Hence, for the relative errors of $\tilde{x}$ and $f(\tilde{x})$ we have

$$\frac{|x - \tilde{x}|}{|x|} \approx 1.001\varepsilon$$

and

$$\frac{|f(x) - f(\tilde{x})|}{|f(x)|} = 10^3\varepsilon + 10^6\varepsilon^2 + \cdots,$$

respectively; that is, independently of the method of calculating f, the relative error is magnified by a large factor (here about 1000).

One calls such problems *ill-conditioned*. (Again, this is only a qualitative notion without very precise meaning.) For an ill-conditioned problem with inexact initial data, *all* numerical methods must give rise to large relative errors. A quantitative measure for the condition of differentiable expressions is the *condition number* κ, the (asymptotic) magnification factor of the relative error. For a concise discussion of the asymptotic behavior of quantities, it is useful to recall the *Landau symbols o* and *O*. In order to characterize the asymptotic behavior of the functions f and g in the neighborhood of $r^* \in \mathbb{R} \cup \{\infty\}$ (a value usually apparent from the context), one writes

$$f(r) = o(g(r)) \quad \text{when} \quad \frac{f(r)}{g(r)} \to 0 \quad \text{as } r \to r^*,$$

and

$$f(r) = O(g(r)) \quad \text{when} \quad \frac{f(r)}{g(r)} \text{ remains bounded as } r \to r^*.$$

In particular,

$$f(r) = o(g(r)), \quad g(r) \text{ bounded} \Rightarrow \lim_{r \to r^*} f(r) = 0,$$

and

$$f(r) = O(g(r)), \quad \lim_{r \to r^*} g(r) = 0 \Rightarrow \lim_{r \to r^*} f(r) = 0.$$

(One reads the symbols as "small oh of" and "big oh of," respectively.)

Now let $\tilde{x}$ be an approximation to x with relative error $\varepsilon = (\tilde{x} - x)/x$, so that $\tilde{x} = (1 + \varepsilon)x$. We expand $f(\tilde{x})$ in a Taylor series about x, to obtain

$$\begin{aligned}
f(\tilde{x}) &= f(x + \varepsilon x) \\
&= f(x) + \varepsilon x f'(x) + O(\varepsilon^2) \\
&= f(x)\left(1 + \frac{x f'(x)}{f(x)}\varepsilon + O(\varepsilon^2)\right).
\end{aligned}$$

We therefore have for the relative error of f,

$$\begin{aligned}
\frac{|f(x) - f(\tilde{x})|}{|f(x)|} &= \left|\frac{x f'(x)}{f(x)}\right| |\varepsilon| + O(\varepsilon^2) \\
&= \kappa |\varepsilon| + O(\varepsilon^2),
\end{aligned}$$

with the (relative) *condition number*

$$\kappa = \left|\frac{x f'(x)}{f(x)}\right|. \tag{4.1}$$

Assuming exact calculation, we can therefore predict approximately the relative error of f at the point x by calculating κ. Neglecting terms of higher order, we have

$$\frac{|f(x) - f(\tilde{x})|}{|f(x)|} \approx \kappa \frac{|x - \tilde{x}|}{|x|}.$$

1.4.2 Examples.

(i) In the previous example $f(x) = 1/(1-x)$, we have at $x = .999$, the values $f(x) \approx 10^3$ and $f'(x) \approx 10^6$; hence, $\kappa \approx .999 * 10^6/1000 \approx 10^3$ predicts the inflation factor correctly.

(ii) The function defined by $f(x) := x^5 - 2x^3$ has the derivative $f'(x) = 5x^4 - 6x^2$. Suppose that $|\tilde{x} - 2| \le r := 10^{-3}$, so that the relative error of $\tilde{x}$ is $r/2$. We estimate the error $|f(\tilde{x}) - f(2)|$ by means of the condition number: For $x = 2$,

$$\kappa = \left| \frac{xf'(x)}{f(x)} \right| = 7,$$

whence

$$\frac{|f(\tilde{x}) - f(2)|}{|f(2)|} \approx \frac{7}{2}r = 0.0035,$$

and $|f(\tilde{x}) - f(2)| \approx 0.0035|f(2)| = 0.056$.

In a qualitative condition analysis, one speaks of *error damping* or of *error magnification*, depending on whether the condition number satisfies $\kappa < 1$ or $\kappa > 1$. *Ill-conditioned problems* are characterized by having very large condition numbers, $\kappa \gg 1$.

From (4.1), it is easy to see when the evaluation of a function of a single variable is ill conditioned. We distinguish several cases.

CASE 1: If $f(x^*) = 0$ and $f'(x^*) \ne 0$ (these conditions characterize a *simple zero x^**), then κ approaches infinity as $x \to x^*$; that is, f is ill conditioned in the neighborhood of a simple zero $x^* \ne 0$.

CASE 2: If there is a positive integer m such that

$$f(x) = (x - x^*)^m g(x) \quad \text{with } g(x^*) \ne 0, \tag{4.2}$$

we say that x^* is a *zero of f of order m* because the factor in front of $g(x)$ can be viewed as the limiting, confluent case $x_i \to x^*$ of a product of m distinct linear factors $(x - x_i)$; if (4.2) holds with a negative integer $m = -s < 0$, then x^* is referred to as a *pole of f of* *order s.* In both cases,

$$f'(x) = m(x - x^*)^{m-1}g(x) + (x - x^*)^m g'(x),$$

so that

$$\kappa = \left| \frac{xf'(x)}{f(x)} \right|$$

$$= |x| \cdot \left| \frac{m}{x - x^*} + \frac{g'(x)}{g(x)} \right|$$

$$= |m| \cdot \left| \frac{x - x^*}{x} \right|^{-1} + o(1).$$

Therefore, as $x \to x^*$,

$$\kappa \to \begin{cases} \infty & \text{if } x^* \neq 0, \\ |m| & \text{if } x^* = 0. \end{cases}$$

For $m = 1$, this agrees with what we obtained in Case 1. In general, we see that in the neighborhood of zeros or poles $x^* \neq 0$, the condition number is large and is nearly inversely proportional to the relative distance of x from x^*. (This explains the bad behavior near the multiple zero in Example 1.1.8.) However, if the zero or pole is at $x^* = 0$, the condition number is approximately equal to the order of the pole or zero, so that f is well conditioned near such zeros and poles.

CASE 3: If $f'(x)$ becomes infinite for $x^* \to 0$ then f is ill-conditioned at x^*. For example, the function defined for $x > 1$ by $f(x) = 1 + \sqrt{x-1}$ has the condition number

$$\kappa = \frac{x}{2(x - 1 + \sqrt{x-1})}$$

and $\kappa \to \infty$ as $x \to 1$.

We now demonstrate with an example the important fact that condition and stability are completely different notions.

1.4.3 Example. Let $f(x)$ be defined by

$$f(x) := \sqrt{x^{-1} - 1} - \sqrt{x^{-1} + 1} \quad (0 < x < 1).$$

(i) *Numerical stability:* For $x \approx 0$, we have $x^{-1} - 1 \approx x^{-1} + 1$; hence, there is cancellation and the expression is unstable; however, the expression is stable for $x \approx 1$.

(ii) *Condition:* We have

$$f'(x) = \frac{-x^{-2}}{2\sqrt{x^{-1}-1}} - \frac{-x^{-2}}{2\sqrt{x^{-1}+1}}$$

$$= \frac{\sqrt{x^{-1}-1} - \sqrt{x^{-1}+1}}{2x^2\sqrt{x^{-1}-1}\sqrt{x^{-1}+1}}.$$

and therefore, for the condition number κ,

$$\kappa = \left| \frac{x \cdot f'(x)}{f(x)} \right|$$

$$= \frac{1}{2\sqrt{1-x^2}}.$$

Thus, for $x \approx 0$, $\kappa \approx \frac{1}{2}$ so that f is well conditioned. However, for $x \to 1$, we have $\kappa \to \infty$ so that f is ill-conditioned. So, the formula for $f(x)$ is unstable but well-conditioned for $x \approx 0$, whereas it is stable but ill conditioned for $x \approx 1$.

We see from this that the two ideas have nothing to do with each other: the *condition of a problem* makes a statement about the *magnification of* **initial** *errors through exact calculation,* whereas the *stability of a formula* refers to the *influence of* **rounding** *errors due to inexact calculation* because of finite precision arithmetic.

It is not difficult to derive a formula similar to (4.1) for the condition number of the evaluation of an expression f in several variables $x_1, \ldots, x_n$. Let $\tilde{x}_i = x_i(1 + \varepsilon_i)$ and $|\varepsilon_i| \le \varepsilon$ $(i = 1, \ldots, n)$. Then one obtains the multidimensional Taylor series

$$f(\tilde{x}_1, \ldots, \tilde{x}_n) = f(x_1, \ldots, x_n) + \sum_{i=1:n} \frac{\partial}{\partial x_i} f(x_1, \ldots, x_n) x_i \varepsilon_i + O(\varepsilon^2).$$

We make use of the derivative

$$f'(x) = \left(\frac{\partial}{\partial x_1} f(x), \ldots, \frac{\partial}{\partial x_n} f(x) \right),$$

to obtain the relative error

$$\frac{|f(x_1, \ldots, x_n) - f(\tilde{x}_1, \ldots, \tilde{x}_n)|}{|f(x_1, \ldots, x_n)|} \le \frac{\sum_{i=1:n} \left| \frac{\partial}{\partial x_i} f(x_1, \ldots, x_n) \right| |x_i|}{|f(x_1, \ldots, x_n)|} |\varepsilon| + O(\varepsilon^2)$$

$$= \frac{|x| \cdot |f'(x)|}{|f(x)|} |\varepsilon| + O(\varepsilon^2),$$

where the multiplication dot denotes the scalar product of two vectors. The resulting error propagation formula for functions of several variables is

$$\frac{|f(x) - f(\tilde{x})|}{|f(x)|} \lesssim \kappa \max_{1 \le i \le n} \frac{|x_i - \tilde{x}_i|}{|x_i|},$$

with the (relative) *condition number*

$$\kappa = |x| \cdot \frac{|f'(x)|}{|f(x)|},$$

looking just as in the one-dimensional case. Again, one expects bad condition only near zeros of f or near singularities of f'.

In the preceding discussion, the higher-order terms in ε were neglected in the calculation of the condition number κ. This means that the error magnification ratio is given closely by the condition number only when the input error ε is sufficiently small. Often, however, it is difficult to tell whether a specified accuracy in the inputs is small enough. Using an additional tool discussed in the next section, it is possible to determine rigorous bounds for the propagation error for a given bound on the input error.

1.5 Interval Arithmetic

Interval arithmetic is a kind of automatic, rigorous error analysis. It serves as an essential tool for mathematically rigorous computer-assisted proofs that involve finite precision calculations. The most conspicuous application is the recent solution by Hales [37] of Kepler's more than 300 year old conjecture that the face-centered cubic lattice is the densest packing of equal spheres in Euclidean three-dimensional space. For other highlights (which require, of course, additional machinery from the applications in question), see Eckmann, Koch, and Wittwer [23], Hass, Hutchings, and Schlafli [42], Mischaikow and Mrozek [62], and Neumaier and Rage [73]. Many algorithms of computational geometry depend on the correct evaluation of the sign of certain numbers for correct performance, and if these are tiny, the correct sign depends on rigorous error estimates (see, e.g., Mehlhorn and Näher [61]).

Independent of rounding issues, interval arithmetic is an important tool in global optimization because of its ability to provide global information over wide intervals, such as bounds on ranges or derivatives, or Lipschitz constants (see Hansen [39] and Kearfott [49]).

The propagation of errors is accounted for in interval arithmetic by taking as given not a particular inexact number, but an interval of machine numbers that contains this number, and critical computations are performed with intervals instead of approximate numbers. The operations of interval arithmetic are defined in such a way that the resulting interval always contains the true result that would be obtained by using exact inputs and exact calculations; by careful rounding, this property is assured, even when all calculations are done with finite precision only.

In the context of interval calculations, it is more natural to express errors as absolute errors, and we replace a number $\tilde{x}$, which has an absolute error $\leq r$, with the interval $[\tilde{x} - r, \tilde{x} + r]$. In the following, M is the space of real numbers or a space of real vectors or matrices, with component-wise inequalities.

1.5.1 Definition. The symbol

$$\mathbf{x} := [\underline{x}, \bar{x}] := \{\tilde{x} \in M \mid \underline{x} \leq \tilde{x} \leq \bar{x}\}$$

denotes a (closed and bounded) interval in M with *lower bound* $\underline{x} \in M$ and *upper bound* $\bar{x} \in M$, $\underline{x} \leq \bar{x}$, and

$$\mathbb{I}M := \{[\underline{x}, \bar{x}] \mid \underline{x}, \bar{x} \in M, \underline{x} \leq \bar{x}\}$$

denotes the set of intervals over M (of *interval vectors* if M is a space of vectors, of *interval matrices* if M is a space of matrices). The *midpoint* of $\mathbf{x}$,

$$\mathrm{mid}\,\mathbf{x} := \check{x} := \frac{1}{2}(\underline{x} + \bar{x})$$

and the *radius* of $\mathbf{x}$,

$$\mathrm{rad}\,\mathbf{x} := \frac{1}{2}(\bar{x} - \underline{x}) \geq 0,$$

allow one to convert the interval representation of an inaccurate number to the absolute error representation,

$$\tilde{x} \in \mathbf{x} \Longleftrightarrow |\tilde{x} - \check{x}| \leq \mathrm{rad}\,\mathbf{x}.$$

Intervals with zero radius (called *point intervals*) contain a single point only, and we always use the *identification*

$$[x, x] \equiv x$$

of point intervals with the element of M they contain.

$$|\mathbf{x}| := \sup\{|\tilde{x}| \mid \tilde{x} \in \mathbf{x}\} = \sup\{\bar{x}, -\underline{x}\}$$

defines the absolute value of $\mathbf{x}$. For a bounded subset S of M,

$$\square S := [\inf S, \sup S]$$

is called the *interval hull* of S; for example, $\square\{1, 2\} = \square\{2, 1\} = [1, 2]$. We also define the relation

$$\mathbf{x} \le \mathbf{y} :\Longleftrightarrow \bar{x} \le \underline{y}$$

on $\mathbb{I}M$, and other relations are defined in a similar way.

1.5.2 Remarks.

(i) For $\mathbf{x} \in \mathbb{IR}$, $|\mathbf{x}| = \max\{\bar{x}, -\underline{x}\}$. However, the maximum may be undefined in the componentwise order of $\mathbb{R}^n$, and, for interval vectors, the supremum takes the componentwise maximum. (MATLAB, however, uses max for the componentwise maximum.)

(ii) The relation $\le$ is not an order relation on $\mathbb{I}M$ because reflexivity fails.

We now extend the definition of operations and elementary functions to intervals over $M = \mathbb{R}$. For $\circ \in O$, we define

$$\mathbf{x} \circ \mathbf{y} := \square\{\tilde{x} \circ \tilde{y} \mid \tilde{x} \in \mathbf{x}, \tilde{y} \in \mathbf{y}\},$$

and for $\varphi \in J$, we define

$$\varphi(\mathbf{x}) := \square\{\varphi(\tilde{x}) \mid \tilde{x} \in \mathbf{x}\}$$

when the right side is defined; that is, excluding cases such as $[1, 2]/[0, 1]$ or $\sqrt{[-1, 1]}$. (There are extensions of interval arithmetic that even give values to such expressions; we do not discuss these here.) Thus, in both cases, the result of the operation is the tightest interval that contains all possible results with inaccurate operands selected from the corresponding interval operands.

1.5.3 Theorem.

(i) *For all intervals,*

$$-\mathbf{x} = [-\bar{x}, -\underline{x}].$$

(ii) *For $\circ \in \{+, -, *, /, \hat{\ }\}$, if $\tilde{x} \circ \tilde{y}$ is defined for all $\tilde{x} \in \mathbf{x}, \tilde{y} \in \mathbf{y}$, we have*

$$\mathbf{x} \circ \mathbf{y} = \square\{\underline{x} \circ \underline{y}, \underline{x} \circ \bar{y}, \bar{x} \circ \underline{y}, \bar{x} \circ \bar{y}\}.$$

In particular,

$$\mathbf{x} + \mathbf{y} = \square\{\underline{x} + \underline{y}, \bar{x} + \bar{y}\},$$
$$\mathbf{x} - \mathbf{y} = \square\{\underline{x} - \bar{y}, \bar{x} - \underline{y}\}.$$

(iii) For monotone $\varphi \in J$,

$$\varphi(\mathbf{x}) = \square\{\varphi(\underline{x}), \varphi(\bar{x})\}.$$

Proof. The proof follows immediately from the monotonicity of $\circ$ and φ. □

For nonmonotone elementary functions and for powers with even integral exponents, one must also look at the values of the local extrema within the interval; thus, for example,

$$\mathbf{x}^2 = \begin{cases} [\underline{x}^2, \bar{x}^2] & \text{if } \mathbf{x} \geq 0, \\ [\bar{x}^2, \underline{x}^2] & \text{if } \mathbf{x} \leq 0, \\ [0, \max\{\underline{x}^2, \bar{x}^2\}] & \text{if } 0 \in \mathbf{x}. \end{cases}$$

Note that the standard functions have well-known extrema, only finitely many in each interval, so that all operations and elementary functions can be calculated for intervals in finitely many steps. Thus we can get an enclosure valid simultaneously for all selections of inaccurate values from the intervals.

1.5.4 Remarks.

(i) Intervals are just a new type of numbers similar to complex numbers (but with different properties). However, it is often convenient to think of an interval as a single inaccurately known real number known to lie within that interval. In this context, it is useful to write narrow intervals in an abbreviated form, in which lower and upper bounds are written over each other and identical figures are given only once. For example,

$$17.4_{58548}^{63751} = [17.458548, 17.463751].$$

(ii) Let $f(x) := x - x$. Then for $\mathbf{x} = [1, 2]$, $f(\mathbf{x}) = [-1, 1]$, and for $\mathbf{x} := [1 - r, 1 + r]$, $f(\mathbf{x}) = [-2r, 2r]$. Interval arithmetic has no memory; it does not "notice" that in $\mathbf{x} - \mathbf{x}$ the same inaccurately known number occurs twice, and thus calculates the result as if coming from two different inaccurately known numbers both lying in $\mathbf{x}$.

(iii) Many rules for calculating with real numbers are no longer valid for intervals. As we have seen, we usually have

$$\mathbf{x} - \mathbf{x} \neq 0,$$

and another example is

$$\mathbf{a}(\mathbf{b} + \mathbf{c}) \neq \mathbf{ab} + \mathbf{ac},$$

except in special cases. So one must be careful in theoretical arguments involving interval arithmetic.

Basic books on interval arithmetic and associated algorithms are those by Moore [63] (introductory), Alefeld and Herzberger [4] (intermediate), and Neumaier [70] (advanced).

Outward Rounding

Let $\mathbf{x} = [\underline{x}, \bar{x}] \in \mathbb{IR}$. If $\underline{x}$ and $\bar{x}$ are not machine numbers, then they must be rounded. In what follows, let $\tilde{\mathbf{x}} = [\underline{\tilde{x}}, \tilde{\bar{x}}]$ be the rounded interval corresponding to $\mathbf{x}$. In order that all elements in $\mathbf{x}$ should also lie in $\tilde{\mathbf{x}}$, $\underline{x}$ must be rounded downward and $\bar{x}$ must be rounded upward; this is called *outward rounding*. Then, $\mathbf{x} \subseteq \tilde{\mathbf{x}}$. The outward rounding is called *optimal* when

$$\tilde{\mathbf{x}} := \cap \{\mathbf{y} = [\underline{y}, \bar{y}] \mid \mathbf{y} \supseteq \mathbf{x}, \ \underline{y}, \bar{y} \text{ are machine numbers}\}.$$

Note that for optimally rounded intervals, the bounds are usually *not* optimally rounded in the sense used for real numbers, but only correctly rounded. Indeed, the lower bound must be rounded to the next smaller machine number even when the closest machine number is larger, and similarly, the upper bound must be rounded to the next larger machine number. This *directed rounding*, which is necessary for optimal outward rounding, is available for the results of $+, -, *, /$ and the square root on all processors conforming to the IEEE standard for floating point arithmetic, defined in [45]; for the power and other elementary functions, the IEEE standard prescribes nothing and one may have to be content with a suboptimal outward rounding, which encloses the exact result, but not necessarily in an optimal way.

For the common programming languages, there are extensions in which rigorously outward rounded interval arithmetic is easily available: the (public domain) INTLAB toolbox [85] for MATLAB, the PASCAL extension PASCAL-XSC [52], the C extension C-XSC [51], and the FORTAN 90 modules FORTRAN-XSC [95] and (public domain) INTERVAL_ARITH-METIC [50].

The following examples illustrate interval operations and outward rounding; we use $B = 10$, $L = 2$, and optimal outward rounding.

$$[1.1, 1.2] + [-2.1, 0.2] = [-1.0, 1.4],$$
$$[1.1, 1.2] - [-2.1, 0.2] = [0.9, 3.3],$$
$$[1.1, 1.2] * [-2.1, 0.2] = [-2.52, 0.24], \text{ rounded: } [-2.6, 0.24],$$
$$[1.1, 1.2]/[-2.1, 0.2] \text{ not defined,}$$
$$[-2.1, 0.2]/[1.1, 1.2] = [-21/11, 2/11], \text{ rounded: } [-2.0, 0.19],$$
$$[-2.1, 0.2]/[1.1, 1000] = [-21/11, 2/11], \text{ rounded: } [-2.0, 0.19],$$
$$[-1.2, -1.1]^2 = [1.21, 1.44], \text{ rounded: } [1.2, 1.5],$$
$$[-2.1, 0.2]^2 = [0, 4.41], \text{ rounded: } [0, 4.5],$$
$$\sqrt{[1.0, 1.5]} = [1.0, 1.22\ldots], \text{ rounded: } [1.0, 1.3].$$

1.5.5 Proposition. *For optimal outward rounding and unbounded exponent range,*

$$\tilde{\mathbf{x}} \subseteq \mathbf{x} \cdot [1 - \varepsilon, 1 + \varepsilon]$$

where $\varepsilon = B^{1-L}$.

Proof. We have $\tilde{\mathbf{x}} = [\underline{\tilde{x}}, \overline{\tilde{x}}]$, in which $\underline{\tilde{x}} \leq \underline{x}$ and $\overline{\tilde{x}} \geq \bar{x}$ when correctly rounded. It follows from the correctness of the rounding that

$$|\underline{\tilde{x}} - \underline{x}| \leq \varepsilon|\underline{x}| \quad \text{and} \quad |\overline{\tilde{x}} - \bar{x}| \leq \varepsilon|\bar{x}|.$$

There are three cases to consider: $\mathbf{x} > 0$, $0 \in \mathbf{x}$, and $\mathbf{x} < 0$.

CASE 1: $\mathbf{x} > 0$ (i.e., $\underline{x} > 0$). By hypothesis, $\bar{x} \geq \underline{x} > 0$, and by outward rounding,

$$\underline{x} - \underline{\tilde{x}} \leq \varepsilon\underline{x} \quad \text{and} \quad \overline{\tilde{x}} - \bar{x} \leq \varepsilon\bar{x},$$

whence

$$\underline{\tilde{x}} \geq \underline{x}(1 - \varepsilon) \quad \text{and} \quad \overline{\tilde{x}} \leq \bar{x}(1 + \varepsilon).$$

So

$$\tilde{\mathbf{x}} = [\underline{\tilde{x}}, \overline{\tilde{x}}] \subseteq [\underline{x}(1 - \varepsilon), \bar{x}(1 + \varepsilon)] = [\underline{x}, \bar{x}][1 - \varepsilon, 1 + \varepsilon],$$

as asserted.

CASE 2: $0 \in \mathbf{x}$ (i.e., $\underline{x} \leq 0 \leq \bar{x}$). By hypothesis, $\underline{\tilde{x}} \leq \underline{x} \leq 0 \leq \bar{x} \leq \bar{\tilde{x}}$, whence we can evaluate the absolute values to obtain

$$-\underline{\tilde{x}} + \underline{x} \leq -\varepsilon \underline{x} \quad \text{and} \quad \bar{\tilde{x}} - \bar{x} \leq \varepsilon \bar{x},$$

giving

$$0 \geq \underline{\tilde{x}} \geq \underline{x}(1 + \varepsilon) \quad \text{and} \quad 0 \leq \bar{\tilde{x}} \leq \bar{x}(1 + \varepsilon).$$

So

$$\tilde{\mathbf{x}} = [\underline{\tilde{x}}, \bar{\tilde{x}}] \subseteq [\underline{x}(1 + \varepsilon), \bar{x}(1 + \varepsilon)] \subseteq [\underline{x}, \bar{x}][1 - \varepsilon, 1 + \varepsilon].$$

CASE 3: $(\mathbf{x} < 0$, i.e., $\bar{x} < 0)$ follows from Case 1 by multiplying with -1. $\quad \square$

Interval Evaluation of Expressions

In arithmetic expressions, one can replace the variables with intervals and evaluate the resulting expressions using interval arithmetic. Different expressions for the same function may produce different interval results; although this already holds for real arguments in finite precision arithmetic, it is much more pronounced here and persists even in exact interval arithmetic. The simplest example is the fact that $\mathbf{x} - \mathbf{x}$ is generally not zero, but only an enclosure of it, with a radius of the order of the radius of the operands. We now generalize this observation to the interval evaluation of arbitrary expressions.

1.5.6 Theorem. *Suppose that the arithmetic expression $f(z_1, \ldots, z_n) \in \mathcal{A}(z_1, \ldots, z_n)$ can be evaluated at $\mathbf{z}_1, \ldots, \mathbf{z}_n \in \mathbb{IR}$, and let*

$$\mathbf{x}_1 \subseteq \mathbf{z}_1, \ldots, \mathbf{x}_n \subseteq \mathbf{z}_n.$$

Then:

(i) f can be evaluated at $\mathbf{x}_1, \ldots, \mathbf{x}_n$ and

$$f(\mathbf{x}_1, \ldots, \mathbf{x}_n) \subseteq f(\mathbf{z}_1, \ldots, \mathbf{z}_n) \quad \text{(inclusion isotonicity)},$$

(ii)

$$\{f(\tilde{z}_1, \ldots, \tilde{z}_n) \mid \tilde{z}_i \in \mathbf{z}_i\} \subseteq f(\mathbf{z}_1, \ldots, \mathbf{z}_n) \quad \text{(range inclusion)}.$$

In (ii), equality holds if each variable occurs only once in f (as, e.g., the single variable z in the expression $\log(\sin(1 - z^3)))$.

(iii) Suppose that in the evaluation of $f(\mathbf{z}_1, \ldots, \mathbf{z}_n)$, the elementary functions $\varphi \in J$ and the power $\char`\^$ are evaluated only on intervals in which they are differentiable. Then, for small

$$r := \max_i \operatorname{rad}(\mathbf{x}_i),$$

we have

$$\operatorname{rad} f(\mathbf{x}_1, \ldots, \mathbf{x}_n) = O(r).$$

We refer to this asymptotic result by saying that naive interval evaluation has a *linear approximation order*.

Before we prove the theorem, we illustrate the statements with some examples.

1.5.7 Example. We mentioned previously that expressions equivalent for real arguments may behave differently for interval arguments. We look closer at the example $f(x) = x^2$, which always gives the range. Indeed, the square is an elementary operation, defined such that this holds. What happens with the equivalent expression $x * x$?

(i) If $f(x) := x * x$, and $\mathbf{x} := [-r, r]$, with $0 < r < 1$, then $f(\mathbf{x}) = [-r, r] \cdot [-r, r] = [-r^2, r^2]$, but the range of values of f is only $\{f(\tilde{x}) \mid \tilde{x} \in \mathbf{x}\} = [0, r^2]$. The reason for the overestimation of the range of values of f is as follows. In the multiplication $\mathbf{x} * \mathbf{x} = \square\{\tilde{x} * \tilde{\tilde{x}} \mid \tilde{x} \in \mathbf{x}, \tilde{\tilde{x}} \in \mathbf{x}\}$, each $\tilde{x}$ is multiplied with every element of $\mathbf{x}$, but in the calculation of the range of values, $\tilde{x}$ is multiplied only with itself, giving a tighter result. Again, this shows the memory-less nature of interval analysis.

(ii) If $f(x) := x * x$ and $\mathbf{x} := [\frac{1}{2} - r, \frac{1}{2} + r]$, with $0 \le r \le \frac{1}{2}$, then

$$f(\mathbf{x}) = \left[\left(\tfrac{1}{2} - r\right)^2, \left(\tfrac{1}{2} + r\right)^2\right] = \{f(\tilde{x}) \mid \tilde{x} \in \mathbf{x}\},$$

that is, the range of values of f is not overestimated, in spite of the fact that the condition from Theorem 1.5.6(ii) that each variable should occur only once is not satisfied (the condition is therefore sufficient, but not necessary). For the radius of $f(\mathbf{x})$, we have $\operatorname{rad} f(\mathbf{x}) = O(r)$ as asserted in (iii) of Theorem 1.5.6.

(iii) If $f(x) := \sqrt{x}$, $\mathbf{z} := [\varepsilon, 1]$ and $\mathbf{x} = [\varepsilon, \varepsilon + 2r]$ $(r \le (1 - \varepsilon)/2)$, then rad $\mathbf{x} = r$ and

$$\text{rad } \sqrt{\mathbf{x}} = \frac{1}{2}(\sqrt{\varepsilon + 2r} - \sqrt{\varepsilon}) = \frac{r}{\sqrt{\varepsilon + 2r} + \sqrt{\varepsilon}} \approx \frac{r}{2\sqrt{\varepsilon}} = O(r)$$

for $r \to 0$. However, the constant hidden in the Landau symbol depends on the choice of $\mathbf{z}$ and becomes unbounded as $\varepsilon \to 0$.

(iv) If $f(x) := \sqrt{x}$, $\mathbf{z} := [0, 1]$, and $\mathbf{x} := [\frac{1}{4}r, \frac{9}{4}r]$, with $0 < r < \frac{4}{9}$, then

$$\sqrt{\mathbf{x}} = \left[\frac{1}{2}\sqrt{r}, \frac{3}{2}\sqrt{r}\right] = \{f(\tilde{x}) \mid \tilde{x} \in \mathbf{x}\},$$

which is to be expected from (ii), and

$$\text{rad } \mathbf{x} = r, \quad \text{rad } \sqrt{\mathbf{x}} = \frac{1}{2}\sqrt{r} \neq O(r),$$

that is, the radius of $\sqrt{\mathbf{x}}$ is considerably greater than is expected from the radius of $\mathbf{x}$, leading to a loss of significant digits. This is due to the fact that $\sqrt{x}$ is not differentiable at $x = 0$; see statement (iii).

Proof of Theorem 1.5.6. Because arithmetic expressions are recursively defined, we proceed by induction on the number of subexpressions. Clearly, it suffices to show (a) that the theorem is valid for constants and variables, and (b) that if the theorem is valid for the expressions g and h, then it is also valid for $-g$, $g \circ h$, and $\varphi(g)$. Now (a) is clear; hence we assume that the theorem is valid for g and h in place of f.

We show that Theorem 1.5.6 is valid for $\varphi(g)$ with $\varphi \in J$. We combine the interval arguments to interval vectors

$$\mathbf{z} := (\mathbf{z}_1, \ldots, \mathbf{z}_n)^T \in \mathbb{IR}^n, \quad \mathbf{x} := (\mathbf{x}_1, \ldots, \mathbf{x}_n)^T \in \mathbb{IR}^n.$$

(i) By definition, and because φ is continuous and $g(\mathbf{z})$ is compact and connected,

$$f(\mathbf{z}) = \varphi(g(\mathbf{z}))$$
$$= \square\{\varphi(\tilde{g}) \mid \tilde{g} \in g(\mathbf{z})\}$$
$$= \{\varphi(\tilde{g}) \mid \tilde{g} \in g(\mathbf{z})\}.$$

In particular, $\varphi(\tilde{g})$ is defined for all $\tilde{g} \in g(\mathbf{z})$. Therefore, because by the inductive hypothesis $g(\mathbf{x}) \subseteq g(\mathbf{z})$, $f(\mathbf{x})$ is also defined, and $f(\mathbf{x}) = \varphi(g(\mathbf{x}))$.

Furthermore,

$$f(\mathbf{x}) = \{\varphi(\tilde{g}) \mid \tilde{g} \in g(\mathbf{x})\}$$
$$\subseteq \{\varphi(\tilde{g}) \mid \tilde{g} \in g(\mathbf{z})\} = f(\mathbf{z}).$$

(ii) For $\mathbf{x} = [\tilde{z}, \tilde{z}]$, (ii) follows immediately from (i). If each variable in f occurs only once, then the same holds for g, and by the inductive hypothesis, $\{g(\tilde{z}) \mid \tilde{z} \in \mathbf{z}\} = g(\mathbf{z})$. From this, it follows that

$$\{f(\tilde{z}) \mid \tilde{z} \in \mathbf{z}\} = \{\varphi(g(\tilde{z})) \mid \tilde{z} \in \mathbf{z}\}$$
$$= \{\varphi(\tilde{g}) \mid \tilde{g} \in g(\mathbf{z})\} = f(\mathbf{z}).$$

(iii) Set $\mathbf{g} := g(\mathbf{x})$. Because over the compact set $\mathbf{g}$, the continuous function φ attains its minimum and maximum, we have

$$f(\mathbf{x}) = \varphi(\mathbf{g}) = [\varphi(\tilde{g}_1), \varphi(\tilde{g}_2)]$$

for suitable $\tilde{g}_1, \tilde{g}_2 \in \mathbf{g}$. Thus by the mean value theorem,

$$\mathrm{rad}\, f(\mathbf{x}) = \frac{1}{2}(\varphi(\tilde{g}_2) - \varphi(\tilde{g}_1)) = \frac{1}{2}\varphi'(\xi)(\tilde{g}_2 - \tilde{g}_1)$$

for some $\xi \in \square\{\tilde{g}_1, \tilde{g}_2\} \subseteq \mathbf{g}$. Let

$$M = \sup_{\xi \in g(\mathbf{z})} |\varphi'(\xi)|.$$

Then

$$\mathrm{rad}\, f(\mathbf{x}) \leq \frac{1}{2}M|\tilde{g}_2 - \tilde{g}_1| \leq M\, \mathrm{rad}\, \mathbf{g} = O(r).$$

The proof for the case $f = g \circ h$ with $\circ \in O$ is similar and is left to the reader. $\square$

The Mean Value Form

The evaluation of $f(\mathbf{x}_1, \ldots, \mathbf{x}_n)$ gives, in general, only an enclosure for the range of values of f. The naive method, in which real numbers are replaced with intervals, often gives pessimistic bounds, especially for intervals of large radius. However, for many numerical problems, an appropriate reformulation of standard solution methods produces interval methods that provide bounds that are provably realistic for narrow intervals. For a thorough treatment, see

Neumaier [70]; in this book, we look at only a few basic techniques for achieving that.

For function evaluation, realistic bounds can be obtained for narrow input intervals using the mean value form, a rigorous version of the linearization method discussed in the previous section.

1.5.8 Theorem. *Let $f(x) = f(x_1, \ldots, x_n) \in \mathcal{A}(x_1, \ldots, x_n)$, and suppose that for the evaluation of $f'(x)$ in $\mathbf{z} \in \mathbb{IR}^n$, the elementary functions and the power are evaluated only on intervals in which they are differentiable. If $\mathbf{x} \subseteq \mathbf{z}$, then*

$$\mathbf{w}^* := \square\{f(\tilde{x}) \mid \tilde{x} \in \mathbf{x}\} \subseteq \mathbf{w} := f(\check{x}) + f'(\mathbf{x})(\mathbf{x} - \check{x}), \tag{5.1}$$

and the overestimation is bounded by

$$0 \leq \text{rad } \mathbf{w} - \text{rad } \mathbf{w}^* \leq 2 \, \text{rad } f'(\mathbf{x}) \, \text{rad } \mathbf{x}. \tag{5.2}$$

1.5.9 Remarks.

(i) Let $\tilde{x}$ be inexactly known with absolute error $\leq r$ and let $\mathbf{x} = [\tilde{x} - r, \tilde{x} + r]$. Statement (5.1) says that the range of values $\mathbf{w}^*$ as well as the required value $f(x)$ of f at x lie in the interval $\mathbf{w}$. Instead of only an approximate bound on the error in the calculation of $f(x)$ as $r \to 0$ (as in Section 1.4), we now have a strict bound on the range of values of f for any fixed value of the radius.

Statement (5.2) says how large the overestimation is. For $r := \max_i \text{rad } \mathbf{x}_i$ tending to zero, the overestimation is $O(r^2)$; that is, the same order of magnitude as the neglected terms in the calculation of the condition number. The order of approximation is therefore *quadratic* if we evaluate f at the midpoint $\check{x} = \text{mid } \mathbf{x}$ and correct it with $f'(\mathbf{x})(\mathbf{x} - \check{x})$. This is in contrast with the linear order of approximation for the interval evaluation $f(\mathbf{x})$.

(ii) The expression $f(\check{x}) + f'(\mathbf{x})(\mathbf{x} - \check{x})$ is called the *mean value form* of f; for a geometrical interpretation, see Figure 1.3. The number

$$q = \max\left(0, 1 - \frac{2 \, \text{rad } f'(\mathbf{x}) \, \text{rad } \mathbf{x}}{\text{rad } \mathbf{w}}\right)$$

is a computable *quality factor* for the enclosure given by the mean value form. Because one can rewrite (5.2) as

$$q \cdot \text{rad } \mathbf{w} \leq \text{rad } \mathbf{w}^* \leq \text{rad } \mathbf{w},$$

a q close to 1 shows that very little overestimation occurred in the

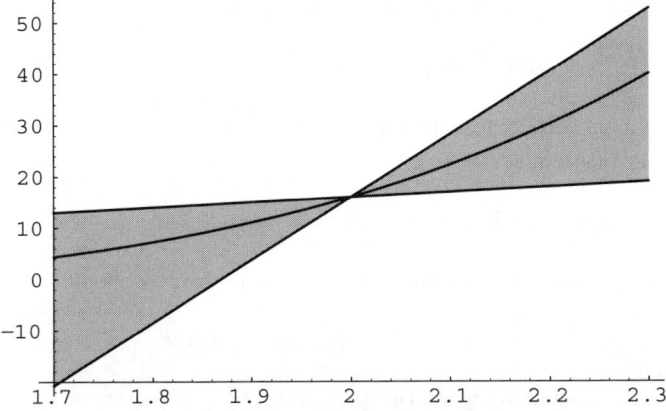

Figure 1.3. Enclosure by the mean value form.

computation of the range. Note that $q = 1 - O(r)$ approaches 1 as the radius of the arguments gets small (cf. Figure 1.4), unless the range $\mathbf{w}^*$ is only $O(r^2)$ (which may happen close to a local minimum).

(iii) For the evaluation of the mean value form in finite precision arithmetic, it is important that the rounding errors in the evaluation of $f(\check{x})$ are taken into account by calculating with a point interval $[\tilde{x}, \tilde{x}]$ instead of with $\check{x}$. However, $\tilde{x}$ itself need not be the exact midpoint of the interval because, as the proof shows, the theorem is valid for any $\tilde{x} \in \mathbf{x}$ instead of $\check{x}$.

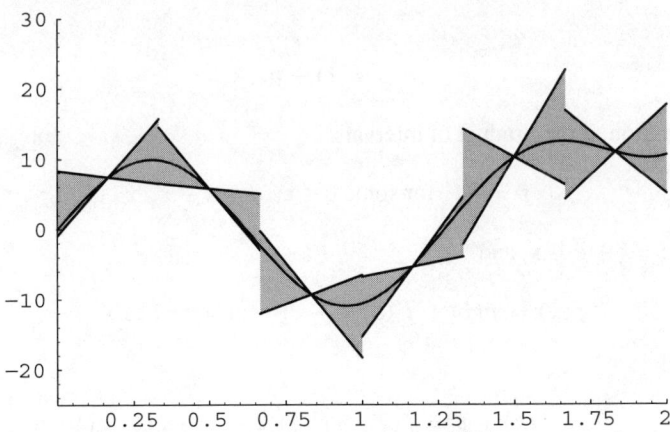

Figure 1.4. Enclosure of $f(x) = x^5 - 2x^3 + 10 \sin 5x$ over $[0, 2]$ with the mean value form.

Proof of Theorem 1.5.8. We choose $\tilde{x} \in \mathbf{x} \subseteq \mathbf{z}$ and we define

$$g(t) := f(\check{x} + t(\tilde{x} - \check{x})) \quad \text{for } t \in [0, 1];$$

$g(t)$ is well defined because the line $\check{x} + t(\tilde{x} - \check{x})$ lies in the interval $\mathbf{x}$. By the mean value theorem

$$f(\tilde{x}) = g(1) = g(0) + g'(\tau) = f(\check{x}) + f'(\xi)(\tilde{x} - \check{x})$$

with $\xi = \check{x} + \tau(\tilde{x} - \check{x})$ for some $\tau \in [0, 1]$. From this, we obtain

$$f(\tilde{x}) \in f(\check{x}) + f'(\mathbf{x})(\mathbf{x} - \check{x}) = \mathbf{w}$$

for all $\tilde{x} \in \mathbf{x}$, whence $\mathbf{w}^* \subseteq \mathbf{w}$. This proves (5.1).

For the overestimation bound (5.2), we first give a heuristic argument. We have

$$f(\tilde{x}) = f(\check{x}) + f'(\xi)(\tilde{x} - \check{x}) \in f(\check{x}) + f'(\mathbf{x})(\mathbf{x} - \check{x}).$$

If one assumes that ξ and $\tilde{x}$ range through the whole interval $\mathbf{x}$, then because of the overestimate of $\tilde{x}$ by $\mathbf{x}$ and of $f'(\xi)$ by $f'(\mathbf{x})$ of $O(r)$, one obtains an overestimate of $O(r^2)$ for the product.

A rigorous proof is a little more technical, and we treat only the one-dimensional case; the general case is similar, but requires more detailed arguments. We write

$$\mathbf{c} := f'(\mathbf{x}), \quad \mathbf{p} := \mathbf{c}(\mathbf{x} - \check{x})$$

so that

$$\mathbf{w} = f(\check{x}) + \mathbf{p}.$$

By definition of the product of intervals,

$$\bar{p} = \tilde{c}\tilde{d} \quad \text{for some } \tilde{c} \in \mathbf{c}, \ \tilde{d} \in \mathbf{x} - \check{x}.$$

Now $\tilde{x} := \check{x} + \tilde{d} \in \mathbf{x}$, and

$$f(\tilde{x}) = f(\check{x}) + f'(\xi)(\tilde{x} - \check{x}) \quad \text{for some } \xi \in \mathbf{x}.$$

Hence

$$\begin{aligned}
\bar{w} &= f(\check{x}) + \bar{p} = f(\tilde{x}) - f'(\xi)(\tilde{x} - \check{x}) + \tilde{c}(\tilde{x} - \check{x}) \\
&= f(\tilde{x}) + (\tilde{c} - f'(\xi))(\tilde{x} - \check{x}) \\
&\in f(\tilde{x}) + (\mathbf{c} - \mathbf{c})(\mathbf{x} - \check{x}).
\end{aligned}$$

Because $f(\tilde{x}) \in \mathbf{w}^*$, we conclude

$$\bar{w} \le \bar{w}^* + (\bar{c} - \underline{c}) \operatorname{rad} \mathbf{x}.$$

Similarly,

$$\underline{w} \ge \underline{w}^* - (\bar{c} - \underline{c}) \operatorname{rad} \mathbf{x},$$

and therefore

$$\begin{aligned}
2 \operatorname{rad} \mathbf{w} &= \bar{w} - \underline{w} \\
&\le \bar{w}^* - \underline{w}^* + 2(\bar{c} - \underline{c}) \operatorname{rad} \mathbf{x} \\
&= 2 \operatorname{rad} \mathbf{w}^* + 4 \operatorname{rad} \mathbf{c} \operatorname{rad} \mathbf{x}.
\end{aligned}$$

This proves the inequality in (5.2), and the asymptotic bound $O(r^2)$ follows from Theorem 1.5.6(iii). □

1.5.10 Examples. We show that the mean value form gives rigorous bounds on the propagation error for inaccurate input data that are realistic when the inaccuracies are small.

(i) We continue the discussion of Example 1.4.2(ii), where we obtained for $f(x) = x^5 - 2x^3$ and $|\tilde{x} - 2| \le 10^{-3}$ the approximate estimate $|f(\tilde{x}) - 16| \approx 0.056$. Because the absolute error of $\tilde{x}$ is $r = 10^{-3}$, an inaccurate $\tilde{x}$ lies in the interval $[2 - r, 2 + r]$. For the calculations, we use finite precision interval arithmetic with $B = 10$ and $L = 5$ and optimal outward rounding.

(a) We bound the error using the mean value form, with $f'(x) = 5x^4 - 6x^2$. From

$$\begin{aligned}
\mathbf{w} &= f(2) + f'([2 - r, 2 + r])[-r, r] \\
&= 16 + [55.815, 56.189][-0.001, 0.001] \\
&= [15.943, 16.057],
\end{aligned}$$

we obtain for the absolute error

$$|f(\tilde{x}) - f(2)| \le 0.057.$$

These are safe bounds, realistic because of Example 1.4.2(ii).

(b) To show the superiority of the mean value form, we also bound the error by means of naive interval evaluation. Here,

$$\mathbf{w}_0 = f([2 - r, 2 + r]) = [15.896, 16.105]$$

whence

$$|f(\tilde{x}) - f(2)| \leq 0.105.$$

These are safe bounds, but there is overestimation by a factor of about 2.

(c) For the same example, we use naive interval arithmetic on the equivalent expression $h(x) = x^3(x^2 - 2)$:

$$\mathbf{w}_1 = h([2 - r, 2 + r]) = [15.944, 16.057]$$

whence

$$|f(\tilde{x}) - f(2)| \leq 0.057.$$

The result here is the same as for the mean value form (and with higher precision calculation it would be even sharper than the latter); but this is due to special circumstances. In all operations, lower bounds are combined with lower bounds only, and upper bounds with upper bounds only. It is not difficult to see that in such a situation, always the exact range is obtained. (For intervals around 1, we no longer have this nice property.)

(ii) Let $f(x) := x^2$. Then $f'(x) = 2x$. For $\mathbf{x} = [-r, r]$, we have $\mathbf{w}^* = [0, r^2]$ and

$$\begin{aligned}
\mathbf{w} &= f(0) + f'([-r, r])([-r, r] - 0) \\
&= 2[-r, r][-r, r] \\
&= 2[-r^2, r^2].
\end{aligned}$$

From this we obtain, for the radii of $\mathbf{w}^*$ and $\mathbf{w}$, rad $\mathbf{w}^* = r^2/2$ and rad $\mathbf{w} = 2r^2$; that is, the mean value form gives rise to an overestimation of the radius of $\mathbf{w}^*$ by a factor of 4. The cause of this is that the range of values $\mathbf{w}^*$ is itself of size $O(r^2)$.

(iii) Let $f(x) := 1/x$. Then $f'(x) = -1/x^2$. For the wide interval $\mathbf{x} = [1, 2]$ (where the asymptotic result of Theorem 1.5.8 is no longer relevant), one has $\mathbf{w}^* = [\frac{1}{2}, 1]$, and

$$\begin{aligned}
\mathbf{w} &= f(1.5) + f'([1, 2])([1, 2] - 1.5) \\
&= \frac{1}{1.5} + \frac{1}{[1, 4]}[-0.5, 0.5] = \left[\frac{1}{6}, \frac{7}{6}\right].
\end{aligned}$$

In this case, the interval evaluation provides better results, namely

$$\mathbf{w}_0 = f([1, 2]) = \frac{1}{[1, 2]} = \left[\tfrac{1}{2}, 1\right],$$

which, by Theorem 1.5.6, is optimal.

When r is not small, it is still true that the mean value form and the interval evaluation provide strict error bounds, but it is no longer possible to make general statements about the size of the overestimation. There may be significant overestimation in the mean value form, and it is not uncommon that for wide intervals, the naive interval evaluation gives better enclosures than the mean value form; a striking example of this is $f(x) = \sin x$.

When the intervals are too wide, both the naive interval evaluation and the mean value form may even break down completely. Indeed, overestimation in intermediate results may lead to a forbidden operation, as in the following example.

1.5.11 Example. The harmless function $f(x) = 1/(1 - x + x^2)$ cannot be evaluated at $\mathbf{x} = [0, 1]$. The evaluation of the denominator (whose range is $[0.75, 1]$) gives the enclosure $1 - [0, 1] + [0, 1] = [0, 2]$, and the subsequent division is undefined. The derivative has the same problem, so that the mean value form is undefined, too. Using as denominator $(x - 1) * x + 1$ reduces the enclosure to $[0, 1]$, but the problem persists.

In this particular example, one can remedy the problem by rewriting f as $f(x) = 1/(0.75 + (x - 0.5)^2)$ and even gets the exact range (by Theorem 1.5.6(ii)), but in more sophisticated examples of the same kind, there is no easy way out.

1.6 Exercises

1. (a) Use MATLAB to plot the functions defined by the following three arithmetical expressions.
    (i) $f(x) := \frac{1}{1+2x} - \frac{1-x}{1+x}$ for $-1 < x \le 1$ and for $|x| \le 10^{-15}$,
    (ii) $f(x) := \sqrt{x + 1/x} - \sqrt{x - 1/x}$ for $1 \le x \le 10$ and for $2 \cdot 10^7 \le x \le 2 \cdot 10^8$,
    (iii) $f(x) := (\tan x - \sin x)/x$ for $0 < x \le 1$ and for $10^{-8} \le x \le 10^{-7}$.
    Observe the strong inaccuracy for the second choices.
   (b) Use MATLAB to print, for the above three functions, the sentences:

```
The answer for f(x) accurate to 6 leading digits is y
The short format answer for f(x) is
y
The long format answer for f(x) is
y
```

but with x and y replaced by the numerical values $x = 0.1111$ and $y = f(0.1111)$. (You must switch the printing mode for the results; use the MAT-LAB functions disp, num2str, format. See also fprintf and sprintf for more flexible printing commands.)

2. Using differential numbers, evaluate the following functions and their derivatives at the points $x_1 = 2$ and $x_2 = 5$. Give the intermediate results for (a) and (b) to (at least) three decimal places. For (c), evaluate as a check an expression for $f'(x)$ at x_1 and x_2. (You may work by hand, using a pocket calculator, or use the automatic differentiation facilities of INTLAB.)

(a) $f(x) := \frac{e^x}{1+x^2}$.

(b) $f(x) := \frac{e^x(x+1)}{\sqrt{x}}$.

(c) $f(x) := \frac{(x^2+3)(x-1)}{x^2(x^2+4)} + \frac{x-1}{x+1}$.

3. Suppose that the polynomial

$$f(x) = \sum_{i=0:n} a_i x^{n-i}$$

is evaluated at the point z using the Horner scheme

$$f_0 := a_0,$$
$$f_i := f_{i-1}z + a_i \quad (i = 1, \ldots, n),$$
$$f(z) := f_n.$$

Show that

$$\sum_{i=0:n-1} f_i x^{n-i-1} = \begin{cases} f'(z) & \text{if } x = z, \\ \frac{f(x)-f(z)}{x-z} & \text{otherwise.} \end{cases}$$

4. Let

$$f(x) = a_0 x^n + a_1 x^{n-1} + \cdots + a_n$$

be a polynomial of degree n. The *complete Horner scheme* is given by

$$f_i^{(0)} := a_i \quad (i = 0, \ldots, n),$$
$$f_j^{(j)} := a_0 \quad (j = 0, \ldots, n),$$
$$f_i^{(j)} := f_{i-1}^{(j)}z + f_{i-1}^{(j-1)} \quad (i = j+1, \ldots, n+1).$$

Show that the polynomials

$$g_j(x) := \sum_{i=j:n} f_i^{(j)} x^{n-i} \quad (j = 0, \ldots, n)$$

satisfy
(a) $g_j(x) = g_j(z) + g_{j+1}(x)(x - z)(j = 0, \ldots, n - 1)$,
(b) $g_0(x) = \sum_{k<j} g_k(z)(x - z)^k + g_j(x)(x - z)^j (j = 1, \ldots, n)$,
(c) $f_{n+1}^{(j+1)} = g_j(z) = \frac{f^{(j)}(z)}{j!} \quad (j = 1, \ldots, n)$.
Hint: For (a), use the preceding exercise; for (c), use the Taylor expansion of $f(x)$.
The relevance of the complete Horner scheme lies in (c), which shows that the scheme may serve to compute scaled higher derivatives $f^{(j)}(z)/j!$ of polynomials.
5. (a) A binary fraction of the form

$$r := (0.a_1 a_2 \ldots a_k)_2$$

with $a_\nu \in \{0, 1\}$ $(\nu = 1, \ldots, k)$ may be represented as

$$r = \sum_{\nu=0:k} a_\nu 0.5^\nu$$

with $a_0 = 0$. Transform the binary fraction

$$(0.101 \; 100 \; 011)_2$$

into a decimal number using the Horner scheme.
(b) Using the complete Horner scheme (see the previous exercise), calculate, for the polynomial

$$p(x) = 4x^4 - 8x^2 + 5x - 1,$$

the values $p(x)$, $p'(x)$, and $p''(x)$ for $x = 2.5$ and for $x = 4/3$. (Use MATLAB or hand calculation. For hand calculation, devise a scheme to arrange the intermediate results so that performing and checking the calculations is easy.)
6. Assuming an unbounded exponent range, prove that, on a computer with an L-digit mantissa and base B, the relative error for optimal rounding is bounded by $\varepsilon = \frac{1}{2} B^{1-L}$.
7. Let $w_0, x \in \mathbb{R}$ with $w_0, x > 0$ and for $i = 0, 1, 2, \ldots$

$$\bar{w}_i := x / w_i^{n-1},$$

$$w_{i+1} := ((n - 1)w_i + \bar{w}_i)/n.$$

(a) Show that, for $w_0 \neq \sqrt[n]{x}$,

$$\bar{w}_1 < \bar{w}_2 < \cdots < \bar{w}_i < \cdots < \sqrt[n]{x} < \cdots < w_i < \cdots < w_2 < w_1$$

and that the sequences $\{w_i\}$ and $\{\bar{w}_i\}$ converge to $w := \sqrt[n]{x}$.

(b) For $x := 4$, $w_0 := 2$, $n := 3, 4$ calculate, in each case, $w_i, \bar{w}_i$ $(i \leq 7)$ and print the values of i, w_i, and $\bar{w}_i$ in long format.

(c) For $x := 4$, $w_0 := 2$, $n := 25, 50, 75, 100$, calculate, in each case, $w_i, \bar{w}_i$ $(i \leq n)$ and print the values of i, w_i, and $\bar{w}_i$ in long format. Use the results to argue that the iteration is unsuitable for the practical calculation of $\sqrt[n]{x}$ for large n.

8. (a) Analyze the behavior of the quotient $f_d(\alpha)/f_s$ of the derivative $f_s = f'(x)$ and the forward difference quotient $f_d(\alpha) = (f(x+\alpha) - f(x))/\alpha$ in exact arithmetic, and in finite precision arithmetic. What do you expect for $\alpha \to 0$? What happens actually?

(b) Write a MATLAB program that tries to discover mistakes in the programming of derivatives. This can be done by computing a table of quotients $f_d(\alpha)/f_s$ for a number of values of $\alpha = 10^{-i}$; say, $i = -10:10$, and checking whether the behavior predicted in (a) actually occurs. Test with some correct and some incorrect pairs of expressions for function and derivative.

(c) How can one reduce the multidimensional case (checking for errors in the gradient) to the previous situation?

9. (How not to calculate e^x.) The value e^x for given $x \in \mathbb{R}$ is to be estimated from the series expansion

$$e^x = \sum_{\nu=0:\infty} \frac{x^\nu}{\nu!}.$$

One calculates, for this purpose, the sums

$$s_n(x) := \sum_{\nu=0:n} \frac{x^\nu}{\nu!}$$

for $n = 0, 1, 2, \ldots$. Obtain a value of n such that $|s_n(x) - e^x| < 10^{-16}$ with $x = -20$ using a remainder estimate. Calculate $u = s_n(-20)$ with MATLAB and compare the values so obtained with the optimally rounded value obtained with the MATLAB command

```
v=exp(-20); disp(v-u);
```

How is the bad result to be explained?

10. (a) Rearrange the arithmetic expressions from Exercise 1 so that the small-est possible loss of precision results if $x \ll 1$ for (i), $x \gg 1$ for (ii), and $0 < |x| \ll 1$ for (iii).

(b) Calculate $f(x)$ from (i) with $x := 10^{-3}$, $B = 10$, and $L = 1, 2, 4, 8$ by hand, and compare the resulting values with those obtained from the rearranged expression.

11. The solutions x_1 and x_2 of a quadratic equation of the form

$$ax^2 + bx + c = 0$$

with $a \neq 0$ are given by

$$x_{1,2} = (-b \pm \sqrt{b^2 - 4ac})/2a. \tag{6.1}$$

Determine, from (6.1), the smaller solution of the equation

$$x = (1 - \alpha x)^2, \quad (\alpha > 0).$$

Why is the formula obtained unstable for small values of α? Derive a bet-ter formula and find five distinct values of α that make the difference in stability between the two formulas evident.

12. Find a stable formula for both solutions of a quadratic equation with com-plex coefficients.

13. *Cardano's formulas.*

(a) Prove that all cubic equations

$$x^3 + ax^2 + bx + c = 0$$

with real coefficients a, b, c may be reduced by the substitution $x = z - s$ to the form

$$z^3 + 3qz - 2r = 0. \tag{6.2}$$

In the following assume $q^3 + r^2 > 0$.

(b) Using the formula $z := p - q/p$, derive a quadratic equation in p^3, and deduce from this an expression for a real solution of (6.2).

(c) Show that there is exactly one real solution. What are the complex solutions?

(d) Show that both solutions of the quadratic equation in (b) give the same real solution.

(e) Determine the real solution in part (b) for $r := \pm 1000$ and $q := 0.01$, choosing for p in turn both solutions of the quadratic equation. For which one is the expression more numerically stable?

(f) For $|r| \ll |q|$, the real solution is small compared with q; its calculation from the difference $p - q/p$ is therefore numerically unstable. Find a stable version.

(g) What happens for $q^3 + r^2 < 0$?

14. The standard deviation $\sigma_n = \sqrt{t_n/n}$ can be calculated for given data x_i ($i = 1, \ldots, n$) either with

$$t_n = \left(\sum_{i=1:n} x_i^2 \right) - \frac{1}{n} \left(\sum_{i=1:n} x_i \right)^2$$

or with the recursion in Section 1.3. Calculate σ_{201} for the values

$$x_i := (499999899 + i)/3000 \quad (i = 1, \ldots, 201)$$

using both methods and compare the two results with the exact value obtained by analytic summation.

Hint: Use $\sum_{i=1:n} i = n(n + 1)/2$ and $\sum_{i=1:n} i^2 = n(n + 1)(2n + 1)/6$.

15. A spring with a suspended mass m oscillates about its position of rest, Hooke's law $K(t) = -D \cdot x(t)$ is assumed. When the force $K_a(t) := A \cdot \cos(\omega_a t)$ acts on the mass m, the equation of motion is

$$\frac{d^2}{dt^2} x(t) + \omega_0^2 x(t) = \frac{A}{m} \cos(\omega_a t)$$

with $\omega_0^2 := D/m$. It has the solution

$$x(t) = -\frac{A/m}{\omega_a^2 - \omega_0^2} \cos(\omega_a t).$$

For which ω_a is this formula ill-conditioned ($t > 0$ and $\omega_0 > 0$ fixed)?

16. The midpoint $\check{x}$ of an interval $\mathbf{x} \in \mathbb{IR}$ can be calculated from

$$\check{x} := (\bar{x} + \underline{x})/2 \tag{6.3}$$

or from

$$\check{x} := \underline{x} + (\bar{x} - \underline{x})/2. \tag{6.4}$$

These formulas give an in general inexact value $\check{x}$ on an L-digit machine with base B and optimal rounding.

(a) For $B = 2$,

$$\check{x} \in \mathbf{x} \tag{6.5}$$

for all $\mathbf{x} \in \mathbb{IR}$ whose end points are machine numbers. Prove this for at least one of the formulas (6.3) or (6.4).

(b) Suppose that $B = 10$. For which formula is (6.5) false? Give a counter-example, and prove that (6.5) holds for the other formula.

(c) Explain purpose and details of the MATLAB code

```
x=input('enter positive x>');
w=max(1,x);wold=2*w;
while w<wold
    wold=w; w=w+(x/w-w)*0.5;
end;
```

What is the limit of w in exact arithmetic? Why does the loop terminate in finite precision arithmetic after finitely many steps?

17. (a) Find intervals $\mathbf{x}, \mathbf{y}, \mathbf{z} \in \mathbb{IR}$ for which the distributive law

$$\mathbf{x} \cdot (\mathbf{y} + \mathbf{z}) = \mathbf{x} \cdot \mathbf{y} + \mathbf{x} \cdot \mathbf{z}$$

is violated.

(b) Show that, for all $\mathbf{x}, \mathbf{y}, \mathbf{z} \in \mathbb{IR}$, the *subdistributive law*

$$\mathbf{x} \cdot (\mathbf{y} + \mathbf{z}) \subseteq \mathbf{x} \cdot \mathbf{y} + \mathbf{x} \cdot \mathbf{z}$$

holds.

18. A camera has a lens with focal length $f = 20$ mm and a film of thickness $2\Delta b$. The distance between the lens and the object is g, and the distance between the lens and the film is b_0. All objects at a distance g are focused sharply if the lens equation

$$\frac{1}{f} = \frac{1}{b} + \frac{1}{g}$$

is satisfied for some $b \in \mathbf{b} = [b_0 - \Delta b, b_0 + \Delta b]$. From the lens equation, one obtains, for a given f and $b := b_0 \pm \Delta b$,

$$g = \frac{1}{1/f - 1/b} = \frac{b \cdot f}{b - f}.$$

A perturbation calculation gives

$$g \approx g(b_0) \pm g'(b_0) \cdot \Delta b.$$

Similarly, one obtains, for a given f and g with $|g - g_0| \leq \Delta g$,

$$b = \frac{1}{1/f - 1/g} = \frac{g \cdot f}{g - f}.$$

(a) How thick must the film be in order that an object at a given distance is sharply focused? For $\mathbf{g} = [3650, 3652], [3650, 4400], [6000, 10000], [10000, 20000]$ (in millimeters), calculate $b_0 = \breve{b}$ and $\Delta b = $ rad $\mathbf{b}$ from $\mathbf{b}$ using the formulas

$$\mathbf{b}_1 = \frac{1}{1/f - 1/\mathbf{g}} \quad \text{and} \quad \mathbf{b}_2 = \frac{\mathbf{g} \cdot f}{\mathbf{g} - f}.$$

Explain the bad results for $\mathbf{b}_2$, and the good results for $\mathbf{b}_1$.

(b) At which distances are objects sharply focused, with a given b_0, for a film thickness of 0.02 mm? Calculate the interval $\mathbf{g}$ using the formulas

$$\mathbf{g}_1 := \frac{1}{1/f - 1/\mathbf{b}} \quad \text{and} \quad \mathbf{g}_2 := \frac{\mathbf{b} \cdot f}{\mathbf{b} - f}$$

with $b_0 = 20.1, 20.05, 20.03, 20.02, 20.01$ mm, and using the approximation

$$\mathbf{g}_3 := g(b_0) + [-1, 1] \cdot g'(b_0) \cdot \Delta b.$$

Compare the results from $\mathbf{g}_1, \mathbf{g}_2$, and $\mathbf{g}_3$. Which of the three results is increasingly in error for decreasing b_0?

19. Write a program for bounding the range of values of $p(x) := (1 - x)^6$ in the interval

$$\mathbf{x} := 1.5 + [-1, 1]/10^i \quad (i = 1, \ldots, 8)$$

(a) by evaluating $p(x)$ on the interval $\mathbf{x}$,

(b) by using the Horner scheme at $\mathbf{z} = \mathbf{x}$ on the equivalent expression

$$p_0(x) = x^6 - 6x^5 + 15x^4 - 20x^3 + 15x^2 - 6x + 1,$$

(c) by using Horner's scheme for the derivative and the mean value form

$$p_0(\breve{x}) + p_0'(\mathbf{x})(\mathbf{x} - \breve{x}).$$

Interpret the results. (You are allowed to use unrounded interval arithmetic if you cannot access directed rounding.)

2

Linear Systems of Equations

In this chapter, we discuss the solution of systems of n linear equations in n variables. At some stage, most of the more advanced problems in scientific calculations require the solution of linear systems; often these systems are very large and their solution is the most time-consuming part of the computation. Therefore, the efficient solution of linear systems and the analysis of the quality of the solution and its dependence on data and rounding errors is one of the central topics in numerical analysis. A fairly complete reference compendium of numerical linear algebra is Golub and van Loan [31]; see also Demmel [17] and Stewart [89]. An exhaustive source for overdetermined linear systems (that we treat in passing only) is Björck [8].

We begin in Section 2.1 by introducing the triangular factorization as a compact form of the well-known Gaussian elimination method for solving linear systems, and show in Section 2.2 how to exploit symmetry with the LDL^T and the Cholesky factorization to reduce work and storage costs. We only touch the savings possible for sparse matrices with many zeros by looking at the simple banded case.

In Section 2.3, we show how to avoid numerical instability by means of pivoting. In order to assess the closeness of a matrix to singularity and the worst case sensitivity to input errors, we look in Section 2.4 at norms and in Section 2.5 at condition numbers. Section 2.6 discusses iterative refinement as a method to increase the accuracy of the computed solution and shows that except for very ill-conditioned systems, one can indeed expect from iterative refinement solutions of nearly optimal accuracy. Finally, Section 2.7 discusses various techniques for a realistic error analysis of the solution of linear systems, including interval techniques for the rigorous enclosure of the solution set of linear systems with uncertain coefficients.

The solution of linear systems of equations lies at the heart of modern computational practice. Therefore, a lot of effort has gone into the development of efficient and reliable implementations of appropriate solution methods. For matrices of moderate size, the state of the art is embodied in the LAPACK [6] subroutine collection. LAPACK routines are freely available from NETLIB [67], an electronic distribution service for public domain numerical software.

If nothing else is said, everything in this chapter remains valid if the set $\mathbb{C}$ of complex numbers is replaced with the set $\mathbb{R}$ of real numbers.

We use the notations $\mathbb{R}^{m \times n}$ and $\mathbb{C}^{m \times n}$ to denote the space of real and complex matrices, respectively, with m rows and n columns; A_{ik} denotes the (i, k)-entry of A. Absolute values and inequalities between vectors or between matrices are interpreted componentwise.

The symbol A^T denotes the *transposed matrix* with

$$(A^T)_{ik} = A_{ki},$$

and A^H denotes the *conjugate transposed matrix* with

$$(A^H)_{ik} = \bar{A}_{ki}.$$

Here, $\bar{\alpha} = a - ib$ denotes the complex conjugate of a complex number $\alpha = a + ib(a, b \in \mathbb{R})$, and not the upper bound of an interval. It is also convenient to write

$$A^{-H} := (A^H)^{-1} = (A^{-1})^H$$

and

$$A^{-T} := (A^T)^{-1} = (A^{-1})^T.$$

A matrix A is called *symmetric* if $A^T = A$, and *Hermitian* if $A^H = A$. For $A \in \mathbb{R}^{m \times n}$ we have $A^H = A^T$, and then symmetry and Hermiticity are synonymous.

The ith row of A is denoted by $A_{i:}$, and the kth column by $A_{:k}$. The matrix

$$I = \begin{pmatrix} 1 & & 0 \\ & \ddots & \\ 0 & & 1 \end{pmatrix},$$

with entries $I_{ii} = 1$ and $I_{ik} = 0$ $(i \neq k)$, is called the *unit matrix*; its size is usually inferred from the context. The columns of I are the *unit vectors* $e^{(k)} = I_{:k}$ $(i = 1, \ldots, n)$ of the space $\mathbb{C}^n$. If A is a nonsingular matrix, then $A^{-1}A = AA^{-1} = I$. The symbols J and e denote matrices and column vectors, all of whose entries are 1.

In MATLAB, the conjugate transpose of a matrix A is denoted by A' (and the transpose by A.'); the ith row and kth column of A are obtained as A(i,:)

and A(:,k), respectively. The dimensions of an $m \times n$ matrix A are found from [m,n]=size(A). The matrices 0 and J of size $m \times n$ are created by zeros(m,n) and ones(m,n); the unit matrix of dimension n by eye(n). Operations on matrices are written in MATLAB as A+B, A-B, A*B ($= A\ B$), inv(A) ($= A^{-1}$), A\B ($= A^{-1}B$), and A/B ($= AB^{-1}$). The componentwise product and quotient of (vectors and) matrices are written as A.*B and A./B, respectively. However, in our pseudo-MATLAB algorithms, we continue to use the notation introduced before for the sake of readability.

2.1 Gaussian Elimination

Suppose that n equations in n unknowns with (real or) complex coefficents are given,

$$A_{11}x_1 + A_{12}x_2 + \cdots + A_{1n}x_n = b_1,$$
$$A_{21}x_1 + A_{22}x_2 + \cdots + A_{2n}x_n = b_2,$$
$$\cdots$$
$$A_{n1}x_1 + A_{n2}x_2 + \cdots + A_{nn}x_n = b_n,$$

in which $A_{ik}, b_i \in \mathbb{C}$ for $i, k = 1, \ldots, n$. We can write this linear system in compact matrix notation as $Ax = b$ with $A = (A_{ik}) \in \mathbb{C}^{n \times n}$ and $x = (x_i)$, $b = (b_i) \in \mathbb{C}^n$. The following theorem is known from linear algebra.

2.1.1 Theorem. *Suppose that $A \in \mathbb{C}^{n \times n}$ and that $b \in \mathbb{C}^n$. The system of equations $Ax = b$ has a unique solution $x \in \mathbb{C}^n$ if and only if A is nonsingular (i.e., if $\det A \neq 0$, equivalently, if $\mathrm{rank}\, A = n$). In this case, the solution x is given by $x = A^{-1}b$. Moreover, the ith component of the solution is given by* Cramer's rule

$$x_i = \frac{\det A^{(i)}}{\det A} = \frac{\sum_{k=1:n}(\mathrm{adj}\, A)_{ik}b_k}{\det A}, \quad i = 1, \ldots, n,$$

where the matrix $A^{(i)}$ is obtained from A by replacing the ith column with b, and the adjoint matrix adj A is formed from suitably signed subdeterminants of A.

From a theoretical point of view, this theorem tells everything about the solution. From a practical point of view, however, solving linear systems of equations is a complex and diverse subject; the actual solution process is influenced by many considerations involving speed, storage, accuracy, and structure. In particular, we see that the numerical calculation of the solution using A^{-1} or Cramer's rule cannot be recommended for $n > 3$. The computational cost in both cases is considerably greater than is necessary; moreover, Cramer's rule is numerically less stable than the other methods.

Solution methods for linear systems fall into two categories: *direct methods*, which provide the answer with finitely many operations; and *iterative methods*, which provide better and better approximations with increasing work. For systems with low or moderate dimensions and for large systems with a band structure, the most efficient algorithms for solving linear systems are direct, generally variants of Gaussian elimination, and we restrict our discussion to these. For sparse systems, see, for example, Duff et al. [22] (direct methods) and Weiss [97] (iterative methods).

Gaussian elimination is a systematic reduction process of general linear systems to those with a triangular coefficient matrix. Although Gaussian elimination currently is generally organized in the form of a triangular factorization, more or less as described as follows, the basic idea is already known from school.

2.1.2 Example. In order to solve a system of equations such as the following:

$$x + y + z = 1,$$
$$2x + 4y + z = 3,$$
$$-x + y - 3z = 5,$$

one begins by rearranging the system such that the coefficients of x in all of the equations save one (e.g., the first) are zero. We achieve this, for example, by subtracting twice the first row from the second row, and adding the first row to the third. We obtain the new equations

$$2y - z = 1,$$
$$2y - 2z = 6.$$

Subtracting the first of these from the second gives

$$-z = 5.$$

It remains to solve a system of equations of triangular form:

$$x + y + z = 1,$$
$$2y - z = 1$$
$$-z = 5$$

and we obtain in reverse order $z = -5$, $y = -2$, $x = 8$.

The purpose of this elimination method is to transform the coefficient matrix A to a matrix R in which only the coefficients in the upper right triangle are distinct from zero, and to solve, instead of $Ax = b$, the triangular system $Rx = y$, where y is the transformed right side. Because the latter is achieved very easily, this is a very useful strategy.

For the compact form of Gaussian elimination, we need two types of triangular matrices. An *upper triangular matrix* has the form

$$
R = \begin{pmatrix}
\times & \times & \times & \times & \times \\
 & \times & \times & \times & \times \\
 & & \times & \times & \times \\
 & & & \times & \times \\
 & & & & \times
\end{pmatrix},
$$

where possible nonzeros, indicated by crosses, appear only in the upper right triangle, that is, where $R_{ik} = 0$ $(i > k)$. (The letter R stands for *right*; there is also a tradition of using the letter U for upper triangular matrices, coding *upper*.) A *lower triangular matrix* has the form

$$
L = \begin{pmatrix}
\times & & & & \\
\times & \times & & & \\
\times & \times & \times & & \\
\times & \times & \times & \times & \\
\times & \times & \times & \times & \times
\end{pmatrix},
$$

where possible nonzeros appear only in the lower left triangle, that is, where $L_{ik} = 0$ $(i < k)$. (The letter L stands for *left* or *lower*.)

A matrix that is both lower and upper triangular, that is, that has the form

$$
D = \begin{pmatrix}
D_{11} & & 0 \\
 & \ddots & \\
0 & & D_{nn}
\end{pmatrix} =: \mathrm{Diag}(D_{11}, \ldots, D_{nn}),
$$

is called a *diagonal matrix*. For a diagonal matrix D, we have $D_{ik} = 0$ $(i \neq k)$.

Triangular Systems of Equations

For an upper triangular matrix $R \in \mathbb{R}^{n \times n}$, we have $Rx = y$ if and only if

$$
y_i = \sum_{k=1:n} R_{ik} x_k = R_{ii} x_i + \sum_{k=i+1:n} R_{ik} x_k.
$$

The following pseudo-MATLAB algorithm solves $Rx = y$ by substituting the already calculated components x_k $(k > i)$ into the ith equation and solving for x_i, for $i = n, n - 1, \ldots, 1$. It also checks whether one of the divisions is forbidden, and then leaves the loop. (In MATLAB, == denotes the logical comparison for equality.) This happens iff one of the R_{ii} vanishes, and because (by induction and development of the determinant by the first column)

$$\det R = R_{11} R_{22} \cdots R_{nn},$$

this is the case iff R is singular.

2.1.3 Algorithm: Solve an Upper Triangular System (slow)

$ier = 0;$
for $i = n : -1 : 1,$
    if $R_{ii} == 0, \quad ier = 1; \quad$ break; end;
    $x_i = (y_i - \sum_{k=i+1:n} R_{ik} x_k)/R_{ii};$
end;
% $ier = 0$: R is nonsingular and $x = R^{-1} y$
% $ier = 1$: R is singular

It is possible to speed up this program by vectorizing the sum using MATLAB's subarray facilities. To display this, we switch from pseudo-MATLAB to true MATLAB.

2.1.4 Algorithm: Solve an Upper Triangular System

```
ier=0;
x=zeros(n,1); % ensures that x will be a column
for i=n:-1:1,
    if R(i,i)==0,  ier=1; break; end;
    x(i)=(y(i)-R(i,i+1:n)*x(i+1:n))/R(i,i);
end;
% ier=0: R is nonsingular and x=R\y
% ier=1: R is singular
```

In the future, we shall frequently use the summation symbol for the sake of visual clarity; it is understood that in programming, such sums are to be converted to vectorized statements whenever possible.

For a lower triangular matrix $L \in \mathbb{R}^{n \times n}$, we have $Ly = b$ if and only if

$$b_i = \sum_{k=1:n} L_{ik} y_k = L_{ii} y_i + \sum_{k=1:i-1} L_{ik} y_k.$$

The following algorithm solves $Ly = b$ by substituting the already calculated components y_k ($k < i$) into the ith equation and solving for y_i for $i = 1, \ldots, n$.

2.1.5 Algorithm: Solve a Lower Triangular System

$ier = 0$;

for $i = 1 : n$,

    if $L(i, i) == 0$, $ier = 1$; break; end;

    $y(i) = (b(i) - L(i, 1 : i - 1) * y(1 : i - 1))/L(i, i)$;

end;

% $ier = 0$: L is nonsingular and $y = L^{-1}b$

% $ier = 1$: L is singular

In case the diagonal entries satisfy $L_{ii} = 1$ for all i, L is called a *unit lower triangular matrix*, and the control structure simplifies because L is automatically nonsingular and no divisions are needed.

Solving a triangular system of linear equations by either forward or back substitution requires $\sum_{i=1:n}(2i - 1) = n^2$ operations (and a few less for a unit diagonal). Note that MATLAB recognizes whether a matrix A is triangular; if it is, it calculates A\b in the previously mentioned cheap way.

The Triangular Factorization

A factorization $A = LR$ into the product of a nonsingular lower triangular matrix L and an upper triangular matrix R is called a *triangular factorization* (or *LR-factorization* or *LU-factorization*) of A; it is called *normalized* if $L_{ii} = 1$ for $i = 1, \ldots, n$. The normalization is no substantial restriction because if L is nonsingular, then $LR = (LD^{-1})(DR)$ is a normalized factorization for the choice $D = \text{Diag}(L)$, where

$$\text{Diag}(A) := \text{Diag}(A_{11}, \ldots, A_{nn}).$$

denotes the *diagonal part* of a matrix $A \in \mathbb{C}^{n \times n}$. If a triangular factorization exists (we show later how to achieve this by *pivoting*), one can solve $Ax = b$ (and also get a simple formula for the determinant $\det A = \det L \cdot \det R$) as follows.

2.1.6 Algorithm: Solve $Ax = b$ without Pivoting

STEP 1: Calculate a normalized triangular factorization LR of A.

STEP 2: Solve $Ly = b$ by forward elimination.

STEP 3: Solve $Rx = y$ by back substitution.

Then $Ax = LRx = Ly = b$ and $\det A = R_{11} R_{22} \cdots R_{nn}$.

This algorithm cannot always work because not every matrix has a triangular factorization.

2.1.7 Example. Let

$$A := \begin{pmatrix} 0 & 1 \\ 1 & 0 \end{pmatrix}$$

$$\overset{?}{=} \begin{pmatrix} L_{11} & 0 \\ L_{21} & L_{22} \end{pmatrix} \begin{pmatrix} R_{11} & R_{12} \\ 0 & R_{22} \end{pmatrix}$$

$$= \begin{pmatrix} L_{11}R_{11} & L_{11}R_{12} \\ L_{21}R_{11} & L_{21}R_{12} + L_{22}R_{22} \end{pmatrix}.$$

If a triangular factorization $A = LR$ existed, then $L_{11}R_{11} = 0$, $L_{11}R_{12} \neq 0$, and $L_{21}R_{11} \neq 0$. This is, however, impossible.

The following theorem gives a complete and constructive answer to the question when a triangular factorization exists. However, this does not yet guarantee numerical stability, and in Section 2.3, we incorporate pivoting steps to improve the stability properties.

2.1.8 Theorem. *A nonsingular matrix $A \in \mathbb{C}^{n \times n}$ has at most one normalized triangular factorization. The triangular factorization exists iff all leading square submatrices $A^{(m)}$ of A, defined by*

$$A^{(m)} := \begin{pmatrix} A_{11} & \dots & A_{1m} \\ \vdots & \ddots & \vdots \\ A_{m1} & \dots & A_{mm} \end{pmatrix} \quad (m = 1, \dots, n),$$

are nonsingular. In this case, the triangular factorization can be calculated recursively from

$$\left. \begin{aligned} R_{ik} &:= A_{ik} - \sum_{j=1:i-1} L_{ij}R_{jk} & (k \geq i), \\ L_{ki} &:= \left(A_{ki} - \sum_{j=1:i-1} L_{kj}R_{ji} \right) \Big/ R_{ii} & (k > i), \end{aligned} \right\} \quad i = 1, \dots, n. \quad (1.1)$$

Proof.

(i) Suppose that a triangular factorization exists. We show that the equations (1.1) are satisfied. Because $A = LR$ is nonsingular, $0 \neq \det A = \det L \cdot \det R = R_{11} \cdots R_{nn}$, so $R_{ii} \neq 0$ $(i = 1, \dots, n)$. Because $L_{ii} = 1$,

$L_{ik} = 0$ for $k > i$ and $R_{ik} = 0$ for $k < i$, it follows that

$$A_{ik} = \sum_{j=1:n} L_{ij} R_{jk} = \sum_{j=1:\min(i,k)} L_{ij} R_{jk} \quad (i, k = 1, \dots, n). \qquad (1.2)$$

Therefore,

$$A_{ik} = \begin{cases} \sum_{j=1:i-1} L_{ij} R_{jk} + R_{ik} & \text{for } k \geq i, \\ \sum_{j=1:k-1} L_{ij} R_{jk} + L_{ik} R_{kk} & \text{for } k < i. \end{cases}$$

Solving the first equation for R_{ik} and the second for L_{ik} gives equations
(1.1). Note that we swapped the indices of L so that there is one outer loop
only.

(ii) By (i), A has a nonsingular triangular factorization iff (1.2) holds with L_{ii},
$R_{ii} \neq 0$ $(i = 1, \dots, n)$. This implies that for all m, $A^{(m)} = L^{(m)} R^{(m)}$ with
$L_{ii}^{(m)} = L_{ii} \neq 0$ $(i = 1, \dots, m)$ and $R_{ii}^{(m)} = R_{ii} \neq 0$ $(i = 1, \dots, m)$.
Therefore, $A^{(m)}$ is nonsingular for $m = 1, \dots, n$. Conversely, if this holds,
then a simple induction argument shows that (1.2) holds, and $L_{ii} = L_{ii}^{(i)} \neq 0$, $R_{ii} = R_{ii}^{(i)} \neq 0$ $(i = 1, \dots, n)$.

(iii) All of the unknown coefficients $R_{ik}(i \leq k)$ and $L_{ki}(i < k)$ can be calculated recursively from (1.1) and are thereby uniquely determined. The
factorization is unique because by (i) each factorization satisfies (1.1).

$\square$

For a few classes of matrices arising frequently in practice, the triangular
factorization always exists. A square matrix A is called an *H-matrix* if there are
diagonal matrices D and D' such that $\|I - DAD'\|_\infty < 1$ (see Section 2.4 for
properties of matrix norms).

We recall that a square matrix A is called *positive definite* if $x^H A x > 0$ for
all $x \neq 0$. In particular, $Ax \neq 0$ if $x \neq 0$; i.e., positive definite matrices are
nonsingular. The matrix A is called *positive semidefinite* if $x^H A x \geq 0$ for all x.
Here, the statements $\alpha > 0$ or $\alpha \geq 0$ include the statement that α is real. Note
that a nonsymmetric matrix A is positive (semi)definite iff its *Hermitian part*
$A_{sym} := \frac{1}{2}(A + A^H)$ has this property.

2.1.9 Corollary. *If A is positive definite or is an H-matrix, then the triangular
factorization of A exists.*

Proof. We check that the submatrices $A^{(i)}$ $(1 \leq i \leq n)$ are nonsingular.

(i) Suppose that $A \in \mathbb{C}^{n \times n}$ is positive definite. For $x \in \mathbb{C}^i$ $(1 \leq i \leq n)$ with
$x \neq 0$, we define a vector $y \in \mathbb{C}^n$ according to $y_j = x_j$ $(1 \leq j \leq i)$ and

$y_j = 0$ ($i < j \leq n$). Then, $x^H A^{(i)} x = y^H A y > 0$. The submatrices $A^{(i)}$ ($1 \leq i \leq n$) are consequently positive definite, and are therefore non-singular.

(ii) Suppose that $A \in \mathbb{C}^{n \times n}$ is an H-matrix; without loss of generality, let A be scaled so that $\|I - A\|_\infty < 1$. Because $\|I - A^{(i)}\|_\infty \leq \|I - A\|_\infty < 1$, each $A^{(i)}$ ($1 \leq i \leq n$) is an H-matrix and is therefore nonsingular.   □

The recursion (1.1) for the calculation of the triangular factorization goes back to Crout. Several alternatives for it exist that compute exactly the same intermediate results but in different orders that may have (dis)advantages on different machines. They are discussed, for example, in Golub and van Loan [31]; there, one can also find a proof of the equivalence of the factorization approach and the reduction process to upper triangular form, outlined in the introductory example.

In anticipation of the need for incorporation of row interchanges when no triangular factorization exists, the following arrangement for the ith recursion step of (1.1) is sensible. One divides the ith step into two stages. In stage (a), R_{ii} is found, and the analogous expressions $A_{ki} - \sum L_{kj} R_{ji}$ ($k = i + 1, \ldots, n$) in the numerators of L_{ik} are calculated as well. In stage (b), the calculation of the R_{ik} ($k = i + 1, \ldots, n$) is done according to (1.1), and the calculation of the L_{ki} ($k = i + 1, \ldots, n$) is completed by dividing the expressions calculated in stage (a) by R_{ii}. This method of calculating the coefficients R_{ii} has the advantage that, when $R_{ii} = 0$, this is immediately known, and unnecessary additional calculations are avoided.

The coefficients A_{ik} ($k = i, \ldots, n$) and A_{ki} ($k = i + 1, \ldots, n$) in (1.1) can be replaced with the corresponding elements of R and L, respectively, because they are no longer needed. Thus, we need no additional storage locations for the matrices R and L. The matrix elements stored in the array A at the conclusion of the ith step correspond to the entries of L, R, and the original A, according to the diagram in Figure 2.1. After n steps, A is completely overwritten with the

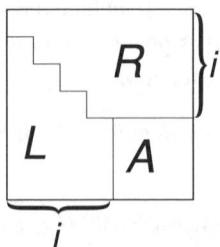

Figure 2.1. The matrices L, R, and A after the ith step.

elements of R and L. The calculation of the triangular factorization therefore proceeds as follows.

2.1.10 Algorithm: Triangular Factorization without Pivoting

for $i = 1 : n$
    for $k = i : n$
        % find R_{ii} and numerators of L_{ki}, $k > i$
$$A_{ki} = A_{ki} - \sum_{j=1:i-1} A_{kj} A_{ji};$$
    end
    if $A_{ii} == 0$, error('no factorization exists'); end;
    for $k = i + 1 : n$
        % find R_{ik}, $k > i$
$$A_{ik} = A_{ik} - \sum_{j=1:i-1} A_{ij} A_{jk};$$
        % complete L_{ki}, $k > i$
$$A_{ki} = A_{ki} / A_{ii};$$
    end
end

The triangular factorization LR of A is obtainable as follows from the elements stored in A after execution of the algorithm:

$$L_{ik} = \begin{cases} A_{ik} & \text{for } k < i, \\ 1 & \text{for } k = i, \\ 0 & \text{for } k > i, \end{cases} \qquad R_{ik} = \begin{cases} A_{ik} & \text{for } k \geq i, \\ 0 & \text{for } k < i. \end{cases}$$

2.1.11 Example. Let the matrix A be given by

$$A = \begin{pmatrix} 2 & 4 & -4 & 2 \\ 4 & 2 & 1 & 1 \\ -4 & 1 & -1 & 3 \\ 2 & 1 & 3 & 1 \end{pmatrix}.$$

STEP 1: In the first stage (a), nothing happens; in the second stage (b), nothing is subtracted from the first row above the diagonal, and the first column is divided by $A_{11} = 2$ below the diagonal. In particular, the first row of R is equal to the first row of A. In the following diagrams, coefficients that result from the subtraction of inner products are marked S and coefficients that result from division are marked with D.

$$
\text{Step 1:} \quad
\overset{\text{(a)}}{\begin{pmatrix}
2^S & 4 & -4 & 2 \\
4^S & 2 & 1 & 1 \\
-4^S & 1 & -1 & 3 \\
2^S & 1 & 3 & 1
\end{pmatrix}}
\qquad
\overset{\text{(b)}}{\begin{pmatrix}
2 & 4^S & -4^S & 2^S \\
2^D & 2 & 1 & 1 \\
-2^D & 1 & -1 & 3 \\
1^D & 1 & 3 & 1
\end{pmatrix}}
$$

STEPS 2 and 3: In stage (a), an inner product is subtracted from the second and third columns, both on and below the diagonal; in stage (b), an inner product is subtracted from the second and third rows above the diagonal, and the second and third columns are divided below the diagonal by the diagonal element.

$$
\text{Step 2:} \quad
\overset{\text{(a)}}{\begin{pmatrix}
2 & 4 & -4 & 2 \\
2 & -6^S & 1 & 1 \\
-2 & 9^S & -1 & 3 \\
1 & -3^S & 3 & 1
\end{pmatrix}}
\qquad
\overset{\text{(b)}}{\begin{pmatrix}
2 & 4 & -4 & 2 \\
2 & -6 & 9^S & -3^S \\
-2 & -\frac{3}{2}^D & -1 & 3 \\
1 & \frac{1}{2}^D & 3 & 1
\end{pmatrix}}
$$

$$
\text{Step 3:} \quad
\overset{\text{(a)}}{\begin{pmatrix}
2 & 4 & -4 & 2 \\
2 & -6 & 9 & -3 \\
-2 & -\frac{3}{2} & \frac{9}{2}^S & 3 \\
1 & \frac{1}{2} & \frac{5}{2}^S & 1
\end{pmatrix}}
\qquad
\overset{\text{(b)}}{\begin{pmatrix}
2 & 4 & -4 & 2 \\
2 & -6 & 9 & -3 \\
-2 & -\frac{3}{2} & \frac{9}{2} & \frac{5}{2}^S \\
1 & \frac{1}{2} & \frac{5}{9}^D & 1
\end{pmatrix}}
$$

STEP 4: In the last step, only A_{nn} remains to be calculated in accordance with (a).

$$
\text{Step 4:} \quad
\overset{\text{(a)}}{\begin{pmatrix}
2 & 4 & -4 & 2 \\
2 & -6 & 9 & -3 \\
-2 & -\frac{3}{2} & \frac{9}{2} & \frac{5}{2} \\
1 & \frac{1}{2} & \frac{5}{9} & -\frac{8}{9}^S
\end{pmatrix}}
$$

The triangular factorization of $A = LR$ has now been carried out; we have

$$
L = \begin{pmatrix}
1 & 0 & 0 & 0 \\
2 & 1 & 0 & 0 \\
-2 & -\frac{1}{3} & 1 & 0 \\
1 & \frac{1}{2} & \frac{5}{9} & 1
\end{pmatrix}
$$

and

$$R = \begin{pmatrix} 2 & 4 & -4 & 2 \\ 0 & -6 & 9 & -3 \\ 0 & 0 & \frac{9}{2} & \frac{5}{2} \\ 0 & 0 & 0 & -\frac{8}{9} \end{pmatrix}.$$

Note that the symmetry of A is reflected in the course of the calculation. Each column below the diagonal, calculated in stage (a), is identical with the row above the diagonal, calculated in stage (b). This implies that for symmetric matrices the computational cost can be nearly halved (cf. Algorithm 2.2.4).

Finding the triangular factorization of a matrix $A \in \mathbb{R}^{n \times n}$ takes

$$\sum_{i=1:n} (i - 1)4(n - i) = \frac{2}{3}n^3 + O(n^2)$$

operations.

Because of the possible division by zero or a small number, the solution of linear systems with a triangular factorization may not be possible or may be numerically unstable. Later, we show that by $O(n^2)$ further operations for implicit scaling and column pivoting, the numerical stability can be improved significantly. The asymptotic cost is unaffected.

Once a factorization of A is available, solving a linear system with any right side takes only $2n^2$ operations because we must solve two triangular systems of equations; thus, the bulk of the work is done in the factorization. It is a very gratifying fact that for the solution of many linear systems with the same coefficient matrix, only the cheap part of the computation must be done repeatedly.

A small number of operations is not the only measure of efficiency of algorithms. Especially for larger problems, further gains can be achieved by making use of *vectorization* or *parallelization* capabilities of modern computers. In many cases, this requires a rearrangement of the order of computations that allow bigger pieces of the data to be handled in a uniform fashion or that reduce the amount of communication between fast and slow storage media. A thorough discussion of these points is given in Golub and van Loan [31].

2.2 Variations on a Theme

For special classes of linear systems, Gaussian elimination can be speeded up by exploiting the structure of the coefficient matrices.

Band Matrices

A $(2m + 1)$-*band matrix* is a square matrix A with $A_{ik} = 0$ for $|i - k| > m$; in the special case $m = 1$, such a matrix is called *tridiagonal*. When Gaussian elimination is applied to band matrices, the band form is preserved. It is therefore sufficient for calculation on a computer to store and calculate only the elements of the bands (of which there are less than $2mn + n$) instead of all the n^2 elements.

Gaussian elimination (without pivot search) for a system of linear equations with a $(2m + 1)$-band matrix A consists of the following two algorithms.

2.2.1 Algorithm: Factorization of $(2m + 1)$-Band Matrices

$ier = 0;$
for $i = 1 : n,$
  $p = \max(i - m, 1);$
  $q = \min(i + m, n);$
  $A_{ii} = A_{ii} - \sum\limits_{j=p:i-1} A_{ij} A_{ji};$
  if $A_{ii} = 0, ier = 1;$ break; end;
  for $k = i + 1 : q,$
    $A_{ki} = A_{ki} - \sum\limits_{j=p:i-1} A_{kj} A_{ji};$
  end;
  for $k = i + 1 : q,$
    $A_{ik} = A_{ik} - \sum\limits_{j=p:i-1} A_{ij} A_{jk};$
    $A_{ki} = A_{ki} / A_{ii};$
  end;
end;

2.2.2 Algorithm: Banded Forward and Back Solve

$ier = 0;$
for $i = 1 : n,$
  $p = \max(i - m, 1);$
  $y_i = b_i - \sum\limits_{j=p:i-1} L_{ij} y_j;$
end;
for $i = n : -1 : 1,$
  if $R_{ii} = 0, ier = 1;$ break; end;
  $q = \min(i + m, n);$
  $x_i = (y_i - \sum\limits_{j=i+1:q} R_{ij} x_j) / R_{ii}.$
end;

The factorization of a $(2m + 1)$-band matrices takes $\frac{2}{3}m^2n + O(mn)$ operations. For small m, the total cost is $O(n)$, the same order of magnitude as that for solving a factored band system. Hence, the advantage of reusing a factorization to solve a second system with the same matrix is not as big as for dense systems.

For the discretization of boundary value problems with ordinary differential equations, we always obtain band matrices with small bandwidth m. For discretized partial differential equations the bandwidth is generally large, but most of the entries within the band vanish. To exploit this, one must resort to more general methods for (large and) *sparse matrices*, that is, matrices containing a high proportion of zeros among its entries (see, e.g., Duff, Erisman and Reid [22]. To explore MATLAB's sparse matrix facilities, start with `help sparfun`.

Symmetry

As seen in Example 2.1.11, the symmetry of a matrix is reflected in its triangular factorization LR. For this reason, it is sufficient to calculate only the lower triangular matrix L.

2.2.3 Proposition. *If the Hermitian matrix A has a nonsingular normalized triangular factorization $A = LR$, then $A = LDL^H$ where $D = \mathrm{Diag}(R)$ is a real diagonal matrix.*

Proof. Let $D := \mathrm{Diag}(R)$. Then

$$LR = A = A^H = R^H L^H = (R^H (D^H)^{-1})(D^H L^H).$$

Because of the uniqueness of the normalized triangular factorization, it follows from this that $R = D^H L^H$ and $\mathrm{Diag}(R) = D^H$. Hence, $D = D^H$, so that D is real and $A = LR = LDL^H$, as asserted. $\square$

The following variant of Algorithm 2.1.10 for real symmetric matrices (so that $LDL^H = LDL^T$) exploits symmetry, using only $\frac{1}{3}n^3 + O(n^2)$ operations, half as much as the nonsymmetric version.

2.2.4 Algorithm: Modified LDL^T Factorization of a real Symmetric Matrix A

$ier = 0;\ \delta = \mathrm{sqrt(eps)} * \mathrm{norm}(A, \mathrm{inf});$
for $i = 1 : n,$

$$piv = A_{ii} - \sum_{j=1:i-1} A_{ij} A_{ji};$$

if $\text{abs}(piv) > \delta, A_{ii} = piv;$

else $ier = 1; A_{ii} = \delta;$

    end

    for $k = i + 1 : n,$

$$A_{ik} = A_{ik} - \sum_{j=1:i-1} A_{ij} A_{jk};$$

$$A_{ki} = A_{ik}/A_{ii};$$

    end

end

%Now A contains the nontrivial elements of L, D and DL^T

% if ier > 0, iterative refinement is advisable

Note that, as for the standard triangular factorization, the LDL^H factorization can be numerically unstable when a divisor becomes tiny, and it fails to exist when a division by zero occurs. We simply replaced tiny pivots by a small threshold value (using the machine accuracy eps in MATLAB) – small enough to cause little perturbation, and large enough to eliminate stability problems. For full accuracy in this case, it is advisable to improve the solutions of linear systems by iterative refinement (see Algorithm 2.6.1). More careful remedies (diagonal pivoting and the use of 2×2 diagonal blocks in D) are possible (see Golub and van Loan [31]).

So far, the uniqueness of the triangular factorization has been attained through the normalization $L_{ii} = 1 (i = 1, \ldots, n)$. However, for Hermitian positive definite matrices, the factors can also be normalized so that $A = LL^H$. This is called a *Cholesky factorization* of A (the Ch is pronounced as a K, cf. [96]), and L is referred to as the *Cholesky factor* of A. The conjugate transpose $R = L^H$ is also referred to as the *Cholesky factor*; then the Cholesky factorization takes the form $A = R^H R$. (MATLAB's chol uses this version.)

2.2.5 Theorem. *If A is Hermitian positive definite, then A has a unique Cholesky factorization $A = LL^H$ in which the diagonal elements L_{ii} $(i = 1, \ldots, n)$ are real and positive.*

Proof. For $x \in \mathbb{C}^i$ $(1 \le i \le n)$ with $x \ne 0$, we define a vector $y \in \mathbb{C}^n$ by $y_j = x_j$ $(1 \le j \le i)$ and $y_j = 0$ $(i < j \le n)$. Then, $x^H A^{(i)} x = y^H A y > 0$. The submatrices $A^{(i)} (1 \le i \le n)$ are consequently positive definite and are therefore nonsingular. Corollary 2.1.9 now implies that there exists a triangular factorization $A = L_0 R_0 = L_0 D L_0^H$. Because A is positive definite, we have for each diagonal entry $D_{ii} = (e^{(i)})^H D e^{(i)} = x^H A x > 0$ for $x = L_0^{-H} e^{(i)}$. Thus we can take square roots and obtain, with $D^{1/2} := \text{Diag}(\sqrt{D_{ii}})$, $L := L_0 D^{1/2}$, the relation $A = L_0 D L_0^H = L_0 D^{1/2}(D^{1/2})^H L_0^H = LL^H$. $\qquad\square$

If A is not positive definite, then it turns out that the square root of some pivot element $p \leq 0$ must be determined, resulting in a failure. However, it is sometimes meaningful in this case to construct instead a so-called *modified Cholesky factorization*. This is obtained by replacing such pivot elements and, to achieve numerical stability, all $p < \delta$, by a small positive number δ. A suitable value is $\delta = \sqrt{\varepsilon}\|A\|$, where ε is the machine accuracy. (In various implementations of modified Cholesky factorizations, the choice of this threshold differs and may vary from row to row. For details of a method especially adapted to optimization, see Schnabel and Eskow [86].) The following algorithm results directly from the equation $A = LL^T$ for real, symmetric A; by careful indexing, it is ensured that the modified Cholesky factorization is overwritten over the upper triangular part of A, and the strict lower triangle remains unchanged.

2.2.6 Algorithm: Modified Cholesky Factorization of a Real Symmetric Matrix A

$\delta = \varepsilon\|A\|$;
if $\delta == 0$, $def = -1$; return; end;
$def = 1$;
for $i = 1 : n$,
$\qquad \Delta = A_{ii} - \displaystyle\sum_{j=1:i-1} A_{ij}^2$;
$\qquad$ if $\Delta < \delta$, $def = 0$; $\Delta = \delta$; end;
$\qquad A_{ii} = \sqrt{\Delta}$;
$\qquad$ for $k = i + 1 : n$,
$$A_{ik} = \left(A_{ik} - \sum_{j=1:i-1} A_{ji} A_{jk} \right) \Big/ A_{ii};$$
$\qquad$ end;
end;
% $def = 1$: A is positive definite; no modification applied
% $def = 0$: A is numerically not positive definite
% $def = -1$: A is the zero matrix

It can be shown that the Cholesky factorization is always numerically stable (see, e.g., Higham [44]).

The computational cost of the Cholesky factorization is essentially (i.e., apart from some square roots) the same as that for the LDL^T factorization. Because in Algorithm 2.2.6 only the upper triangle of A is used and changed, one could halve the storage location requirement by storing the strict upper triangle of A in a vector of length $n(n-1)/2$ and the diagonal of A (or its inverse) in a vector of length n, and adapting the corresponding index calculations. Alternatively, one can keep a copy of the diagonal of the original A in a vector a; then, one has both A and its Cholesky factor available for later use. This is useful for iterative refinement (see Section 2.6).

Normal Equations

An overdetermined linear system, with more equations than variables, is solvable only in special cases. However, approximation problems frequently lead to the problem of finding a vector x such that $Ax \approx b$ for some vector b, where the dimension of b is larger than that of x.

A typical case is that the data vector is supposed to be generated by a stochastic model $y = Ax + \varepsilon$, where ε is a noise vector with well-defined statistical properties. If the components of ε are uncorrelated, random variables with mean zero and constant variance, the *Gauss-Markov theorem* (see, e.g., Björck [8]) asserts that the best linear unbiased estimator (BLUE) for the parameter vector x can be found by minimizing the 2-norm $\|y - Ax\|_2$ of the residual. Because then

$$\|y - Ax\|_2^2 = \|\varepsilon\|_2^2 = \sum_i \varepsilon_i^2$$

is also minimal, this way of finding a vector x such that $Ax \approx y$ is called the *method of least squares*, and any vector x minimizing $\|y - Ax\|_2$ is called a *least squares solution* of $Ax \approx y$. The least squares solution can be found by solving the so-called *normal equations* (2.1).

2.2.7 Theorem. *Suppose that $A \in \mathbb{C}^{m \times n}$, $b \in \mathbb{C}^m$, and $A^H A$ is nonsingular. Then $\|b - Ax\|_2$ assumes its minimum value exactly at the solution x^* of the normal equations*

$$A^H A x^* = A^H b. \tag{2.1}$$

Proof. We put $r := b - Ax$ and $r^* := b - Ax^*$. Then (2.1) implies that $A^H r^* = 0$, whence also $(r^* - r)^H r^* = (A(x - x^*))^H r^* = (x - x^*)^H A^H r^* = 0$. It follows from this that

$$\|r\|_2^2 = r^H r + r^{*H}(r^* - r) = r^{*H} r^* - (r^* - r)^H r$$
$$= r^{*H} r^* + (r^* - r)^H(r^* - r) = \|r^*\|_2^2 + \|r^* - r\|_2^2.$$

Therefore, $\|r\|_2^2 \geq \|r^*\|_2^2$, and equality holds precisely when $r^* = r$. In this case, however,

$$A^H A(x - x^*) = A^H(r - r^*) = 0;$$

and because of the regularity of $A^H A$, the minimum is attained just for $x^* = x$. □

Thus, the normal equations are obtained from the approximate equations by multiplying with the transposed coefficient matrix. The coefficient matrix $A^H A$ of the normal equations is Hermitian because $(A^H A)^H = A^H (A^H)^H = A^H A$, and positive semidefinite because $x^H A^H A x = \|Ax\|_2^2 \geq 0$ for all x. If A has rank n, then $A^H A$ is positive definite because $x^H A^H A x = 0$ implies $Ax = 0$, and hence $x = 0$.

In cases where the coefficient matrix $A^H A$ is well-conditioned (such as for the approximation of functions by splines discussed in Section 3.4), the normal equations can be used directly to obtain least squares solutions, by forming a Cholesky factorization $A^H A = LL^H$, and solving the corresponding triangular systems of equations. However, frequently the normal equations are much more ill-conditioned than the underlying minimization problem; the condition number essentially squares. Thus for a problem in which half of the significant figures would be lost because of the condition of the least squares problem, one would obtain no significant figures by solving the normal equations.

2.2.8 Example. In curve fitting, the approximation function is often represented as a linear combination of basis functions. When these basis functions have a similar behavior, the resulting equations are usually ill conditioned.

For example, let us try to compute with $B = 10$, $L = 5$, and optimal rounding the best straight line through the three points $(s_i, t_i) = (3.32, 4.32), (3.33, 4.33),$ $(3.34, 4.34)$. Of course, the true solution is the line $s = t + 1$. The formula $s = x_1 t + x_2$ leads to the overdetermined system $Ax = b$, where

$$A = \begin{pmatrix} 3.32 & 1 \\ 3.33 & 1 \\ 3.34 & 1 \end{pmatrix}, \quad b = \begin{pmatrix} 4.32 \\ 4.33 \\ 4.34 \end{pmatrix}.$$

For the normal equations (2.1), one obtains

$$A^H A \approx \begin{pmatrix} 33.267 & 9.99 \\ 9.99 & 3 \end{pmatrix}, \quad A^H b \approx \begin{pmatrix} 43.257 \\ 12.99 \end{pmatrix}.$$

The Cholesky factorization gives $A^H A \approx L L^T$ with

$$L \approx \begin{pmatrix} 5.7678 & 0 \\ 1.7320 & 0.014142 \end{pmatrix}$$

and solving the system of equations gives

$$L^{-1} A^H b \approx \begin{pmatrix} 7.4997 \\ 0.001 \end{pmatrix}$$

and from this one obtains the (completely false) "solution"

$$x \approx (L^H)^{-1} L^{-1} A^H b \approx \begin{pmatrix} 1.2790 \\ 0.070817 \end{pmatrix}.$$

Indeed, a computation of the condition number $\text{cond}_\infty(A^H A) \approx 3.1 \cdot 10^6$ (see Section 2.5) reveals that with the accuracy used for the computations, one could have expected no useful results.

Note that we should have fared much better by representing the line as a linear combination $s = x_1(t - 3.33) + x_2$; one moral of the example is that one should select good basis functions before fitting data. (A good choice is usually one in which different basis functions differ in the number and distribution of the zeros within the interval of interest.)

The right way to solve moderately ill-conditioned least squares problems is by means of a so-called *orthogonal factorization* (or *QR-factorization*) of A into a product $A = QR$ of a unitary matrix Q and an upper triangular matrix R. Here, a matrix Q is called *unitary* if $Q^H Q = I$. A unitary matrix with real entries is called *orthogonal*, and in this case, $Q^T Q = I$.

From an orthogonal factorization, the least squares solution of $Ax \approx y$ can be computed by solving the triangular system

$$Rx = Q^H b. \tag{2.2}$$

Indeed, because $Q^H Q = I$, $A^H A = R^H Q^H Q R = R^H R$, so that R is the (transposed) Cholesky factor of A^H, and the solution x^* of (2.2) satisfies the normal equations because

$$A^H A x^* = R^H R x^* = R^H Q^H b = A^H b.$$

For the stable computation of orthogonal factorizations, we refer to the books mentioned in the introduction to this chapter. MATLAB provides the orthogonal factorization via the routine qr, but it automatically produces a stable least squares solution of $Ax \approx b$ by x=A\b, so that the casual user need not to know the details.

For strongly ill-conditioned least squares problems, even the solution of (2.2) produces meaningless results. Remarkably, it is still possible to get sensible solutions of $Ax \approx b$, provided one has some qualitative information about the desired solution. The key to a good solution is called *regularization* and involves the so-called *singular-value decomposition* (SVD) of A, a factorization into a product $A = U \Sigma V^H$ of two unitary matrices U and V and a diagonal matrix Σ. For details about the SVD, we refer again to the books mentioned in the introduction to this chapter; for details about regularization, refer to Hansen [40] and Neumaier [71].

Matrix Inversion

Occasionally, one must compute a matrix inverse explicitly. (However, as we shall see, it is inadvisable to solve $Ax = b$ by computing first A^{-1} and then $x = A^{-1}b$!)

To avoid later repetition, we discuss immediately the numerically stable case using a permuted factorization, obtained by multiplying the matrix by a nonsingular (permutation) matrix P (see Algorithm 2.3.7). From a factorization $PA = LR$, it is not difficult to compute an explicit inverse of A.

If we introduce the matrices $\hat{R} := R^{-1}$ and $\hat{A} := R^{-1}L^{-1} = \hat{R}L^{-1}$, we can split the computation of the inverse matrix $A^{-1} = R^{-1}L^{-1}P = \hat{A}P$ into three steps.

2.2.9 Algorithm: Matrix Inversion

STEP 1: Compute a permuted triangular factorization $PA = LR$ (see Algorithm 2.3.7).
STEP 2: Determine $\hat{R}$ from the equation $\hat{R}R = I$.
STEP 3: Determine $\hat{A}$ from the equation $\hat{A}L = \hat{R}$.
STEP 4: Form the product $A^{-1} = \hat{A}P$.

The details for how to proceed in steps 2 and 3 can be obtained as in Theorem 2.1.8 by expressing the matrix equations in components,

$$\hat{R}_{ii}R_{ii} = 1,$$

$$\sum_{j=i:k} \hat{R}_{ij}R_{jk} = 0 \quad \text{if } i < k,$$

$$\sum_{j=k:n} \hat{A}_{ij}L_{jk} = \hat{R}_{ik},$$

and solving for the unknown variables in an appropriate order (see Exercise 14). For the calculation of an explicit inverse A^{-1}, we need $2n^3 + O(n^2)$ operations (see Exercise 14); that is, three times as many as for the triangular factorization. For banded matrices, where the inverse is full, this ratio is even more pessimistic.

In practice, the computation of the inverse should be avoided whenever possible. The reason is that, given a factorization, the solution of the resulting two triangular systems of equations takes only $2n^2 + O(n)$ operations. This is essentially the same amount as needed for the multiplication $A^{-1}b$ when an explicit inverse is available. However, because the inverse is much more expensive than the factorization, there is no incentive to compute it at all, except in circumstances where the entries of A^{-1} are of independent interest. In other applications where matrix inverses are used, it is usually possible (and profitable) to avoid the explicit inverse, too.

2.3 Rounding Errors, Equilibration, and Pivot Search

We now consider techniques that ensure that the solution of linear systems is computed in a numerically stable way.

Rounding Errors and Equilibration

In finite precision arithmetic, the triangular factorization is subject to rounding errors. We first present a short and very simplified argument that serves to identify the main source of errors; a detailed error analysis is given at the end of this section (Theorem 2.3.9).

For simplicity, we suppose first that a rounding error appears only in a single multiplication $L_{ij}R_{jk}$; say, for $i = p$, $j = q$, $k = r$. In this case, we have

$$\tilde{R}_{pr} = A_{pr} - \sum_{\substack{j=1:p-1 \\ j \neq q}} L_{pj}R_{jr} - L_{pq}R_{qr}(1 - \tilde{\varepsilon}_{pr})$$

$$= A_{pr} + L_{pq}R_{qr}\tilde{\varepsilon}_{pr} - \sum_{j=1:p-1} L_{pj}R_{jr},$$

where $|\tilde{\varepsilon}_{pr}| \leq B^{1-L}$. If we define the matrix $\tilde{A}$ obtained by changing in A the (p, r)-entry to

$$\tilde{A}_{pr} := A_{pr} + L_{pq}R_{qr}\tilde{\varepsilon}_{pr},$$

the triangular factorization $L\tilde{R}$ (in which $\tilde{R}_{ik} := R_{ik}$ for $(i, k) \neq (p, r)$) that we obtain instead of LR may be interpreted as the exact triangular factorization of the matrix $\tilde{A}$. If $|L_{pq}R_{qr}| \gg A_{pr}$; then this corresponds to a large relative perturbation $|\tilde{A}_{pr} - A_{pr}|/|A_{pr}|$ of A_{pr}, and the algorithm is unstable.

If rounding errors arise from other arithmetic operations in addition to multiplication, one finds similar expressions. Instability always arises if the magnitude of a term $L_{ij}R_{jk}$ is very large compared with the magnitudes of the elements of A. In particular, if a divisor element R_{ii} has a very small magnitude, then the magnitudes of the corresponding elements L_{ki} $(k > i)$ are large, causing instability. For a few classes of matrices (positive definite matrices and H-matrices; cf. Corollary 2.1.9), it can be shown that no instability can arise. For general matrices, we must ensure stability by avoiding not only vanishing, but also small divisors R_{ii}. This is achieved by row (or sometimes column) permutations.

One calls the row permuted into the ith row the ith *pivoting row* and the resulting divisors R_{ii} the *pivot elements*; the search for a suitable divisor is called the *pivot search*. The most popular pivot search is *column pivoting*, also called *partial pivoting*, which usually achieves stability with only $O(n^2)$ additional comparisons. (The stability of Gaussian elimination for balanced matrices may be further improved by so-called complete pivoting, which allows row and column permutations. Here, the element of greatest magnitude in the remainder of the matrix becomes the pivot element. However, the cost is considerably higher; $O(n^3)$ comparisons are necessary.)

In column pivoting, one searches the remainder of the column under consideration for the element of greatest magnitude as the pivot element. Because in the calculation of the $L_{ki}(k > i)$ the pivot element is a divisor, it follows that, with column pivoting, $|L_{ki}| \leq 1$. This implies that

$$|R_{ik}| \leq |A_{ik}| + \sum_{j=1:i-1} |R_{jk}|,$$

so that by an easy induction argument, all R_{ik} are bounded by the expression $K_n \max_{i,k} |A_{ik}|$, where the factor K_n can, in unusual cases, be exponential in n, but typically varies only slowly with n. Therefore, if A is *balanced* (i.e., if all nonzero entries of A have a similar magnitude), then we typically have numerical stability.

To balance a matrix with entries of widely different magnitudes, one usually uses some form of scaling. In compact notation, scaling of a matrix A corresponds to the multiplication of A by some diagonal matrix D; left multiplication by D corresponds to *row scaling* (i.e., multiplication of each row by the corresponding diagonal element), and right multiplication by D corresponds to *column scaling* (i.e., multiplication of each column by the corresponding diagonal element). For example, if

$$A = \begin{pmatrix} a & b \\ c & d \end{pmatrix}, \quad D = \begin{pmatrix} e & 0 \\ 0 & f \end{pmatrix}$$

then

$$DA = \begin{pmatrix} ea & eb \\ fc & fd \end{pmatrix}, \quad AD = \begin{pmatrix} ea & fb \\ ec & fd \end{pmatrix}.$$

If one uses row scaling only, the most natural strategy for balancing A is *row equilibration* (i.e., scaling each row such that the absolute values of its entries sum to 1). This is achieved for the choice

$$D_{ii} := \left(\sum_{k=1:n} |A_{ik}| \right)^{-1}. \tag{3.1}$$

An optimality property for this choice among all *row* scalings (but not among two-sided scalings) is proved in Theorem 2.5.4. (Note that (3.1) is well defined unless the ith row of A vanishes identically; in this case, we may take w.l.o.g. $D_{ii} = 1$.) In the following MATLAB program, the vector d contains the diagonal entries of D.

2.3.1 Algorithm: Row Equilibration

```
d=ones(n,1);
for i=1:n,
   dd=sum(abs(A(i,:)));
   if dd>0,
      d(i)=1/dd;
      A(i,:)=d(i)*A(i,:);
   end;
end;
```

Although it usually works well, scaling by row equilibration only is sometimes not the most appropriate way to balance a matrix.

2.3.2 Example. Rice [82] described the difficulties in scaling the matrix

$$A = \begin{pmatrix} 1 & 10^{+20} & 10^{+10} & 1 \\ 10^{+20} & 10^{+20} & 1 & 10^{+40} \\ 10^{+10} & 1 & 10^{+40} & 10^{+50} \\ 1 & 10^{+40} & 10^{+50} & 1 \end{pmatrix}.$$

to a well-behaved matrix; row equilibration and column equilibration emphasize very different components of the matrix, and it is not at all clear which elements should be regarded as large or small. However, using the scaling matrices

$$D = \text{diag}(1, 10^{-15}, 10^{-20}, 10^{-25}),$$
$$D' = \text{diag}(10^{-5}, 10^{-20}, 10^{-25}, 10^{-30})$$

and the permutation matrix (cf. below)

$$P = \begin{pmatrix} 0 & 1 & 0 & 0 \\ 1 & 0 & 0 & 0 \\ 0 & 0 & 0 & 1 \\ 0 & 0 & 1 & 0 \end{pmatrix},$$

we get the strongly diagonally dominant matrix

$$PDAD' = \begin{pmatrix} 1 & 10^{-15} & 10^{-40} & 10^{-5} \\ 10^{-5} & 1 & 10^{-15} & 10^{-30} \\ 10^{-30} & 10^{-5} & 1 & 10^{-55} \\ 10^{-15} & 10^{-40} & 10^{-5} & 1 \end{pmatrix},$$

which is perfectly scaled for Gaussian elimination. (For a general method to scale and permute a matrix to so-called *I-matrix form*, where all diagonal entries are 1 and the other entries have absolute value of at most 1, see Olschowka and Neumaier [76]).

Row Interchanges

We now consider what to do when no triangular factorization exists because some divisor R_{ii} is zero. One usually proceeds by what is called *column pivoting* (i.e., performing suitable row permutations).

In linear algebra, permutations are written in compact notation by means of permutation matrices. A matrix P is called a *permutation matrix* if, in each row and column of P, the number 1 occurs exactly once and all other elements are equal to 0. The multiplication of a matrix A with P on the left corresponds to a permutation of rows; multiplication with P on the right corresponds to a permutation of columns. For example, the permutation matrix

$$P_1 = \begin{pmatrix} 1 & 0 \\ 0 & 1 \end{pmatrix} = I$$

leaves $A \in \mathbb{C}^{2 \times 2}$ unchanged; the only other 2×2 permutation matrix is

$$P_2 = \begin{pmatrix} 0 & 1 \\ 1 & 0 \end{pmatrix},$$

and we have

$$A = \begin{pmatrix} a & b \\ c & d \end{pmatrix} \implies P_2 A = \begin{pmatrix} c & d \\ a & b \end{pmatrix}, \quad A P_2 = \begin{pmatrix} b & a \\ d & c \end{pmatrix}.$$

A *symmetric* permutation of rows and columns that moves diagonal entries to diagonal entries is given by replacing A with PAP^T; they may be useful, too, because they preserve symmetry. Permutation matrices form a group (i.e., products and inverses of permutation matrices are again permutation matrices). This allows one to store permutation matrices in terms of a product representation by *transpositions*, which are simple permutations that only interchange two rows (or columns), by storing information about the indices involved.

2.3.3 Example. We replace, in the matrix A of Example 2.1.11, the element $A_{33} = -1$ with $A_{33} = -11/2$. The calculation is then unchanged up to and including Step (2b), and we obtain the following diagrams.

$$
\begin{matrix} & & (2b) & \\ \begin{pmatrix} 2 & 4 & -4 & 2 \\ 2 & -6 & 9 & -3 \\ -2 & -\frac{3}{2} & -\frac{11}{2} & 3 \\ 1 & \frac{1}{2} & 3 & 1 \end{pmatrix} \end{matrix}
\qquad
\begin{matrix} & & (3a) & \\ \begin{pmatrix} 2 & 4 & -4 & 2 \\ 2 & -6 & 9 & -3 \\ -2 & -\frac{3}{2} & 0^S & 3 \\ 1 & \frac{1}{2} & \frac{5}{2}^S & 1 \end{pmatrix} \end{matrix}
$$

In the next step, division by the element $R_{33} = A_{33}$ should occur, among other things. In this example, however, $A_{33} = 0$ (i.e., the division cannot be performed). This makes an additional step necessary in the previously mentioned algorithm: we must exchange the third row with the fourth row. This corresponds to a row permutation of the initial matrix A.

If one forms the triangular factorization of the matrix PA, then one obtains just the corresponding permuted diagrams for A, including Step (3a). Because the third diagonal element A_{33} is different from zero, one can continue the process.

$$
\begin{matrix} & & \text{Permutation} & \\ \begin{pmatrix} 2 & 4 & -4 & 2 \\ 2 & -6 & 9 & -3 \\ 1 & \frac{1}{2} & \frac{5}{2} & 1 \\ -2 & -\frac{3}{2} & 0 & 3 \end{pmatrix} \end{matrix}
\qquad
\begin{matrix} & & (3b) & \\ \begin{pmatrix} 2 & 4 & -4 & 2 \\ 2 & -6 & 9 & -3 \\ 1 & \frac{1}{2} & \frac{5}{2} & \frac{1}{2}^S \\ -2 & -\frac{3}{2} & 0^D & 3 \end{pmatrix} \end{matrix}
$$

$$\begin{array}{c}\text{(4a)}\\\begin{pmatrix}2 & 4 & -4 & 2\\ 2 & -6 & 9 & -3\\ 1 & \frac{1}{2} & \frac{5}{2} & \frac{1}{2}\\ -2 & -\frac{3}{2} & 0 & \frac{5}{2}{}^S\end{pmatrix}\end{array}$$

After Step (4a), the triangular factorization of PA with

$$P = \begin{pmatrix}1 & 0 & 0 & 0\\ 0 & 1 & 0 & 0\\ 0 & 0 & 0 & 1\\ 0 & 0 & 1 & 0\end{pmatrix}$$

is overwritten over A.

What happens if all elements in the remaining column are equal to zero? To construct an example for this case, we replace the elements $A_{33} = -1$ and $A_{43} = 3$ with $A_{33} = -11/2$ and $A_{43} = 1/2$ and obtain the following diagrams

$$\begin{array}{cc}\text{(2b)} & \text{(3a)}\\\begin{pmatrix}2 & 4 & -4 & 2\\ 2 & -6 & 9 & -3\\ -2 & -\frac{3}{2} & -\frac{11}{2} & 3\\ 1 & \frac{1}{2} & \frac{1}{2} & 1\end{pmatrix} & \begin{pmatrix}2 & 4 & -4 & 2\\ 2 & -6 & 9 & -3\\ -2 & -\frac{3}{2} & 0^S & 3\\ 1 & \frac{1}{2} & 0^S & 1\end{pmatrix}\end{array}$$

Again, we obtain a zero on the diagonal; but this cannot be removed from the diagonal by an exchange of the third and fourth rows. If we simply continue, then the division in Step (3b) leads to $L_{43} = 0/0$ (i.e., any value can be taken for L_{43}). For simplicity, we set $L_{43} = 0$ and continue the recursion.

$$\begin{array}{cc}\text{(3b)} & \text{(4a)}\\\begin{pmatrix}2 & 4 & -4 & 2\\ 2 & -6 & 9 & -3\\ -1 & -\frac{3}{2} & 0 & \frac{5}{2}{}^S\\ 1 & \frac{1}{2} & 0^D & 1\end{pmatrix} & \begin{pmatrix}2 & 4 & -4 & 2\\ 2 & -6 & 9 & -3\\ -1 & -\frac{3}{2} & 0 & \frac{5}{2}\\ 1 & \frac{1}{2} & 0 & \frac{1}{2}{}^S\end{pmatrix}\end{array}$$

Indeed, we can carry out the factorization process formally, obtaining, however, $R_{33} = 0$ (i.e., R, and therefore A, is singular).

It is easy to see that the permuted triangular factorization can be constructed for arbitrary matrices A in exactly the same fashion as illustrated in the examples. Therefore, the following theorem holds.

2.3.4 Theorem. *For each $n \times n$ matrix A there is at least one permutation matrix P such that the permuted matrix PA has a normalized triangular factorization $PA = LR$.*

Noting that $\det P \det A = \det(PA) = \det(LR) = \det L \det R = \det R$ and using the permuted system $PAx = Pb$ in place of $Ax = b$, we obtain the following final version of Algorithm 2.1.6.

2.3.5 Algorithm: Solve $Ax = b$ with Pivoting

STEP 1: Calculate a permuted normalized triangular factorization $PA = LR$.
STEP 2: Solve $Ly = Pb$.
STEP 3: If R is nonsingular, solve $Rx = y$.

The resulting vector x is the solution of $Ax = b$. Moreover, $\det A = \det R / \det P$.

2.3.6 Remarks.

(i) If several systems of equations with the same coefficient matrix A are to be solved, then only steps 2 and 3 must be repeated. Thus we save a considerable amount of computing time.

(ii) P is never formed explicitly, but is represented as a product of single row exchanges (transpositions); the number of the row exchanged with the ith row in the ith step is stored as the ith component of a *pivot vector*.

(iii) As a product of many factors, $\det R$ is prone to overflow or underflow, unless special precautions are taken to represent it in the form $a \cdot M^k$ for some big number M, and updating k occasionally. The determinant $\det P$ is either $+1$ or -1, depending on whether an even or an odd number of exchanges were carried out.

(iv) $\log |\det A| = \log |\det R|$ is needed in many statistical applications involving maximum likelihood estimation.

Pivot Search with Implicit Scaling

As we have seen, scaling is important to obtain an equilibrated matrix for which column pivoting may be applied sensibly. In practice, it is customary to avoid explicit scaling. Instead, one uses for the actual factorization the original matrix and performs the scaling only implicitly to determine the pivot element of largest absolute value. If we would start the factorization with the scaled matrix DAD' in place of A, the factors L and R would scale to DLD^{-1} (because of normalization) and DRD'. Thus, R scales as A. Hence, in the ith step, the pivot row index j would be selected by the condition

$$D_{jj}A_{ji}D'_{ii} \geq D_{kk}A_{ki}D'_{ii} \quad \text{for all } k \geq i,$$

One sees that in this condition, the D'_{ii} cancel and only the row scaling affects the choice of the pivot row. Writing $d_k := D_{kk}$, we find the selection rule

$$\text{Choose } j \geq i \text{ such that } d_j A_{ji} = \max\{d_k A_{ki} \mid k \geq i\}.$$

This selection rule can now be used together with the unscaled A, and one obtains the following algorithm.

2.3.7 Algorithm: Permuted Triangular Factorization with Implicit Row Equilibration and Column Pivoting

```
% find scale factors
d=ones(n,1);
for i=1:n,
  dd=sum(abs(A(i,:)));
  if dd>0, d(i)=1/dd; end;
end;
% main loop
for i=1:n,
  % find possible pivot elements
  for k=i:n,
    A(k,i)=A(k,i)- A(k,1:i-1)*A(1:i-1,i);
  end;
  % find and save index of pivoting row
  [val,j]=max(d(i:n).*abs(A(i:n,i)));j=i-1+j;p(i)=j;
  % interchange rows i and j
  if j>i,
    A([i,j],:)=A([j,i],:);
    d(j)=d(i); % d(i) no longer needed
  end;
  for k=i+1:n,
    % find R(i,k)
    A(i,k)=A(i,k)-A(i,1:i-1)*A(1:i-1,k);
    % complete L(k,i)
    if A(i,i) ~= 0, A(k,i)=A(k,i)/A(i,i); end;
  end;
end;
```

Note that is necessary to store the pivot indices in a vector p because this information is needed for the solution of a triangular system $Ly = Pb$; the original right side b must be permuted as well.

2.3.8 Algorithm: Solving a Factored Linear System

% forward elimination
for $i = 1 : n$,
  $j = p_i$; if $j > i$, $b([i, j]) = b([j, i])$; end;
  $y_i = b_i - \sum\limits_{k=1:i-1} A_{ik}y_k$;
end;
% back substitution
$ier = 0$;
for $i = n : -1 : 1$,
  if $A_{ii} == 0$, $ier = 1$; break; end;
  $x_i = (y_i - \sum\limits_{k=i+1:n} A_{ik}x_k)/A_{ii}$;
end;
% $ier = 0$: A is nonsingular and $x = A^{-1}b$
% $ier = 1$: A is singular

Rounding Error Analysis for Gaussian Elimination

Rounding errors imply that any computed approximate solution $\tilde{x}$ is in general inaccurate. Already, rounding the components of the exact x^* gives $\tilde{x}$ with $\tilde{x}_k = x_k^*(1 + \varepsilon_k)$, $|\varepsilon_k| \leq \varepsilon$ (ε the machine precision), and the *residual* $b - A\tilde{x}$, instead of being zero, can be as large as the upper bound in

$$|b - A\tilde{x}| = |A(x^* - \tilde{x})| \leq |A||x^* - \tilde{x}| \leq \varepsilon|A|\,|x^*|.$$

We now show that in the execution of Gaussian elimination, the residual usually is not much bigger than the unavoidable bound just derived. Theorem 2.5.6 then implies that rounding errors have the same effect as a small perturbation of the right side b. We may assume w.l.o.g. that the matrix is already permuted in such a way that no further permutations are needed in the pivot search.

2.3.9 Theorem. *Let $\tilde{x}$ be an approximation to the solution of the system of linear equations $Ax^* = b$ with $A \in \mathbb{C}^{n \times n}$ and $b \in \mathbb{C}^n$, calculated in finite-precision arithmetic by means of Gaussian elimination without pivoting. Let ε be the machine precision. If the computed triangular matrices are denoted by*

$\tilde{L}$ and $\tilde{R}$ and $5n\varepsilon \le 1$, *then*

$$|b - A\tilde{x}| \le 5n\varepsilon|\tilde{L}||\tilde{R}||\tilde{x}|.$$

Usually (after partial pivoting), the entries of $|\tilde{L}||\tilde{R}|$ have the same order of magnitude as A; then, the errors in Gaussian elimination with partial pivoting are comparable to those made in forming products Ax. However (cf. Exercise 11), occasionally unreasonable pivot growth may occur such that $|\tilde{L}||\tilde{R}|$ has much bigger entries than A. In such cases, either *complete pivoting* (involving row and column interchanges to select the absolutely largest pivot element) or an alternative factorization such as the QR-factorization, which are provably stable without exception, should be used.

To prove the theorem, we must know how the expressions of the form

$$b_m := \left(c - \sum_{i=1:m-1} a_i b_i\right)\Big/ a_m, \quad a_m \ne 0 \qquad (3.2)$$

that occur in the recurrence for the triangular factorization and the solution of the triangular systems are computed. We assume that the rounding errors in these expressions are equivalent to those obtained if these expressions are calculated with the following algorithm.

2.3.10 Algorithm: Computation of (3.2)

$s_0 = c$;
for $j = 1 : m - 1$,
    $s_j = s_{j-1} - a_j b_j$;
end;
$b_m = s_{m-1}/a_m$;

Proof of Theorem 2.3.9. We proceed in 5 steps.

STEP 1: Let $\varepsilon_i := i\varepsilon/(1 - i\varepsilon)(i < 5n)$. Then

$$|\alpha| \le \varepsilon_i, |\eta| \le \varepsilon \Longrightarrow (1+\alpha)(1+\eta)^{\pm 1} = 1+\beta \text{ with } |\beta| \le \varepsilon_{i+1}, \quad (3.3)$$

because for the positive sign,

$$|\beta| = |\eta + \alpha(1 + \eta)| \le \varepsilon + \varepsilon_i(1 + \varepsilon) = \frac{(i + 1)\varepsilon}{1 - i\varepsilon} \le \varepsilon_{i+1},$$

and for the negative sign,

$$|\beta| = \left| \frac{\alpha - \eta}{1 + \eta} \right| \leq \frac{\varepsilon_i + \varepsilon}{1 - \varepsilon} = \frac{(i+1)\varepsilon - i\varepsilon^2}{1 - (i+1)\varepsilon + i\varepsilon^2} \leq \varepsilon_{i+1}.$$

STEP 2: We show that with the initial data $\tilde{a}_i$, $\tilde{b}_i$ $(i = 1, \ldots, m-1)$ and $\tilde{c}$, the calculated values $\tilde{s}_j$ satisfy the relation

$$\tilde{c} = \sum_{i=1:j} \tilde{a}_i \tilde{b}_i (1 + \alpha_i) + \tilde{s}_j (1 + \beta_j) \quad (j = 0, 1, \ldots, m-1) \quad (3.4)$$

for suitable α_i and β_i with magnitude $\leq \varepsilon_i$. This clearly holds for $j = 0$ with $\beta_0 := 0$. Suppose that (3.4) holds for some $j < m - 1$. Because the rounding is correct,

$$\tilde{s}_{j+1} = (\tilde{s}_j - \tilde{a}_{j+1} \tilde{b}_{j+1} (1 + \mu))(1 + \nu) \quad \text{with } |\mu|, |\nu| \leq \varepsilon.$$

Solving for $\tilde{s}_j$ and multiplying by $(1 + \beta_j)$ gives, because of (3.3),

$$\begin{aligned} \tilde{s}_j (1 + \beta_j) &= \tilde{s}_{j+1} (1 + \beta_j)(1 + \nu)^{-1} + \tilde{a}_{j+1} \tilde{b}_{j+1} (1 + \beta_j)(1 + \mu) \\ &= \tilde{s}_{j+1} (1 + \beta_{j+1}) + \tilde{a}_{j+1} \tilde{b}_{j+1} (1 + \alpha_{j+1}) \end{aligned}$$

with $|\alpha_{j+1}|, |\beta_{j+1}| \leq \varepsilon_{j+1}$. From the inductive hypothesis (3.4), we obtain

$$\begin{aligned} \tilde{c} &= \sum_{i=1:j} \tilde{a}_i \tilde{b}_i (1 + \alpha_i) + \tilde{s}_j (1 + \beta_j) \\ &= \sum_{i=1:j+1} \tilde{a}_i \tilde{b}_i (1 + \alpha_i) + \tilde{s}_{j+1} (1 + \beta_{j+1}). \end{aligned}$$

STEP 3: From $\tilde{b}_m = (\tilde{s}_{m-1}/\tilde{a}_m)(1 + \eta)$, $|\eta| \leq \varepsilon$, and (3.4), it follows that

$$\begin{aligned} \tilde{c} &= \sum_{i=1:m-1} \tilde{a}_i \tilde{b}_i (1 + \alpha_i) + \tilde{a}_m \tilde{b}_m (1 + \beta_{m-1})(1 + \eta)^{-1} \\ &= \sum_{i=1:m} \tilde{a}_i \tilde{b}_i (1 + \alpha_i), \end{aligned}$$

whence, by (3.3), $|\alpha_m| \leq \varepsilon_m$; we therefore obtain

$$\left| \tilde{c} - \sum_{i=1:m} \tilde{a}_i \tilde{b}_i \right| = \left| \sum_{i=1:m} \tilde{a}_i \tilde{b}_i \alpha_i \right| \leq \varepsilon_m \sum_{i=1:m} |\tilde{a}_i| |\tilde{b}_i|. \quad (3.5)$$

STEP 4: Applying (3.5) to the recursive formulae of Crout (without pivoting) gives, because $\varepsilon_i \leq \varepsilon_n$ for $i \leq n$,

$$\left| A_{ik} - \sum_{j=1:i} \tilde{L}_{ij} \tilde{R}_{jk} \right| \leq \varepsilon_n \sum_{j=1:i} |\tilde{L}_{ij}||\tilde{R}_{jk}|,$$

$$\left| A_{ki} - \sum_{j=1:i} \tilde{L}_{kj} \tilde{R}_{ji} \right| \leq \varepsilon_n \sum_{j=1:i} |\tilde{L}_{kj}||\tilde{R}_{ji}|$$

if $i \leq k$, so for the computed triangular factors $\tilde{L}$ and $\tilde{R}$ of A,

$$|A - \tilde{L}\tilde{R}| \leq \varepsilon_n |\tilde{L}||\tilde{R}|.$$

Similarly, for the computed solutions $\tilde{y}, \tilde{x}$ of the triangular systems $\tilde{L}y = b, \tilde{R}x = \tilde{y}$,

$$\left| b_i - \sum_{j=1:i} \tilde{L}_{ij} \tilde{y}_j \right| \leq \varepsilon_n \sum_{j=1:i} |\tilde{L}_{ij}||\tilde{y}_j|,$$

$$\left| \tilde{y}_i - \sum_{j=1:n} \tilde{R}_{ij} \tilde{x}_j \right| \leq \varepsilon_n \sum_{j=1:n} |\tilde{R}_{ij}||\tilde{x}_j|,$$

and we obtain

$$|b - \tilde{L}\tilde{y}| \leq \varepsilon_n |\tilde{L}||\tilde{y}|,$$

$$|\tilde{y} - \tilde{R}\tilde{x}| \leq \varepsilon_n |\tilde{R}||\tilde{x}|.$$

STEP 5: Using steps 3 and 4, we can estimate the residual. We have

$$
\begin{aligned}
|b - A\tilde{x}| &= |b - \tilde{L}\tilde{y} - \tilde{L}(\tilde{R}\tilde{x} - \tilde{y}) + (\tilde{L}\tilde{R} - A)\tilde{x}| \\
&\leq |b - \tilde{L}\tilde{y}| + |\tilde{L}||\tilde{R}\tilde{x} - \tilde{y}| + |\tilde{L}\tilde{R} - A||\tilde{x}| \\
&\leq \varepsilon_n(|\tilde{L}||\tilde{y}| + |\tilde{L}||\tilde{R}||\tilde{x}| + |\tilde{L}||\tilde{R}||\tilde{x}|) \\
&\leq \varepsilon_n(|\tilde{L}||\tilde{y} - \tilde{R}\tilde{x}| + 3|\tilde{L}||\tilde{R}||\tilde{x}|) \\
&\leq \varepsilon_n(\varepsilon_n + 3)|\tilde{L}||\tilde{R}||\tilde{x}|.
\end{aligned}
$$

By hypothesis, $5n\varepsilon < 1$, so $\varepsilon_n = n\varepsilon/(1 - n\varepsilon) \leq \frac{5}{4}n\varepsilon < 1$, whence the theorem is proved.

Rounding error analyses like these are available for almost all numerical algorithms that are widely used in practice. A comprehensive treatment of error analyses for numerical linear algebra is in Higham [44]; see also Stummel and Hainer [92]. Generally, these analyses confirm rigorously what can be argued much simpler without rigor, using the rules of thumb given in Chapter 1. Therefore, later chapters discuss stability in this more elementary way only.

2.4 Vector and Matrix Norms

Estimating the effect of a small residual on the accuracy of the solution is a problem independent of rounding errors and can be treated in terms of what is called *perturbation theory*. For this, we need new tools; in particular, the concept of a norm.

A mapping $\| \cdot \| : \mathbb{C}^n \to \mathbb{R}$ is called a *(vector) norm* if, for all $x, y \in \mathbb{C}^n$:

(i) $\|x\| \geq 0$ with equality iff $x = 0$,
(ii) $\|\alpha x\| = |\alpha| \|x\|$ for $\alpha \in \mathbb{C}$,
(iii) $\|x + y\| \leq \|x\| + \|y\|$.

The most important vector norms are the *Euclidean norm*

$$\|x\|_2 := \sqrt{\sum_i |x_i|^2} = \sqrt{x^H x},$$

the *maximum norm*

$$\|x\|_\infty := \max_{i=1,\dots,n} |x_i|,$$

and the *sum norm*

$$\|x\|_1 := \sum_{i=1:n} |x_i|.$$

In a finite-dimensional vector space, all norms are equivalent, in the sense that any ε-neighborhood contains a suitable ε'-neighborhood and is contained in a suitable ε''-neighborhood (cf. Exercise 16). However, for practical purposes, different norms may yield quite different measures of "size." The previously mentioned norms are appropriate if the objects they measure have components of roughly the same order of magnitude.

The norm of a sum or difference of vectors can be bounded from below as follows:

$$\|x \pm y\| \geq \|x\| - \|y\|$$

and

$$\|x \pm y\| \geq \|y\| - \|x\|$$

because, for example,

$$
\begin{aligned}
\|x\| - \|y\| &= \|x \pm y \mp y\| - \|y\| \\
&\leq \|x \pm y\| + \|y\| - \|y\| \\
&= \|x \pm y\|.
\end{aligned}
$$

A metric in $\mathbb{C}^n$ is defined through $d(x, y) = \|x - y\|$. (Restricted to real 3-space $\mathbb{R}^3$, the "ball" consisting of all x with $\|x - x_0\| \leq \varepsilon$ is, for $\|\cdot\|_2$, $\|\cdot\|_\infty$, and $\|\cdot\|_1$ a sphere-shaped, a cube-shaped, and an octahedral-shaped ε-neighborhood of x_0, respectively.)

The *matrix norm* belonging to a vector norm is defined by

$$
\|A\| := \sup\{\|Ax\| \mid \|x\| = 1\}.
$$

The relations

$$
\|Ax\| \leq \|A\|\|x\| \quad \text{for all } A \in \mathbb{C}^{n \times n}, \ x \in \mathbb{C}^n,
$$

$$
\|AB\| \leq \|A\|\|B\| \quad \text{for all } A, \ B \in \mathbb{C}^{n \times n},
$$

$$
\|\alpha A\| = |\alpha|\|A\| \quad \text{for all } A \in \mathbb{C}^{n \times n}, \ \alpha \in \mathbb{C},
$$

$$
\|I\| = 1
$$

can be proved easily from the definition.

The following matrix norms belong to the vector norms $\|\cdot\|_2$, $\|\cdot\|_\infty$, and $\|\cdot\|_1$.

$$
\begin{aligned}
\|A\|_2 &= \sup\{\sqrt{x^H A^H A x} \mid x^H x = 1\} \\
&= \sqrt{\text{maximal eigenvalue of } A^H A};
\end{aligned}
$$

is called the *spectral norm*; by definition, for a unitary matrix A, $\|A\|_2 = \|A^{-1}\|_2 = 1$.

$$
\|A\|_\infty = \max\left\{\sum_k |A_{ik}| \mid i = 1, \ldots, n\right\}
$$

is called the *row sum norm*, and

$$
\|A\|_1 = \max\left\{\sum_i |A_{ik}| \mid k = 1, \ldots, n\right\}
$$

the *column sum norm*. We shall derive only the formula for the row sum norm.

We have

$$\|A\|_\infty = \sup\{\|Ax\|_\infty \,|\, \|x\|_\infty = 1\}$$

$$= \sup_{|x_k| \le 1} \max_i \left| \sum_k A_{ik} x_k \right|$$

$$\le \max_i \sum_k |A_{ik}|,$$

and equality holds because the maximum is attained for $x_k = |A_{ik}|/A_{ik}$.

Obviously, the calculation of the spectral norm is more difficult than the calculation of the row sum norm or the column sum norm, but there are some useful, easily computable upper bounds.

2.4.1 Proposition. *If $A \in \mathbb{C}^{n \times n}$, then*

(i) $\|A\|_2 \le \sqrt{\|A\|_1 \|A\|_\infty}$, *with equality, e.g., when A is diagonal.*

(ii) $\|A\|_2 \le \|A\|_F := \sqrt{\sum_{i,k} |A_{ik}|^2}$; *the equality sign holds if and only if $A = xy^H$ with $x, y \in \mathbb{C}^n$ (i.e., the rank of A is at most 1). $\|A\|_F$ is called the* Frobenius norm *(or* Schur norm*) of A.*

(iii) *If A is Hermitian then $\|A\|_2 \le \|A\|$ for every matrix norm $\|\cdot\|$.*

Proof. We use the fact that for any matrix norm $\|\cdot\|$ and for each eigenvalue λ of a matrix A we have $|\lambda| \le \|A\|$, because for a corresponding eigenvector x we have

$$|\lambda| = \|\lambda x\|/\|x\| = \|Ax\|/\|x\| \le \|A\|. \tag{4.1}$$

(i) Let $\lambda_{\max}$ be the eigenvalue of $A^H A$ of maximum absolute value. Then

$$\|A\|_2^2 = |\lambda_{\max}| \le \|A^H A\|_\infty \le \|A^H\|_\infty \|A\|_\infty = \|A\|_1 \cdot \|A\|_\infty.$$

(ii) By definition,

$$\|A\|_2 = \sup_{\|x\|_2 = 1} \sqrt{(Ax)^H (Ax)}$$

and the supremum is attained for some $x = \hat{x}$ because of the compactness of the unit ball. By the Cauchy-Schwarz inequality,

$$(A_{i:}\hat{x})^2 \le \|A_{i:}\|_2^2 \|\hat{x}\|_2^2 = \|A_{i:}\|_2^2 \quad (i = 1, \ldots, n)$$

and therefore

$$\|A\|_2^2 = (A\hat{x})^H (A\hat{x}) = \sum_i |A_{i:}\hat{x}|^2 \le \sum_i \|A_{i:}\|_2^2 = \sum_{i,k} |A_{ik}|^2.$$

Equality holds when equality holds in the Cauchy-Schwarz inequality (i.e., when for all i the vectors $(A_{i:})^T$ and $\hat{x}$ are linearly dependent). This is equivalent to saying that the rank of the matrix A is ≤ 1. Thus A may be written as an *outer product*

$$A = (\bar{y}_1 x, \bar{y}_2 x, \ldots, \bar{y}_n x) = xy^H.$$

(iii) If λ_i are the eigenvalues of the Hermitian matrix A, then $A^H A = A^2$ has the eigenvalues λ_i^2. Let λ_{max} be the eigenvalue of A of maximum absolute value. Then, by (4.1),

$$\|A\|_2 = \sqrt{\lambda_{max}^2} = |\lambda_{max}| \le \|A\|. \qquad \square$$

Norms may serve to prove the nonsingularity of matrices close to the identity.

2.4.2 Proposition. *If $\|I - A\| \le \beta < 1$, then A is nonsingular and*

$$\|A^{-1}\| \le 1/(1 - \beta).$$

Proof. Suppose that $Ax = 0$. Then

$$\|x\| = \|x - Ax\| = \|(I - A)x\| \le \|I - A\| \cdot \|x\| \le \beta \|x\|.$$

Because $\beta < 1$ we conclude that $\|x\| = 0$, hence $x = 0$. Thus $Ax = 0$ has only the trivial solution, so A is nonsingular. Furthermore, it follows from

$$\|A^{-1}\| \le \|A^{-1} - I\| + \|I\| \le \|A^{-1}\| \cdot \|I - A\| + \|I\| \le \beta \|A^{-1}\| + 1$$

that $\|A^{-1}\| \le 1/(1 - \beta)$. $\qquad \square$

M-Matrices and H-Matrices

There are important classes of matrices which satisfy a generalization of the criterion in Proposition 2.4.2. We call A an *H-matrix* if diagonal matrices D_1 and D_2 exist such that $\|I - D_1 A D_2\|_\infty < 1$. By the proposition, an H-matrix is nonsingular. *Diagonally dominant matrices* are matrices in which the weak row sum criterion

$$|A_{ii}| \geq \sum_{k \neq i} |A_{ik}| \quad \text{for } i = 1, \ldots, n \tag{4.2}$$

holds. Diagonally dominant matrices are H-matrices if strict inequality holds in (4.2) for all i. (To see this, use $D_1 = \text{Diag}(A)^{-1}$ and $D_2 = I$.) In fact, strict inequality in (4.2) for one i suffices if, in addition, A happens to be "irreducible"; see, e.g., Varga [94].

M-matrices are H-matrices with the sign distribution

$$A_{ii} \geq 0, \quad A_{ik} \leq 0 \quad \text{for } i \neq k;$$

there are many other equivalent definitions of M-matrices. Each M-matrix A is nonsingular and satisfies $A^{-1} \geq 0$ (see Exercise 19). This inequality mirrors the fact that M-matrices arise naturally as discretization of inverse positive elliptic differential operators; but they arise also in other context (e.g., input-output models in economics).

Monotone Norms

The *absolute value* of a vector or a matrix is defined by replacing all components with their absolute value; for example,

$$\left| \begin{pmatrix} 1 & -1 \\ -2 & 5 \end{pmatrix} \right| = \begin{pmatrix} 1 & 1 \\ 2 & 5 \end{pmatrix},$$

For $x, y \in \mathbb{C}^n$ and $A, B \in \mathbb{C}^{n \times n}$, one easily sees that

$$|x \pm y| \leq |x| + |y|, \quad |Ax| \leq |A||x|,$$
$$|A \pm B| \leq |A| + |B|, \quad |AB| \leq |A||B|.$$

A norm is called *monotone* if

$$|x| \leq y \implies \|x\| = \||x|\| \leq \|y\|.$$

All vector norms considered above are monotone. The analogous result

$$|A| \leq B \implies \|A\| = \||A|\| \leq \|B\|$$

holds for the matrix norms $\| \cdot \|_1$ and $\| \cdot \|_\infty$, but for $\| \cdot \|_2$ we only have

$$|A| \leq B \implies \|A\|_2 \leq \||A|\|_2 \leq \|B\|_2.$$

2.4.3 Proposition. *A vector norm is monotone iff, for all diagonal matrices D, the corresponding matrix norm satisfies*

$$\|D\| = \max\{|D_{ii}| \,|\, i = 1, \ldots, n\}. \tag{4.3}$$

Proof. Let $\|\cdot\|$ be a monotone vector norm. Because $|D_{ii}| \|e^{(i)}\| = \|D_{ii}e^{(i)}\| = \|De^{(i)}\| \le \|D\| \|e^{(i)}\|$, we have $|D_{ii}| \le \|D\|$, hence $\|D\| \ge \max |D_{ii}| =: \alpha$. Because $|Dx| \le \alpha|x|$ for all x, monotony implies $\|Dx\| = \||Dx|\| \le \|\alpha|x|\| = \alpha\|x\|$, hence $\|D\| = \sup \|Dx\|/\|x\| \le \alpha$. Thus $\|D\| = \alpha$, and (4.3) holds.

Conversely, suppose (4.3) holds for all diagonal D. We consider the special diagonal matrix $D = \mathrm{Diag}(\mathrm{sign}(x_1), \ldots, \mathrm{sign}(x_n))$, where

$$\mathrm{sign}(x) = \begin{cases} x/|x| & \text{if } x \ne 0, \\ 1 & \text{if } x = 0. \end{cases}$$

Then $\|D\| = \|D^{-1}\| = 1$ and $x = D|x|$, hence $\|x\| = \|D|x|\| \le \|D\| \||x|\| = \||x|\| = \|D^{-1}x\| \le \|D^{-1}\| \|x\| = \|x\|$. Thus we must have equality throughout; hence $\|x\| = \||x|\|$. Now suppose $|x| \le y$. We take $D = \mathrm{Diag}(|x_1|/y_1, \ldots, |x_n|/y_n)$ and find $\|D\| \le 1$, $|x| = Dy$, hence $\||x|\| = \|Dy\| \le \|D\| \|y\| \le \|y\|$. Hence, the norm is monotone. □

2.5 Condition Numbers and Data Perturbations

As is well known, a matrix A is nonsingular if and only if $\det A \ne 0$. Although one can compute the determinant efficiently by a triangular factorization, checking whether it is zero is difficult to do numerically. Indeed, rounding errors in the numerical calculation of $\det A$ imply that the calculated value of the determinant is almost always different from zero, even when the exact value is $\det A = 0$. Thus the unique solvability of $Ax = b$ would be predicted from an inaccurate determinant even when in reality no solution or infinitely many solutions exist.

The size of the determinant is not even an appropriate measure for closeness to singularity! For example, $\det(\alpha A) = \alpha^n \det A$, so that multiplying a 50×50 matrix by a harmless factor of 3 increases the determinant by a factor of $3^n > 10^{23}$.

For the analysis of the numerical solution of a system of linear algebraic equations, however, it is important to know "how far from singular" a given matrix is. We can then estimate how much error is allowed in order to still obtain a useful result (i.e., in order to ensure that no singular matrix is found). A useful measure is the *condition number* of A; it depends on the norm used and is defined by

$$\mathrm{cond}(A) := \|A^{-1}\| \cdot \|A\|$$

for nonsingular matrices A; we add the norm index (e.g., $\mathrm{cond}_\infty(A)$) if we refer to a definite norm. For singular matrices, the condition number is set to ∞. Unlike for the determinant,

$$\mathrm{cond}(\alpha A) = \mathrm{cond}(A) \quad \text{for } \alpha \in \mathbb{C}.$$

The condition number indeed deserves its name because, as we show now, it is a measure of the sensitivity of the solution x^* of a linear system $Ax^* = b$ to changes in the right side b.

2.5.1 Proposition. *If $Ax^* = b$ then, for arbitrary $\tilde{x}$,*

$$\|x^* - \tilde{x}\| \le \|A^{-1}\| \cdot \|b - A\tilde{x}\| \tag{5.1}$$

and

$$\frac{\|x^* - \tilde{x}\|}{\|x^*\|} \le \mathrm{cond}(A) \frac{\|b - A\tilde{x}\|}{\|b\|}. \tag{5.2}$$

Proof. (5.1) follows from

$$\|x^* - \tilde{x}\| = \|A^{-1}(b - A\tilde{x})\| \le \|A^{-1}\| \cdot \|b - A\tilde{x}\|,$$

and (5.2) then from

$$\|b\| = \|Ax^*\| \le \|A\| \cdot \|x^*\|. \qquad \square$$

The condition number also gives a lower bound on the distance from singularity (For a related upper bound, see Rump [83]):

2.5.2 Proposition. *If A is nonsingular, and B is a singular matrix with*

$$|B - A| \le \delta |A|, \tag{5.3}$$

then

$$\delta \ge 1/\mathrm{cond}(D_1 A D_2)$$

for all nonsingular diagonal matrices D_1, D_2.

Note that (5.3) specifies a componentwise relative error of at most δ; in particular, zeros are unperturbed.

Proof. Because (5.3) is scaling invariant, it suffices to consider the case where $D_1 = D_2 = I$. Because B is singular, then so is $A^{-1}B$, so, by Proposition 2.4.2, $\|I - A^{-1}B\| \geq 1$, and we obtain

$$1 \leq \|I - A^{-1}B\|$$
$$= \|A^{-1}(A - B)\| \leq \|A^{-1}\|\|B - A\|$$
$$\leq \|A^{-1}\|\|\,\delta|A|\,\| = \delta\mathrm{cond}(A),$$

whence $\delta \geq 1/\mathrm{cond}(A)$. $\qquad\square$

In particular, a matrix that has a small condition number cannot be close to a singular matrix. The converse is not true because the condition number depends strongly on the scaling of the matrix, in a way illustrated by the following example.

2.5.3 Example. If

$$A := \begin{pmatrix} 0.005 & 1 \\ 1 & 1 \end{pmatrix}$$

then

$$A^{-1} = \frac{1}{0.995}\begin{pmatrix} -1 & 1 \\ 1 & -0.005 \end{pmatrix},$$

$\|A\|_\infty = 2$, $\|A^{-1}\|_\infty = 2/0.995$, and $\mathrm{cond}_\infty(A) \approx 4$. Now

$$D := \begin{pmatrix} 200 & 0 \\ 0 & 1 \end{pmatrix} \Rightarrow DA = \begin{pmatrix} 1 & 200 \\ 1 & 1 \end{pmatrix},$$

whence

$$(DA)^{-1} = \frac{1}{199}\begin{pmatrix} -1 & 200 \\ 1 & -1 \end{pmatrix},$$

$\|DA\|_\infty = 201$, $\|(DA)^{-1}\|_\infty = 201/199$, and $\mathrm{cond}_\infty(DA) \approx 200$. So the condition number is increased by a factor of about 50 through scaling.

The following theorem (due to van der Sluis [93]) shows that equilibration generally improves the condition number; in particular, the condition number of the equilibrated matrix gives a better bound in Proposition 2.5.2.

2.5.4 Theorem. *If D runs through the set of nonsingular diagonal matrices, then* $\text{cond}_\infty(DA)$ *is minimal for*

$$D_{ii} = 1 \Big/ \sum_k |A_{ik}| \quad (i = 1, \dots, n), \tag{5.4}$$

that is, if

$$\sum_k |(DA)_{ik}| = 1 \quad \text{for } i = 1, \dots, n.$$

Proof. Without loss of generality, let A be scaled so that $\sum_k |A_{ik}| = 1$ for all i. If D is an arbitrary nonsingular diagonal matrix, then

$$\|DA\|_\infty = \max_i \left\{ |D_{ii}| \sum_k |A_{ik}| \right\} = \max_i |D_{ii}| = \|D\|_\infty;$$

so

$$\begin{aligned}
\text{cond}_\infty(A) &= \|A^{-1}\|_\infty \|A\|_\infty = \|(DA)^{-1}D\|_\infty \\
&\leq \|(DA)^{-1}\|_\infty \|D\|_\infty = \|(DA)^{-1}\|_\infty \|DA\|_\infty \\
&= \text{cond}_\infty(DA). \qquad \square
\end{aligned}$$

To compute the condition number, one must usually have the inverse of A explicitly, which is unduly expensive in many cases. If A is an H-matrix, we can use Proposition 2.4.2 to get an upper bound that is often (but not always) quite reasonable.

If one is content with an approximate value, one may proceed as follows, using the row-sum norm $\|A^{-1}\|_\infty$. Let s_i be the sum of the magnitudes of the elements in the ith row of A^{-1}. Then

$$\|A^{-1}\|_\infty = \max\{s_i \mid i = 1, \dots, n\} = s_{i_0}$$

for at least one index i_0. The ith row of A^{-1} is

$$f^{(i)T} := e^{(i)T} A^{-1}.$$

If a triangular factorization $PA = LR$ is given, then $f^{(i)}$ is calculated as the solution of the transposed system

$$A^T f^{(i)} = e^{(i)}.$$

We solve $R^T y = e^{(i)}$, $L^T z = y$, and set $f^{(i)} = P^T z$. If we were able to guess the correct index $i = i_0$, then we would have simply $s_i = \|A^{-1}\|_\infty$; for an arbitrary index i, we obtain a lower bound s_i for $\|A^{-1}\|_\infty$. If d is an arbitrary vector with $|d| = e$, where e is the all-one vector, then

$$|(A^{-1}d)_i| = e^{(i)^T}|A^{-1}d| \le e^{(i)^T}|A^{-1}| \cdot |d| = \left|f^{(i)}\right|^T e = s_i.$$

The calculation of $|(A^{-1}d)_i|$ for all i is less expensive than the calculation of all s_i, because only a single linear system must be solved, and not all the $f^{(i)}$. The index of the absolutely largest component of $A^{-1}d$ would therefore seem to be a cheap and good substitute for the optimal index i_0.

Among the possibilities for d, the choice $d := \text{sign}(A^{-T}e)$ (interpreted componentwise) proves to be especially favorable. For this choice, we even obtain the exact value $s_{i_0} = \|A^{-1}\|_\infty$ whenever A^{-1} has columnwise constant sign! Indeed, the ith component a_i of $a := A^{-T}e$ then has the sign of the ith column of A^{-1}, and because $d_i = \text{sign}(a_i)$, the ith component of $A^{-1}d$ satisfies the relation

$$|(A^{-1}d)_i| = e^{(i)^T}|A^{-1}d| = e^{(i)^T}|A^{-1}| \cdot |d| = s_i.$$

2.5.5 Algorithm: Condition Estimation

STEP 1: Solve $A^T a = e$ and set $d = \text{sign}(a)$.
STEP 2: Solve $Ac = d$ and find i^* such that $|c_{i^*}| = \max_i |c_i|$.
STEP 3: Solve $A^T f = e^{(i^*)}$ and determine $s = |f|^T e$.

Then $\|A^{-1}\|_\infty \ge s$ and often equality holds. Thus s serves as an estimate of $\|A^{-1}\|_\infty$.

Indeed, for random matrices A with uniformly distributed $A_{ik} \in [-1, 1]$, equality holds in 60–80% of the cases and $\|A^{-1}\|_\infty \le 3s$ in more than 99% of the cases. (However, for specially constructed matrices, the estimate may be arbitrarily bad.) Given an LR factorization, only $O(n^2)$ additional operations are necessary, so the estimation effort is small compared to the factorization work.

Data Perturbations

Often the coefficients of a system of equations are only inexactly known and we are interested in the influence of this inexactness on the solution. In view of Proposition 2.5.1, we discuss only the influence on the residuals, which is somewhat easier to determine. The following basic result is due to Oettli and Prager [75].

2.5.6 Theorem. *Let $A \in \mathbb{C}^{n \times n}$ and $b, \tilde{x} \in \mathbb{C}^n$. Then for arbitrary nonnegative matrices $\Delta A \in \mathbb{R}^{n \times n}$ and nonnegative vectors $\Delta b \in \mathbb{R}^n$, the following statements are equivalent:*

(i) There exists $\tilde{A} \in \mathbb{C}^{n \times n}$ and $\tilde{b} \in \mathbb{C}^n$ with $\tilde{A}\tilde{x} = \tilde{b}$, $|A - \tilde{A}| \leq \Delta A$, and $|\tilde{b} - b| \leq \Delta b$.

(ii) The residual satisfies the inequality

$$|b - A\tilde{x}| \leq \Delta b + \Delta A |\tilde{x}|.$$

Proof. It follows from (i) that

$$|b - A\tilde{x}| = |b - \tilde{b} + (\tilde{A} - A)\tilde{x}| \leq |b - \tilde{b}| + |\tilde{A} - A| \cdot |\tilde{x}| \leq \Delta b + \Delta A |\tilde{x}|;$$

that is, (ii) holds.

Conversely, if (ii) holds, then we define q_i as the ith component of the residual divided by the ith component of the right side of (ii); that is,

$$q_i := \frac{b_i - \sum_k A_{ik}\tilde{x}_k}{(\Delta b)_i + \sum_k (\Delta A)_{ik}|\tilde{x}|_k}.$$

By assumption, $|q_i| \leq 1$. We define $\tilde{A} \in \mathbb{C}^{n \times n}$ and $\tilde{b} \in \mathbb{C}^n$ by

$$\tilde{A}_{ik} := A_{ik} + q_i(\Delta A)_{ik}/\mathrm{sign}(\tilde{x}_k), \ (i, k = 1, \dots, n),$$
$$\tilde{b}_i := b_i - q_i(\Delta b)_i, \ (i = 1, \dots, n),$$

with the complex sign

$$\mathrm{sign}(x) = \begin{cases} x/|x| & \text{if } x \neq 0, \\ 1 & \text{if } x = 0. \end{cases}$$

(However, note that MATLAB evaluates `sign(0)` as 0.) Then $|\tilde{A} - A| \leq \Delta A$, $|\tilde{b} - b| \leq \Delta b$, and for $i = 1, \dots, n$,

$$\begin{aligned}(\tilde{b} - \tilde{A}\tilde{x})_i &= b_i - q_i(\Delta b)_i - \sum (A_{ik} + q_i(\Delta A)_{ik}/\mathrm{sign}(\tilde{x}_k))\tilde{x}_k \\ &= \left(b_i - \sum A_{ik}\tilde{x}_k\right) - q_i\left((\Delta b)_i + \sum (\Delta A)_{ik}|\tilde{x}_k|\right) \\ &= 0,\end{aligned}$$

so $\tilde{A}\tilde{x} = \tilde{b}$. Thus (i) is proved. $\qquad\square$

2.5.7 Remarks.

(i) The implication (ii) $\Rightarrow$ (i) in Theorem 2.5.6 says that an approximation $\tilde{x}$ to the solution x^* of $Ax^* = b$ with small residual can be represented as the solution of a slightly perturbed system $\tilde{A}\tilde{x} = \tilde{b}$. (This argument is a typical *backward error analysis*: Instead of the more difficult task of showing how data perturbations affect the solutions, one analyzes whether the approximations obtained can be interpreted as an exact solution of a nearby system.)

In particular, if the coefficients of the system $\tilde{A}\tilde{x} = \tilde{b}$ have a relative error of at most ε per component, then $|A - \tilde{A}| \le \varepsilon |A|$ and $|\tilde{b} - b| \le \varepsilon |b|$. An approximation $\tilde{x}$ to $x^* = A^{-1}b$ can then be regarded as acceptable if the relation $|b - A\tilde{x}| \le \varepsilon(|b| + |A| \cdot |\tilde{x}|)$ holds (i.e., if $\varepsilon \ge \varepsilon_0$), where

$$\varepsilon_0 := \max_i \frac{|b_i - \sum_k A_{ik}\tilde{x}_k|}{|b_i| + \sum_k |A_{ik}\tilde{x}_k|}. \tag{5.5}$$

(Here the natural interpretation for $0/0$ is zero.) ε_0 is a scaling invariant *quality factor* for computed approximations to systems of linear equations. It is always between 0 and 1, and it is small iff the error in $\tilde{x}$ can be explained by small relative errors in the original data.

(ii) If we exchange the quantities distinguished by $\sim$ and $*$, we can interpret the implication (i) $\Rightarrow$ (ii) as follows. The solution x^* of a system of linear equations $Ax^* = b$ approximates the solution $\tilde{x}$ of a slightly perturbed system $\tilde{A}\tilde{x} = \tilde{b}$ with $\tilde{A} \approx A$ and $\tilde{b} \approx b$ in the sense that the residual $\tilde{b} - \tilde{A}x^*$ remains small. As seen in Proposition 2.5.1, the absolute error $x^* - \tilde{x}$ can be a factor of $\|A^{-1}\|$ greater than the residual, and the relative error can be a factor of cond(A) greater than the relative residual. As the condition number for function evaluation in Section 1.4, the number cond(A) is therefore interpretable as the maximal magnification factor of the relative error.

As an application of Theorem 2.5.6, we prove an easily calculable upper bound on the distance to a singular matrix.

2.5.8 Proposition. *Let $\tilde{x} \ne 0$, and let*

$$\delta := \max_i \frac{|\sum_k A_{ik}\tilde{x}_k|}{\sum_k |A_{ik}\tilde{x}_k|}. \tag{5.6}$$

Then there is a singular matrix $\tilde{A}$ with $|\tilde{A} - A| \le \delta |A|$.

Proof. We have $|A\tilde{x}| \leq \delta |A| \cdot |\tilde{x}|$. By Theorem 2.5.6 with $b = \Delta b = 0$, there exists $\tilde{A}$ with $\tilde{A}\tilde{x} = 0$ and $|\tilde{A} - A| \leq \delta |A|$. Because $\tilde{x} \neq 0$, it follows that $\tilde{A}$ is singular. $\qquad\square$

To make δ small, a sensible choice for $\tilde{x}$ is a solution of

$$R\tilde{x} = e', \qquad (|e'_1| = \cdots = |e'_n| = 1) \tag{5.7}$$

where $PA = LR$ is a normalized triangular factorization that has been determined by using a column pivot search, and the signs of the e'_i are chosen during back substitution such that the $|\tilde{x}_i|$ are as large as possible. If A is close to a singular matrix, then $|A\tilde{x}|$ remains bounded, whereas, typically the solution of (5.7) determined in this way becomes very large and δ becomes tiny.

2.6 Iterative Refinement

The accuracy of the solution that has been calculated with the aid of Gaussian elimination can be improved by using the method of iterative refinement.

The equality $PA = LR$ is only approximate because of rounding error. It follows from this that the calculated "solution" $\tilde{x}$ differs, in general, from the "exact" solution $x^* = A^{-1}b$. The error vector $\delta^* := x^* - \tilde{x}$ satisfies

$$A\delta^* = Ax^* - A\tilde{x} = b - A\tilde{x};$$

therefore δ^* is the solution of the system of equations

$$A\delta^* = b - A\tilde{x}. \tag{6.1}$$

Thus we can calculate the error vector by solving a system of equations with the same matrix, so that no new triangular factorization is necessary. The right side $\tilde{r} := b - A\tilde{x}$ is called the *residual* corresponding to the approximation $\tilde{x}$; the residual corresponding to the exact solution x^* is zero. In an attempt to obtain a better approximation than $\tilde{x}$, we solve the system of equations (6.1) for δ^* and obtain $x^* = \tilde{x} + \delta^*$.

Because instead of the absolute error δ^* only an approximation $\tilde{\delta}$ can be calculated, we repeat this method with $\tilde{\tilde{x}} := \tilde{x} + \tilde{\delta}$ until $\|\tilde{\delta}\|$ does not become appreciably smaller – by a factor of 2, say. This serves as a convenient *termination criterion* for the iteration.

2.6.1 Algorithm: Iterative Refinement

STEP 1: Compute a permuted triangular factorization $PA = LR$, and initialize $x^0 = 0, r^1 = b, l = 1$.

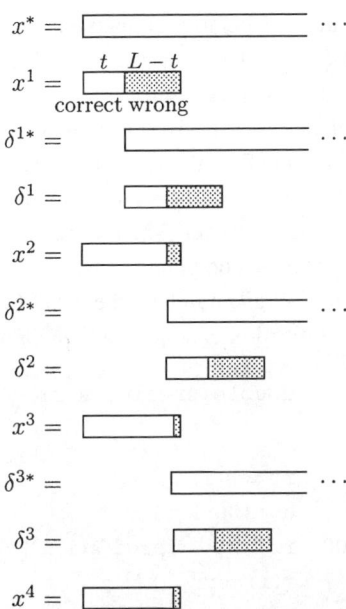

Figure 2.2. Iterative refinement (shaded digits are inaccurate).

STEP 2: Solve $LR\delta^l = Pr^l$ for δ^l and put $x^l = x^{l-1} + \delta^l$.

STEP 3: If $l > 1$ and $\|\delta^l\|_\infty \geq \frac{1}{2}\|\delta^{l-1}\|_\infty$: stop with $x^* = x^l$ as best approximation found. Otherwise, update $l = l + 1$.

STEP 4: Calculate $r^l = b - Ax^{l-1}$ (if possible in double precision) and continue with step 2.

Figure 2.2 illustrates how the iterative refinement algorithm improves the accuracy of a calculated solution. The solution $\tilde{x} = x^1$, calculated by means of Gaussian elimination, typically agrees with some leading digits of x^* (cf. the first two rows of the figure). Similarly, the calculated error $\tilde{\delta} = \delta^1$ agrees with some leading digits of the true error δ^{1*}; because δ is several orders of magnitudes smaller than x^*, the new error δ^{2*} is much smaller than δ^{1*}, and so on. In the particular example drawn, the calculated error δ^3 satisfies $\|\delta^3\|_\infty \geq \frac{1}{2}\|\delta^2\|_\infty$ and hence is not appreciably smaller than δ^2; we have reached the limiting accuracy. In this way, we can obtain nearly L valid digits for the calculated solution $\tilde{x}^*$, although the initial approximation was rather inaccurate.

2.6.2 Remarks.

(i) The error $\|x^* - \tilde{x}^*\|_\infty$ in Algorithm 2.6.1 almost always has an order of magnitude $\|\delta^l\|_\infty$, where l is the last index used; but this cannot be proved because one has no control over the rounding error.

(ii) For the implementation of iterative refinement, we need both A and the factorization LR. Because A is overwritten by LR, a copy of A must be made before computing the factorization.

(iii) Because the size of the error δ^* is determined by the residual through (6.1), it is essential to calculate the residual as accurately as possible. In MATLAB, double precision accuracy is used for all calculations, and nothing special must be done. However, when working in single precision, one can exploit the fact that most computers construct, for the multiplication of two single precision (L-digit) numbers the exact ($2L$-digit) product, and then round this to L digits. For example, in FORTRAN77, the instructions

```
         double precision rr
         ⋮
         rr = b(i)
         do 100 k = 1,n
100      rr = rr − dprod(A(i, k), x(k))
         r(i) = rr
```

produce the double precision value of the ith component of the residual $r = b - Ax$, with A and b stored in single precision only. The final rounding of rr to a single precision number stored in $r(i)$ is harmless because it has a small relative error.

2.6.3 Examples.

(i) The system $Ax = b$ with

$$A = \begin{pmatrix} 372 & 241 & -613 \\ -573 & 63 & 511 \\ 377 & -484 & 107 \end{pmatrix}, \quad b = \begin{pmatrix} 210 \\ -281 \\ 170 \end{pmatrix}$$

is to be solved. Single precision Gaussian elimination with iterative refinement using double precision residuals gives

x^0	x^1	x^2	x^3
17.46114540	17.46091557	17.46099734	17.46101689
16.99352741	16.99329805	16.99337959	16.99339914
16.93472481	16.93449497	16.93457675	16.93459606.

Comparing x^2 with x^3, one would expect that x^3 is accurate to six places. The quality index (5.5) has the value $\varepsilon_0 = .61_{10} - 8$ and indicates that

x^3 can be regarded as the exact solution of a system $\tilde{A}x^3 = \tilde{b}$ in which $\tilde{A}$ and $\tilde{b}$ are obtained from A and b by rounding to the next machine number.

Now let us consider the scaled system $(10^{-3}A)x = 10^{-3}b$. The given data are now no longer exactly representable in the (base 2) computer, and the resulting data errors have the following effect on the computed iterates.

$\tilde{x}^1$	$\tilde{x}_2$	$\tilde{x}^3$	$\tilde{x}^4$
17.46083331	17.46096158	17.46093845	17.46093869
16.99321566	16.99334383	16.99332047	16.99332070
16.93441272	16.93454123	16.93451762	16.93451786

One sees that, compared with the unscaled system, the seventh places of the last iterations are affected; we see also that the agreement of $\tilde{x}^3$ and $\tilde{x}^4$ up to the eighth place indicates a higher accuracy than is actually available. If one supposes that the coefficients of A and b are accurate only to three of the eight given figures, one must reckon with a loss of another $8 - 3 = 5$ figures in the result; then, only $6 - 5 = 1$ figures are correct. This indicates a nearly singular problem; we therefore estimate the distance to singularity from Proposition 2.5.8 and see that our prognosis of the accuracy was too optimistic: With the given choice of $\tilde{x}$, we obtain a singular matrix within a relative error $\delta = 0.002$. From this, it follows that an absolute error of 0.5 in all of the coefficients < 250 (of the original problem) leads to a possibly singular matrix, and one must doubt whether the absolute accuracy of 0.5 in the coefficients guarantees a nonsingular problem.

A more precise inspection of the data, which is still possible in this 3×3 problem, shows that if one replaces A_{21} with -572.5 and A_{23} with 510.5 then A becomes singular, since the sum of the columns becomes zero. Therefore the given data are actually too imprecise to furnish sensible results; the relative error of this particular singular matrix is $< 0.5/511 < 10^{-4}$, a factor of about 20 smaller than the upper bound from Proposition 2.5.8.

(ii) The system $Ax = b$ with

$$A = \begin{pmatrix} 372 & 241 & -125 \\ -573 & 63 & 182 \\ 377 & -484 & 437 \end{pmatrix}, \quad b = \begin{pmatrix} 155 \\ -946 \\ 38 \end{pmatrix}$$

is to be solved. Gaussian elimination with iterative refinement gives

$$
\begin{array}{cc}
x^0 & x^1 \\
-0.255\,228\,35 & -0.255\,228\,37 \\
2.811\,513\,04 & 2.811\,513\,04 \\
3.421\,037\,59 & 3.421\,037\,53,
\end{array}
$$

so that we expect that x^1 is accurate to eight places. If we assume that the coefficients (measured in millimetres) are accurate only to within an absolute tolerance of 0.5, we expect that $8 - 5 = 3$ places of x^1 are still reliable. The calculated bound $\delta_0 = 0.76$ from Proposition 2.5.8 indicates no particular problem.

Limiting Accuracy of Iterative Refinement

For singular or almost singular matrices, iterative refinement does not converge. To prove convergence and obtain a formula for the limiting accuracy of the iterative refinement algorithm 2.6.1 in the presence of rounding error, we suppose that the approximations (denoted by ~) satisfy for $l = 0, 1, \ldots$ the following inequalities (i) – (iv):

(i) $\|\tilde{r}^l - A\tilde{\delta}^l\|_\infty \le \varepsilon_0 \|\tilde{\delta}^l\|_\infty$ (this is a test for the calculation of δ^l; by Theorem 2.3.9, this is certainly satisfied with $\varepsilon_0 = 5n\varepsilon\|\tilde{L}\|_\infty\|\tilde{R}\|_\infty$);

(ii) $\|\tilde{x}^{l-1} + \tilde{\delta}^l - \tilde{x}^l\|_\infty \le \varepsilon\|\tilde{x}^l\|_\infty$ (rounding error due to the addition of the correction);

(iii) $\|b - A\tilde{x}^{l-1} - \tilde{r}^l\|_\infty \le \varepsilon_r$ (rounding error due to the calculation of the residual);

(iv) $\|A^{-1}\|_\infty\varepsilon_0 \le \frac{1}{4}$ (a quantitative nonsingularity requirement).

Assuming that

$$
\|\tilde{L}\|_\infty\|\tilde{R}\|_\infty \approx \|\tilde{L}\tilde{R}\|_\infty \approx \|A\|_\infty
$$

and using (i), requirement (iv) is approximately equivalent to the requirement $20n\varepsilon\,\mathrm{cond}_\infty(A) < 1$.

2.6.4 Theorem. *Suppose that (i) – (iv) hold for $l = 0, 1, \ldots$ and let $\tilde{x}^*$ be the approximation for x^* computed by iterative refinement. Then the error satisfies the relation*

$$
\|x^* - \tilde{x}^*\|_\infty \le 5\varepsilon^* \quad \text{where } \varepsilon^* = \|A^{-1}\|_\infty\varepsilon_r + \varepsilon\|\tilde{x}^*\|_\infty.
$$

Proof. From $x^* = A^{-1}b$ it follows that

$$
\begin{aligned}
\|x^* - \tilde{x}^{l-1} - \delta^l\|_\infty &= \|A^{-1}(\tilde{r}^l - A\tilde{\delta}^l + b - A\tilde{x}^{l-1} - \tilde{r}^l)\|_\infty \\
&\leq \|A^{-1}\|_\infty (\|\tilde{r}^l - A\tilde{\delta}^l\|_\infty + \|b - A\tilde{x}^{l-1} - \tilde{r}^l\|_\infty) \\
&\leq \|A^{-1}\|_\infty (\varepsilon_0\|\tilde{\delta}^l\|_\infty + \varepsilon_r) \text{ by (i) and (iii),} \\
&\leq \tfrac{1}{4}\|\tilde{\delta}^l\|_\infty + \|A^{-1}\|_\infty \varepsilon_r \text{ by (iv).}
\end{aligned}
$$

Now let l be the index such that $\tilde{x}^{l-1} = \tilde{x}^*$. Because the termination criterion must have been passed, we have $\|\tilde{\delta}^{l-1}\|_\infty \leq 2\|\tilde{\delta}^l\|_\infty$. Therefore

$$
\begin{aligned}
\|x^* - \tilde{x}^*\|_\infty &= \|(x^* - \tilde{x}^{l-1} - \tilde{\delta}^{l-1}) + (\tilde{x}^{l-1} + \tilde{\delta}^{l-1} - \tilde{x}^{l-1})\|_\infty \\
&\leq \|x^* - \tilde{x}^{l-1} - \tilde{\delta}^{l-1}\|_\infty + \varepsilon\|\tilde{x}^{l-1}\|_\infty, \text{ by (ii)} \\
&\leq \tfrac{1}{4}\|\tilde{\delta}^{l-1}\|_\infty + \|A^{-1}\|_\infty \varepsilon_r + \varepsilon\|\tilde{x}^*\|_\infty,
\end{aligned}
$$

whence

$$
\|x^* - \tilde{x}^*\|_\infty \leq \frac{1}{2}\|\tilde{\delta}^l\|_\infty + \varepsilon^*. \tag{6.2}
$$

From this, it follows that

$$
\begin{aligned}
\|\tilde{\delta}^l\|_\infty &= \|(x^* - \tilde{x}^*) - (x^* - \tilde{x}^{l-1} - \tilde{\delta}^l)\|_\infty \\
&\leq \|x^* - \tilde{x}^*\|_\infty + \|x^* - \tilde{x}^{l-1} - \tilde{\delta}^l\|_\infty \\
&\leq \tfrac{1}{2}\|\tilde{\delta}^l\|_\infty + \varepsilon^* + \tfrac{1}{4}\|\tilde{\delta}^l\|_\infty + \varepsilon^* \text{ by (i).}
\end{aligned}
$$

So $\|\tilde{\delta}^l\|_\infty \leq 8\varepsilon^*$, and substituting this into (6.2), $\|x^* - \tilde{x}^*\| \leq 5\varepsilon^*$. □

2.6.5 Remarks.

(i) The factor ε^* that occurs in the limiting accuracy estimate results from putting together the unavoidable rounding error $\varepsilon\|\tilde{x}^*\|_\infty$ and the error $\|A^{-1}\|_\infty \varepsilon_r$ caused by the inexact residual. The double precision calculation of the residual that has already been mentioned is therefore important for the reduction of ε_r because $\|A^{-1}\|_\infty$ can be comparatively large.

(ii) In the proof, the final δ^l is smaller than $8\varepsilon^*$ in norm, so that (as already mentioned) $\|\delta^l\|_\infty$ and $\|x^* - \tilde{x}^*\|_\infty$ have the same order of magnitude generally. However, no lower bound for $\|\delta^l\|_\infty$ can be obtained from the previous analysis, so that it is conceivable that $\|\delta^l\|_\infty$ is very much smaller than $\|x^* - \tilde{x}^*\|_\infty$.

(iii) Under the same hypotheses,

$$
\|x^* - \tilde{x}^1\|_\infty \leq \|A^{-1}\|_\infty \varepsilon_0\|\tilde{x}^1\|_\infty \leq \frac{1}{4}\|\tilde{x}^1\|_\infty
$$

for Gaussian elimination without iterative refinement if $\tilde{x}^1 = \tilde{\delta}^0$. For well-conditioned matrices $(\text{cond}(A) \approx \|A^{-1}\|_\infty$ for equilibrated matrices), Gaussian elimination without iterative refinement also gives good results; for (not too) ill-conditioned matrices, only the order of magnitude is correct. However, even then, if we obtain only one valid decimal without iterative refinement, then iterative refinement with sufficiently accurate calculation of the residuals provides good results. Of course, we then need about as many iterative steps as valid decimals. (Why?)

2.7 Error Bounds for Solutions of Linear Systems

Proposition 2.5.1 gives (fairly rough) error bounds for an approximation $\tilde{x}$ to the solution of a linear system $Ax = b$. The bounds cannot be improved in general (e.g., inequality (5.1) is best possible because $\|A^{-1}\| = \sup\{\|A^{-1}\tilde{r}\| \mid \|\tilde{r}\| = 1\}$). In particular, harmless looking approximations with a tiny residual can give rise to large errors if $\text{cond}(A)$ is large (or rather, if A is nearly singular).

2.7.1 Example. In order to illustrate this kind of effect, we consider an almost singular equilibrated matrix A. We obtain such a matrix if, for example, we change the singular matrix $\begin{pmatrix} 1 & 1 \\ 1 & 1 \end{pmatrix}$ by approximately 0.1% to $A := \begin{pmatrix} 1.001 & 0.999 \\ 0.999 & 1.001 \end{pmatrix}$. (Changing the matrix by a smaller relative error gives similar but more pronounced results.)

The solution of the system of equations $Ax^* = b$ with the right side $b := \begin{pmatrix} 2 \\ 2 \end{pmatrix}$ is $x^* = \begin{pmatrix} 1 \\ 1 \end{pmatrix}$. For the approximate solution $\tilde{x} = \begin{pmatrix} 1.001 \\ 1.001 \end{pmatrix}$, we calculate the residual $\tilde{r} = \begin{pmatrix} -0.001 \\ -0.002 \end{pmatrix}$ and the error $\delta^* = \begin{pmatrix} -0.001 \\ -0.001 \end{pmatrix}$. Because $A^{-1} = \begin{pmatrix} 250.25 & -249.75 \\ -249.75 & 250.25 \end{pmatrix}$, we have $\|A^{-1}\|_\infty = 500$. Hence the bound (5.1) gives $\|\delta^*\|_\infty \leq 1$, and the error in (5.1) (and in (5.2)) is overestimated by a factor of 1000. The totally wrong approximation $\tilde{x} = \begin{pmatrix} 2 \\ 0 \end{pmatrix}$ has a residual $\tilde{r} = \begin{pmatrix} -0.002 \\ 0.002 \end{pmatrix}$ of the same order, but the error is $\delta^* = \begin{pmatrix} -1 \\ 1 \end{pmatrix}$, and this time both bounds (5.1) and (5.2) are tight, without any overestimation.

We see from this example that for nearly singular matrices we cannot conclude from a small residual that the error is small. However, for equilibrated matrices, one can conclude from a large residual that the absolute error has at least the same order of magnitude,

$$\|\tilde{r}\| = \|A\delta^*\| \leq \|A\| \cdot \|\delta^*\| \approx \|\delta^*\|,$$

and it may be much larger when the matrix is ill-conditioned (which, for equilibrated matrices, is the same as saying that $\|A^{-1}\|$ is large).

From Theorem 2.3.9, we may deduce the heuristic but useful rule of thumb that for the computed solution $\tilde{x}$ one must expect a loss of accuracy of approximately $\log_{10} \text{cond}(A)$ decimal places (relative to the machine precision). This follows from the theorem if we make the additional assumption that

$$\| |\tilde{L}| |\tilde{R}| \| \approx \| \tilde{L}\tilde{R} \| \approx \| A \|$$

because then

$$\begin{aligned}
\| \tilde{x} - x^* \| &= \| A^{-1}(b - A\tilde{x}) \| \\
&\leq \| A^{-1} \| 5n\varepsilon \| |\tilde{L}| |\tilde{R}| \| \| \tilde{x} \| \\
&\approx 5n\varepsilon \, \text{cond}(A) \| \tilde{x} \|.
\end{aligned}$$

If we omit the factor $5n$ in the approximate equality in order to compensate for the overestimation of the "worst-case-analysis," then there remains a loss of $\log_{10} \text{cond}(A)$ decimal places.

Realistic Error Bounds

As seen previously, the use of $\| A^{-1} \|$ is often too crude for obtaining realistic error bounds. In particular the bound (5.1) is usually much too bad if $\tilde{x}$ is determined by iterative refinement. To get error bounds that can be assessed as to how realistic they are, one must invest more work.

An approximate but generally quite reliable estimate of the error (and at the same time an improved approximation) can be obtained with one step of iterative refinement, from the order of magnitude $\| \delta^0 \|$ of the first correction δ^0 (cf. Figure 2.2).

Rigorous and realistic error bounds can be obtained from a good approximation C for A^{-1} (called a *preconditioner* and computed, e.g., using Algorithm 2.2.9), as follows. Because $CA \approx I$, the norm $\| I - CA \|$ is small and the following theorem, due to Aird and Lynch [3], is applicable.

2.7.2 Theorem. *If $\| I - CA \| \leq \beta < 1$ then A is nonsingular, and for an arbitrary approximation $\tilde{x}$ of $x^* = A^{-1}b$,*

$$\frac{\| C(b - A\tilde{x}) \|}{1 + \beta} \leq \| x^* - \tilde{x} \| \leq \frac{\| C(b - A\tilde{x}) \|}{1 - \beta}. \tag{7.1}$$

Proof. By Proposition 2.4.2, the matrix CA is nonsingular. Because $0 \neq \det(CA) = \det C \det A$, the matrix A is nonsingular, too. From $A\delta^* = \tilde{r}$, one finds that

$$\delta^* = C\tilde{r} + (I - CA)\delta^*,$$

$$\|\delta^*\| \le \|C\tilde{r}\| + \|I - CA\| \, \|\delta^*\| \le \|C\tilde{r}\| + \beta\|\delta^*\|,$$

$$\|\delta^*\| \ge \|C\tilde{r}\| - \|I - CA\| \, \|\delta^*\| \ge \|C\tilde{r}\| - \beta\|\delta^*\|,$$

whence the assertion follows.   □

2.7.3 Remarks.

(i) The error bound (7.1) is realistic for small β because the actual error is overestimated by a factor of at most $q := (1 + \beta)/(1 - \beta)$; e.g., $q < 1.23$ for $\beta < 0.1$.

(ii) In order to reduce the effects of rounding error one should calculate the residual $b - A\tilde{x}$ in double precision. Indeed, inaccuracies in C or β affect the bound (7.1) much less than errors in the residual.

(iii) To evaluate the bound (7.1), we need $2n^3 + O(n^2)$ operations for the calculation of LR and C, $4n^3$ operations for the multiplication CA (an operation real $\circ$ interval involves two operations with reals!), and $O(n^2)$ for the remainder of the calculation. This means that altogether, $6n^3 + O(n^2)$ operations are necessary for the calculation of a realistic bound for the error of $\tilde{x}$. Comparing this with the $\frac{2}{3}n^3 + O(n^2)$ operations needed for the calculation of $\tilde{x}$ by Gaussian elimination, we see that the calculation of realistic and guaranteed error bounds requires about nine times the cost for solving the system approximatively.

Systems of Interval Equations

For many calculations with inexact data, the data are known to lie in specified intervals. In this case, we can use interval arithmetic to compute error bounds simultaneously with the solution instead of estimating the error from an analysis of the error propogation.

In the following, we work with a fixed interval matrix $\mathbf{A} \in \mathbb{IR}^{n \times n}$ and a fixed interval vector $\mathbf{b} \in \mathbb{IR}^n$. We want to find a good enclosure for the solution of $\tilde{A}\tilde{x}^* = \tilde{b}$, where it is only known that $\tilde{A} \in \mathbf{A}$ and $\tilde{b} \in \mathbf{b}$. Clearly, $\tilde{x}^*$ lies in the *solution set*

$$\Sigma(\mathbf{A}, \mathbf{b}) := \{\tilde{x}^* \mid \tilde{A}\tilde{x}^* = \tilde{b} \text{ for some } \tilde{A} \in \mathbf{A}, \tilde{b} \in \mathbf{b}\}.$$

The solution set is typically star-shaped. The calculation of the solution set is

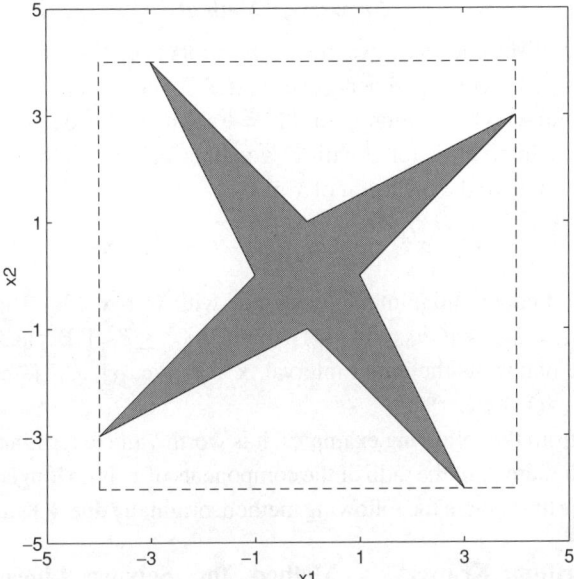

Figure 2.3. Star-shaped solution set.

quite expensive; for systems of dimension n, it can be shown that the star has up to 2^n spikes. We content ourselves with the computation of the interval hull $\square\Sigma(\mathbf{A}, \mathbf{b})$ of the solution set.

2.7.4 Example. The linear system of interval equations with

$$\mathbf{A} := \begin{pmatrix} [2, 2] & [-2, 1] \\ [-1, 2] & [2, 4] \end{pmatrix}, \quad \mathbf{b} := \begin{pmatrix} [-2, 2] \\ [-2, 2] \end{pmatrix}$$

has the solution set drawn in Figure 2.3. The hull of the solution set is $\square\Sigma(\mathbf{A}, \mathbf{b}) = ([-4, 4], [-4, 4])^T$, drawn with dashed lines. (To construct the exact solution set is a nontrivial task, even for 2×2 systems!)

Methods for determining the hull of the solution set are available (see, e.g., Neumaier [70, Chapter 6]) but must still in the worst case have an exponential effort. Indeed, the problem belongs to the class of NP-hard problems (Garey and Johnson [27]) for which it is a long-standing conjecture that no polynomial algorithms exists. However, there are several methods that give at least a reasonable inclusion of the solution set with a cost of $O(n^3)$.

Krawczyk's Method

To obtain a realistic enclosure, we imitate the error estimation of Aird and Lynch (Theorem 2.7.2). Let the preconditioner $C \in \mathbb{R}^{n \times n}$ be an approximation of the inverse of mid($\mathbf{A}$). The solution $\tilde{x}^*$ of $\tilde{A}\tilde{x}^* = \tilde{b}$ satisfies $\tilde{x}^* = C\tilde{b} + (I - C\tilde{A})\tilde{x}^*$. If we know an interval vector $\mathbf{x}^l$ with $\tilde{x}^* \in \mathbf{x}^l$, then also $\tilde{x}^* \in C\mathbf{b} + (I - C\mathbf{A})\mathbf{x}^l$ and we can improve the inclusion of $\tilde{x}^*$:

$$\tilde{x}^* \in \mathbf{x}^{l+1} := (C\mathbf{b} + (I - C\mathbf{A})\mathbf{x}^l) \cap \mathbf{x}^l.$$

How can we find an initial interval vector $\mathbf{x}^0$ with $\tilde{x}^* \in \mathbf{x}^0$? By Theorem 2.7.2 (for $x^0 = 0$), $\|\tilde{x}^*\|_\infty \leq \|C\tilde{b}\|_\infty/(1-\beta)$ if $\|I - C\tilde{A}\|_\infty \leq \beta < 1$. Because $\|C\tilde{b}\|_\infty \leq \|C\mathbf{b}\|_\infty$, we define, as the initial interval, $\mathbf{x}^0 := ([-\alpha, \alpha], \ldots, [-\alpha, \alpha])^T$ with $\alpha := \|C\mathbf{b}\|_\infty/(1 - \beta)$.

As seen from the following examples, it is worthwhile to terminate the iteration when the sum σ_l of the radii of the components of x^l is no longer rapidly decreasing. We thus obtain the following method, originally due to Krawczyk [53].

2.7.5 Algorithm: Krawczyk's Method for Solving Linear Interval Equations

$ier = 0$;
$\mathbf{a} = C * \mathbf{b}$;
$\mathbf{E} = I - C * \mathbf{A}$;
% To save space, $\mathbf{b}$ and $\mathbf{A}$ could be overwritten with $\mathbf{a}$ and $\mathbf{E}$.
$\beta = \max_i \sum_k |\mathbf{E}_{ik}|$; % must be rounded upwards
if $\beta \geq 1$, $ier = 1$; return; end;
$\alpha = \|\mathbf{a}\|_\infty/(1 - \beta)$;
$\mathbf{x} = ([-\alpha, \alpha], \ldots, [-\alpha, \alpha])^T$;
$\sigma_{old} = \inf$; $\sigma = \sum_k \text{rad}(\mathbf{x}_k)$; $fac = (1 + \beta)/2$;
while $\sigma < fac * \sigma_{old}$,
    $\mathbf{x} = (\mathbf{a} + \mathbf{Ex}) \cap \mathbf{x}$;
    $\sigma_{old} = \sigma$; $\sigma = \sum_k \text{rad}(\mathbf{x}_k)$;
end;

If we denote by $\mathbf{x}^l$ the box $\mathbf{x}$ after l iterations, then by construction, $\tilde{x}^* \in \mathbf{x}^l$ for all $l \geq 0$, and $\mathbf{x}^0 \supseteq \mathbf{x}^1 \supseteq \mathbf{x}^2 \ldots$.

2.7.6 Example. In order to make rigorous the error estimates computed in Examples 2.6.3(i)–(ii), we apply Krawczyk's method to the systems of linear equations discussed there. (To enable a comparison with the results in Examples 2.6.3, the calculations were performed in single precision.)

(i) In the first example, when all coefficients are treated as point intervals, $\beta = .000050545 < 1$, $\alpha = 51.392014$, and

x^1	x^2	$x^3 = x^4$
17.4^{63751}_{58548}	17.46^{1262}_{0690}	17.46^{1262}_{0690}
16.99^{5543}_{1516}	16.993^{578}_{005}	16.993^{578}_{006}
16.93^{6372}_{3080}	16.934^{851}_{214}	16.934^{851}_{214}
$\sigma_1 = .006260$	$\sigma_2 = .000891$	$\sigma_3 = .000891$

One can see that x^3 (and already x^2) have five correct figures. If one again supposes that there is an uncertainly of 0.5 in each coefficient, then all of the coefficients are genuine intervals; $A_{11} = [371.5, 372.5]$, and so on. Now, in Krawczyk's method, $\beta = 3.8025804 > 1$ suggesting numerical singularity of the matrix. No enclosure can be calculated.

(ii) In the second example, with point intervals we get the values $\beta = .000\,000\,309$, $\alpha = 6.4877797$, and

x^1	$x^2 = x^3$
-0.255228^6_1	-0.255228^5_3
2.81151^{34}_{27}	2.81151^{32}_{29}
3.42103^{83}_{69}	3.421037^7_3
$\sigma_1 = .00000125$	$\sigma_2 = .00000031$

with six correct figures for x^2. With interval coefficients corresponding to an uncertainty of 0.5 in each coefficient, we find $\beta = .01166754$, $\alpha = 6.5725773$, and (outwardly rounded at the fourth figure after the decimal point)

x^1	x^2	x^3	x^4
-0.2^{802}_{303}	-0.2^{645}_{460}	-0.2^{644}_{461}	-0.2^{644}_{461}
$2.^{8741}_{7489}$	$2.^{8347}_{7883}$	$2.^{8344}_{7887}$	$2.^{8344}_{7887}$
$3.^{5017}_{3404}$	$3.^{4509}_{3912}$	$3.^{4505}_{3916}$	$3.^{4504}_{3916}$
$\sigma_1 = .1681$	$\sigma_2 = .0622$	$\sigma_3 = .0613$	$\sigma_4 = .0613$

with just under two correct figures for x^4 (and already for x^2). The safe inclusions of Krawczyk's method are therefore about one decimal place more pessimistic here than the precision that was estimated for Gaussian elimination with iterative refinement.

We mention without proof the following statement about the quality of the limit $\mathbf{x}^\infty$, which obviously exists. A proof can be found in Neumaier [70].

2.7.7 Theorem. *The limit* $\mathbf{x}^{\infty}$ *of Krawczyk's iteration has the quadratic approximation property:*

$$\text{rad}\,\mathbf{A},\,\text{rad}\,\mathbf{b} = O(\varepsilon) \;\Rightarrow\; 0 \leq \text{rad}\,\mathbf{x}^{\infty} - \text{rad}\,\square\Sigma(\mathbf{A},\mathbf{b}) = O(\varepsilon^2).$$

Thus if the radii of input data $\mathbf{A}$ and $\mathbf{b}$ are of order of magnitude $O(\varepsilon)$, then the difference between the radius of the solution of Krawczyk's iteration $\mathbf{x}^{\infty}$ and the radius of the hull of the solution set is of order of magnitude $O(\varepsilon^2)$. The method therefore provides realistic bounds if $\mathbf{A}$ and $\mathbf{b}$ have small radii (compare with the discussion of the mean value form in Remark 1.5.9).

Even sharper results can be obtained with $O(n^3)$ operations by a more involved method discovered by Hansen and Bliek; see Neumaier [72].

Interval Gaussian Elimination

For special classes of matrices, especially for M-matrices, for diagonally dominant matrices, and for tridiagonal matrices, realistic error bounds can also be calculated without using Krawczyk's method (and so without knowing an approximate inverse), by performing Gaussian elimination in interval arithmetic. The recursions for the triangular factorization and for the solution of the triangular systems of equations are simply calculated using interval arithmetic. Because of the inclusion property of interval arithmetic (Theorem 1.5.6), this gives an inclusion of the solution set. For matrices of the special form mentioned, the quality of the inclusion is very good – in the case of M-matrices $\mathbf{A}$, one even obtains for many right sides $\mathbf{b}$ the precise hull $\square\Sigma(\mathbf{A},\mathbf{b})$:

2.7.8 Theorem. *If* $\mathbf{A}$ *is an M-matrix and* $\mathbf{b} \geq 0$ *or* $\mathbf{b} \leq 0$ *or* $0 \in \mathbf{b}$ *then interval Gaussian elimination gives the interval hull* $\square\Sigma(\mathbf{A},\mathbf{b})$ *of the solution set.*

The proof, which can be found in Neumaier [70], is based on the fact that in these cases the smallest and the largest elements of the hull belong to the solution set.

2.7.9 Example. The interval system of linear equations with

$$\mathbf{A} := \begin{pmatrix} [2,4] & [-2,0] \\ [-1,0] & [2,4] \end{pmatrix}, \quad \mathbf{b} := \begin{pmatrix} [-2,2] \\ [-2,2] \end{pmatrix}$$

satisfies the hypotheses of Theorem 2.20. The solution set (filled) and its interval hull (dashed) are drawn in Figure 2.4.

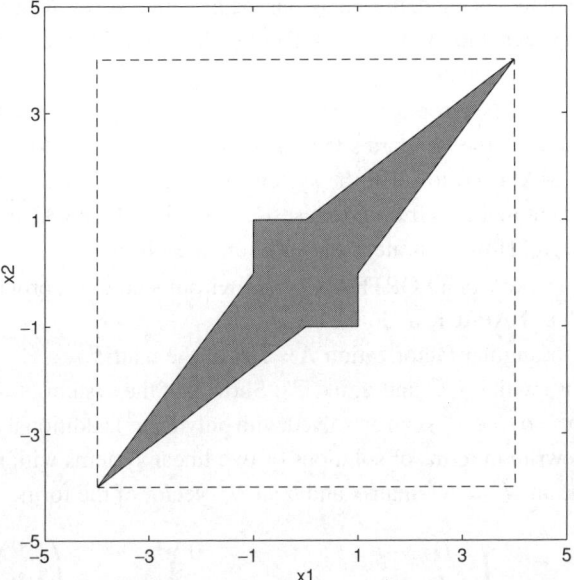

Figure 2.4. Solution set (filled) and interval hull of solution.

Note that, although interval Gaussian elimination can, in principle, be tried for arbitrary interval systems of linear equations, it can be recommended only for the classes of matrices mentioned previously (and certain other matrices, e.g., 2×2 matrices). In many other cases, the radii of the intervals can become so large in the course of the calculation (they may grow exponentially fast with increasing dimension) that at some stage, all candidates for pivot elements contain zero.

2.8 Exercises

1. The two systems of linear equations

$$Ax = b^{(l)}, \quad l = 1, 2 \tag{8.1}$$

with

$$A := \begin{pmatrix} 1 & 2 & -1 & 0 \\ 3 & 4 & 0 & 1 \\ 0 & 2 & 5 & 4 \\ 1 & 2 & 3 & 4 \end{pmatrix}, \quad b^{(1)} := \begin{pmatrix} 2 \\ 15 \\ 35 \\ 30 \end{pmatrix}, \quad b^{(2)} := \begin{pmatrix} -3 \\ -3 \\ 0 \\ -3 \end{pmatrix}$$

are given.

(a) Calculate, by hand, the triangular factorization $A = LR$; then calculate the exact solutions $x^{(l)}$ ($l = 1, 2$) of (8.1) by solving the corresponding triangular systems.

(b) Calculate the errors $\tilde{x}^{(l)} - x^{(l)}$ of the approximations $\tilde{x}^{(l)} = A \backslash b^{(l)}$ computed with the standard solution $x = A \backslash b$ of $Ax = b$ in MATLAB.

2. (a) Use MATLAB to solve the systems of equations $Ax = b$ with the coefficient matrices from Exercises 1, 9, 14, 26, 31, and 32 and right side $b := Ae$. How accurate is the solution in each case?

(b) If you work in a FORTRAN or C environment, use a program library such as LAPACK to do the same.

3. Given a triangular factorization $A = LR$ of the matrix $A \in \mathbb{C}^{n \times n}$. Let $B = A + \alpha u v^T$ with $\alpha \in \mathbb{C}$ and $u, v \in \mathbb{C}^n$. Show that the system of linear equations $Bx = b$ ($b \in \mathbb{C}^n$) can be solved with only $O(n^2)$ additional operations. *Hint*: Rewrite in terms of solutions of two linear systems with matrix A.

4. Let A be an $N^2 \times N^2$-matrix and b an N^2-vector of the form

$$
A := \begin{pmatrix}
D & -I & & & 0 \\
-I & D & -I & & \\
& \ddots & \ddots & \ddots & \\
& & & & -I \\
0 & & & -I & D
\end{pmatrix}, \quad
b := \begin{pmatrix}
b^{(1)} \\
b^{(2)} \\
b^{(2)} \\
\vdots \\
b^{(2)} \\
b^{(1)}
\end{pmatrix}
$$

where the $N \times N$-matrix D is given by

$$
D := \begin{pmatrix}
4 & -1 & & & 0 \\
-1 & 4 & -1 & & \\
& \ddots & \ddots & \ddots & \\
& & & & -1 \\
0 & & & -1 & 4
\end{pmatrix},
$$

I is the $N \times N$ unit matrix, and the N-vectors $b^{(1)}$ and $b^{(2)}$ are given by $b^{(1)} := (2, 1, 1, \ldots, 2)^T$ and $b^{(2)} := (1, 0, 0, \ldots, 0, 1)^T$, with $N \in \{4, 8, 12, 16, 20, \ldots\}$. Matrices such as A result when discretizing boundary value problems for elliptical partial differential equations.

Solve the system of equations $Ax = b$ using the sparse matrix features of MATLAB. Make a list of run times for different N. Suitable values depend on your hardware.

How many different techniques for reordering of sparse matrices are available in MATLAB? Use the MATLAB command spy to visualize the nonzero pattern of the matrix A before and after reordering of the rows and columns.

Compare computational effort and run time for solving the system in various reorderings to the results obtained from a calculation with matrix A stored as a full matrix.

Hints: Have a look at the Sparse item in the demo menu Matrices. You may also try help sparfun, help sparse, or help full.

5. Show that a square matrix A has a Cholesky factorization (possibly with singular L) if and only if it is positive semidefinite.

6. For the data $(x_i, y_i) = (-2, 0.5), (-1, 0.5), (0, 2), (1, 3.5), (2, 3.5)$, determine a straight line $f(x) = \alpha + \beta x$ such that $\sum_i (y_i - f(x_i))^2$ is minimal.

7. (a) Let P be a permutation matrix. Show that if A is Hermitian positive definite or is an H-matrix, then the same holds for PAP^H.

(b) A *monomial matrix* is a matrix A such that exactly one element in each row and column is not equal to zero. Show that A is square, and one can express each monomial matrix A as the product of a nonsingular diagonal matrix and a permutation matrix in both forms $A = PD$ or $A = D'P$.

8. (a) Show that one can realize steps 2 and 3 in Algorithm 2.2.9 for matrix inversion using $\frac{4}{3}n^3 + O(n^2)$ operations only. (Assume $P = I$, the unit matrix.)

(b) Use the algorithm to calculate (by hand) the inverse of the matrix of Exercise 1, starting with the factorization computed there.

9. (a) Show that the matrix

$$A := \begin{pmatrix} 2 & 2 & 1 \\ 1 & 1 & 1 \\ 3 & 2 & 1 \end{pmatrix}$$

is invertible but has no triangular factorization.

(b) Give a permutation matrix P such that PA has a nonsingular triangular factorization LR.

10. Show that column pivoting guarantees that the entries of L have absolute value ≤ 1, but those of R need not be bounded.

Hint: Consider matrices with $A_{ii} = A_{in} = 1$, $A_{ik} = -\beta$ if $i > k$, and $A_{ik} = 0$ otherwise.

11. (a) Find an explicit formula for the triangular factorization of the block
      matrix

$$
A = \begin{pmatrix}
I & 0 & \cdots & 0 & I \\
-B & I & & \mathbf{0} & 0 \\
 & -B & \ddots & & \vdots \\
 & & \ddots & \ddots & 0 \\
\mathbf{0} & & & -B & I
\end{pmatrix}
$$

with unit diagonal blocks and subdiagonal blocks $-B \in \mathbb{R}^{d \times d}$.

(b) Show that column pivoting produces no row interchanges if all entries
    of B have absolute value <1.

(c) Show that choosing for B a matrix with constant entries β $(d^{-1} < \beta < 1)$,
    some entries of the upper triangular factor R grow exponentially with
    the dimension of A.

(d) Check the validity of (a) and (b) with a MATLAB program, using
    MATLAB's `lu`.

Note that matrices of a similar form arise naturally in multiple shooting
methods for 2-point boundary value problems (see Wright [99]).

12. Write a MATLAB program for Gaussian elimination with column pivoting
for tridiagonal matrices. Show that storage and work count can be kept of
the order $O(n)$.

13. How does pivoting affect the band structure?

(a) For the permuted triangular factorization of a $(2m + 1)$-band matrix,
    show that L has $m + 1$ nonzero bands and R has $2m + 1$ nonzero bands.

(b) Asymptotically, how many operations are required to compute the per-
    muted triangular factorization of a $(2m + 1)$-band matrix?

(c) How many more operations (in %) are required for the solution of an
    additional system of linear equations with the same coefficient matrix
    and a different right side, with or without pivoting? (Two cases; keep
    only the leading term in the operation counts!) How does this compare
    with the relative cost of solving a dense system with an additional right
    side?

(d) Write an algorithm that solves a linear system with a banded matrix in
    a numerically stable way. Assume that the square matrix A is stored
    in a rectangular array AB such that $A_{ik} = AB(i, m + 1 + i - k)$ if
    $|k - i| \leq m$, and $A_{ik} = 0$ otherwise.

14. (a) In step 4 of Algorithm 2.2.9, the interchanges making up the permu-
      tation matrix must be applied to the columns, and in reverse order.
      Why?

(b) Write a MATLAB program that realizes the algorithm (make use of Exercise 8), and apply it to compute the inverse of the matrix

$$\begin{pmatrix} 0 & 1 & 1 \\ 2 & 3 & 4 \\ 5 & 5 & 6 \end{pmatrix}.$$

Compare with MATLAB's built in routine `inv`.

15. For $i = 1, \ldots, n$ let $\tilde{x}^{(i)}$ be the computed solution (in finite precision arithmetic) obtained by Gaussian elimination from the system

$$A^T x^{(i)} = e^{(i)}$$

in which $e^{(i)}$ is the ith unit vector.

(a) Show that the computed approximation $\tilde{C}$ to the inverse A^{-1} calculated in finite precision arithmetic as in Exercise 14 coincides with the matrix $(\tilde{x}^{(1)}, \ldots, \tilde{x}^{(n)})^T$.

(b) Derive from Theorem 2.3.9 an upper bound for the residual $|I - \tilde{C}A|$.

16. (a) Show that all norms on $\mathbb{C}^n$ are equivalent; that is, if $\| \cdot \|$ and $\| \cdot \|'$ are two arbitrary norms on $\mathbb{C}^n$, then there are always constants $0 < c, d \in \mathbb{R}$ such that for all $x \in \mathbb{C}^n$

$$c\|x\| \leq \|x\|' \leq d\|x\|.$$

Hint: It suffices to prove this for $\| \cdot \| = \| \cdot \|_\infty$. Now use that continuous functions on the cube surface $\|x\|_\infty = 1$ attain their extrema.

(b) What are the optimal such constants (largest possible c, smallest possible d) for pairs of p-norms ($p = 1, 2, \infty$)?

17. Show that the matrix norm belonging to an arbitrary vector norm has the following properties for all $\alpha \in \mathbb{C}$:

(a) $\|Ax\| \leq \|A\| \cdot \|x\|$

(b) $\|A\| \geq 0;\ \|A\| = 0 \iff A = 0$

(c) $\|\alpha A\| = |\alpha| \cdot \|A\|$

(d) $\|A + B\| \leq \|A\| + \|B\|$

(e) $\|AB\| \leq \|A\| \cdot \|B\|$

(f) $|A| \leq B \Rightarrow \|A\| \leq \| \, |A| \, \| \leq \|B\|$ if the norm is monotone

(g) The implication

$$|A| \leq B \Rightarrow \|A\| = \|\,|A|\,\| \leq \|B\|$$

holds for the matrix norms $\| \cdot \|_1$ and $\| \cdot \|_\infty$, but not in general for $\| \cdot \|_2$.

18. Prove the following bounds for the inverse of a matrix A:

(a) If $\|Ax\| \geq \gamma \|x\|$ for all $x \in \mathbb{C}^n$ with $\gamma > 0$ and $\|\cdot\|$ an arbitrary vector norm then A^{-1} exists and $\|A^{-1}\| \leq \gamma^{-1}$ for the matrix norm belonging to the vector norm $\|\cdot\|$.

(b) If $x^H A x \geq \gamma \|x\|_2^2$ for all $x \in \mathbb{C}^n$ with $\gamma > 0$, then $\|A^{-1}\|_2 \leq \gamma^{-1}$.

(c) If the denominator on the right side of the following three expressions is positive, then

$$\|A^{-1}\|_\infty \leq 1/\min_i \left(|A_{ii}| - \sum_{k \neq i} |A_{ik}| \right),$$

$$\|A^{-1}\|_1 \leq 1/\min_k \left(|A_{kk}| - \sum_{i \neq k} |A_{ik}| \right),$$

$$\|A^{-1}\|_2 \leq 1/\min_i \left(A_{ii} - \frac{1}{2} \sum_{k \neq i} |A_{ik} + A_{ki}| \right).$$

Hints: For arbitrary $x, y \in \mathbb{C}^n$, $|x^H y| \leq \|x\|_2 \|y\|_2$. For arbitrary $u, v \in \mathbb{C}$, $|uv| \leq \frac{1}{2}|u|^2 + \frac{1}{2}|v|^2$.

19. A matrix A is called an *M-matrix* if

$$A_{ii} \geq 0, \quad A_{ik} \leq 0 \quad \text{for } i \neq k;$$

and one of the following equivalent conditions holds:

(i) There are diagonal matrices D_1 and D_2 such that $\|I - D_1 A D_2\| < 1$ (i.e., A is an H-matrix).

(ii) There is a vector $u > 0$ with $Au > 0$.

(iii) $Ax \geq 0 \Rightarrow x \geq 0$.

(iv) A is nonsingular and $A^{-1} \geq 0$.

Show that for matrices with the above sign distribution, these conditions are indeed equivalent.

20. (a) Let $A \in \mathbb{C}^{n \times n}$ be nonpositive outside the diagonal, $A_{ik} \leq 0$ for $i \neq k$. Then A is an M-matrix iff there are vectors $u, v > 0$ such that $Au \geq v$.

(b) If A' and A'' are M-matrices and $A' \leq A \leq A''$ (componentwise) then A is an M-matrix.

Hint: Use Exercise 19. Show first that $C = (A'')^{-1} B$ is an M-matrix; then use $B^{-1} = (A''C)^{-1}$.

21. A lot of high-quality public domain software is freely available on the

World Wide Web (WWW). The distributors of MATLAB maintain a page

http://www.mathworks.com/support/ftp/

with links to user-contributed software in MATLAB.

(a) Find out what is available about linear algebra, and find a routine for the LDL^T factorization. Generate a random normalized lower triangular matrix L and a diagonal matrix D with diagonal entries of both signs, and check whether the routine reconstructs L and D from the product LDL^T.

(b) If you are interested what one can do about linear systems with very ill-conditioned matrices, get the regularization tools, described in Hansen [40], from

http://www.imm.dtu.dk/ pch/Regutools/regutools.html

It takes a while to explore this interesting package!

22. Let x^* be the solution of the system of linear equations

$$\frac{1}{3}x_1 + \frac{1}{7}x_2 = \frac{10}{21},$$
$$\frac{1}{4}x_1 + \frac{29}{280}x_2 = \frac{99}{280}. \tag{8.2}$$

(a) Calculate the condition number of the coefficient matrix for the row-sum norm.

(b) An approximation to the system (8.2) may be expressed in the form

$$0.333x_1 + 0.143x_2 = 0.477,$$
$$0.250x_1 + 0.104x_2 = 0.353 \tag{8.3}$$

by representing the coefficients A_{ik} with three places after the decimal point. Let the solution of (8.3) be $\tilde{x}$. Calculate the relative error $\|x^* - \tilde{x}\|/\|x^*\|$ of the *exact* solutions x^* and $\tilde{x}$ of (8.2) and (8.3), respectively (calculating by hand with rational numbers).

23. Use MATLAB's cond to compute the condition numbers of the matrices A with entries $A_{ik} = f_k(x_i)$ ($k = 1, \ldots, n$ and $i = 1, \ldots, m$; $x_i = (i - 1)/(m - 1)$ with $m = 10$ and $n = 5, 10, 20$) in the following polynomial bases:

(a) $f_k(x) = x^{k-1}$,

(b) $f_k(x) = (x - 1/2)^{k-1}$,

(c) $f_k(x) = T_k(2x - 1)$, where $T_k(x)$ is the kth Chebyshev polynomial (see the proof of Proposition 3.1.14),

(d) $f_k(x) = \prod_{j=m-k, j \text{ even}}^{m-1+k} (x - x_{j/2})$.

Plot in each case some of the basis functions. (Only the bases with small condition numbers are suitable for use in least squares data fitting.)

24. A matrix Q is called *unitary* if $Q^H Q = I$. Show that

$$\text{cond}_2(Q) = 1,$$

$$\text{cond}_2(A) = \text{cond}_2(QA) = \text{cond}_2(AQ) = \text{cond}_2(Q^H AQ)$$

for all unitary matrices $Q \in \mathbb{C}^{n \times n}$ and any invertible matrix $A \in \mathbb{C}^{n \times n}$.

25. (a) Show that for the matrix norm corresponding to any monotone norm,

$$\|A\| \geq \max |A_{ii}|,$$

and, if A is triangular,

$$\text{cond}(A) \geq \max \left| \frac{A_{ii}}{A_{kk}} \right|.$$

(b) Show that, for the spectral norm, $\text{cond}_2(A^T) = \text{cond}_2(A)$.

(c) Show that the condition number of a positive definite matrix A with Cholesky factorization $A = LL^T$ satisfies

$$\text{cond}_2(A) = \text{cond}_2(L)^2 \geq \max \left| \frac{L_{ii}}{L_{kk}} \right|^2.$$

26. (a) Calculate (by hand) an estimate c of $\text{cond}(A)$ from Algorithm 2.5.5 for

$$A := \begin{pmatrix} 0 & -1 & 1 \\ -3 & 3 & 0 \\ -3 & 7 & -1 \end{pmatrix}.$$

Is this estimated value exact, that is, is $c = \|A\|_\infty \|A^{-1}\|_\infty$?

(b) Give a 2×2 matrix A for which the estimated value c is not exact, that is, for which $s < \|A^{-1}\|_\infty$.

27. (a) Does the MATLAB operation x = A\b for solving $Ax = b$ include iterative refinement? Find out by implementing an iterative refinement method based on this operation, and test it with *Hilbert matrices A* with entries $A_{ik} = 1/(i + k - 1)$ and all-one right side $b = e$.

(b) For Hilbert matrices of dimension $n = 5, 10, 15, 20$, after how many steps of iterative refinement is the termination criterion satisfied?

(c) Confirm that Hilbert matrices of dimension n are increasingly ill conditioned for increasing n by calculating their condition number for the dimensions $n = 5, 10, 15, 20$. How does this explain the results of (b)?

28. The two-point boundary value problem

$$-y''(x) + \frac{k(k-1)}{(1-x)^2} y(x) = 0, \quad y(0) = 1, \quad y(1) = 0$$

can be approximated by the linear tridiagonal system of linear equations $Ty = b$ with

$$T = \begin{pmatrix} 2+q_1 & -1 & & & 0 \\ -1 & 2+q_2 & \ddots & & \\ & \ddots & \ddots & \ddots & \\ & & \ddots & \ddots & -1 \\ 0 & & & -1 & 2+q_n \end{pmatrix}, \quad b = e^{(1)}$$

with $q_i = k(k-1)/(n+1-i)^2$. Then $y(i/(n+1)) \approx y_i$.

Solve this system for $k = 0.5$ with $n = 2^s - 1$ for several s, using Exercise 12 and (at least) one step of iterative refinement to obtain an approximation $\tilde{y}^{(s)}$ for the solution vector. Print the approximate values at the points $x = 0.25, 0.5,$ and 0.75 with and without iterative refinement, and compare with the solution $y(x) = (1-x)^k$ of the boundary value problem. Then do the same for $k = 4$.

29. (a) Write a MATLAB program for the iterative refinement of approximate solutions of least squares problems. Update the approximate solution x^l in the lth step of the method by $x^{l+1} = x^l + s^l$, where s^l be the solution of $R^T R s^l = A^T (b - A x^l)$ and L is a Cholesky factor of the normal equations (found with MATLAB's chol).

(b) Give a qualitative argument for the limit accuracy that can be obtained by this way of iterative refinement.

30. The following technique may be used to assess heuristically the accuracy of linear or nonlinear systems solvers or eigenvalues calculation routines without doing any formal error analysis. (It does *not* work for problems involving approximations other than rounding errors, as in numerical differentiation, integration, or solving differential equations.)

Solve the systems $(kA)x^{(k)} = kb$ for $k = 1, 3, 5$ numerically, and derive heuristic error estimates $|x^* - \bar{x}| \approx \Delta x$ for the solution x^* of $Ax^* = b$, where

$$\bar{x} = \frac{1}{3}\left(x^{(1)} + x^{(3)} + x^{(5)}\right),$$

$$\Delta x_i = \sqrt{\frac{1}{2}\left(\left(x_i^{(1)} - \bar{x}_i\right)^2 + \left(x_i^{(3)} - \bar{x}_i\right)^2 + \left(x_i^{(5)} - \bar{x}_i\right)^2\right)}.$$

Compare with the true error for a system with known x^*, constructed by choosing random integral A and x^* and $b = Ax^*$. Explain! How reliable is the heuristics?

31. Given the system of linear equations $Ax = b$ with

$$A := \begin{pmatrix} 0.051 & -0.153 & 0 \\ -0.153 & -0.737 & -0.598 \\ 0 & -0.598 & -0.299 \end{pmatrix}, \quad b := \begin{pmatrix} 1 \\ 3 \\ 3 \end{pmatrix}.$$

(a) Give a matrix $\tilde{A}$ and a vector $\tilde{b}$ such that $\tilde{A}\tilde{x} = \tilde{b}$ where

$$\tilde{x} := \begin{pmatrix} 4280 \\ 1420 \\ -2850 \end{pmatrix},$$

$|A - \tilde{A}| \le \varepsilon_0 |A|$, and $|b - \tilde{b}| \le \varepsilon_0 |b|$, with smallest possible relative error ε_0.

(b) Use $\tilde{x}$ to determine a singular matrix $\tilde{A}$ close to A, and compute the "estimated singularity distance" (5.6).

32. (a) Using the built in solver `verifylss` of INTLAB to solve linear interval equations with the four symmetric interval coefficient matrices A defined by either

$$A := \begin{pmatrix} 36.1 & & & & & \\ -63.4 & 14.7 & & & \text{symm.} & \\ 33.1 & -88.5 & 56.9 & & & \\ -75.2 & 21.6 & -14.0 & 36.3 & & \\ 75.8 & -22.4 & 15.2 & -39.0 & 44.1 & \\ -27.4 & 83.3 & -58.3 & 15.4 & -17.0 & 69.4 \end{pmatrix}$$

or

$$A := \begin{pmatrix} d & c & c & \cdots & c \\ c & d & c & \cdots & c \\ \vdots & & \ddots & & \vdots \\ c & \cdots & & c & d \end{pmatrix}$$

with dimension $n = 16$ and c, d chosen from the following:

	c	d
(a)	1.25	−3.34
(b)	1.25	3.34
(c)	−1.26	18.9

Interpret the matrices either as thin interval matrices or as interval matrices obtained by treating each number as only accurate to the number of digits shown. Use as right side $b = [15.65, 15.75]e$ when the matrix is thin, and $b = e = (1, \ldots, 1)^T$ otherwise.

Is the system solvable in each of the eight cases?

(b) Implement Krawczyk's algorithm in INTLAB and compare the accuracy resulting in each iteration with that of the built-in solver.

3

Interpolation and Numerical Differentiation

In this chapter, we discuss the problem of finding a "nice" function of a single variable that has given values at specified points. This is the so-called *interpolation problem*. It is important both as a theoretical tool for the derivation and analysis of other numerical algorithms (e.g., finding zeros of functions, numerical integration, solving differential equations) and as a means to approximate functions known only at a finite set of points.

Because a continuous function is not uniquely defined by a finite number of function values, one must specify in advance a class of interpolation functions with good approximation properties. The simplest class, polynomials, has considerable theoretical importance; they can approximate any continuous functions in a bounded interval with arbitrarily small error. However, they often perform poorly when used to match many function values at specific (e.g., equally spaced) points over a wide interval because they tend to produce large spurious oscillations near the ends of the interval in which the data are given. Hence polynomials are used only over narrow intervals, and large intervals are split into smaller ones on which polynomials perform well. This results in interpolation by piecewise polynomials, and indeed, the most widely used interpolation schemes today are based on so-called splines – that is, piecewise polynomials with strong smoothness properties.

In Section 3.1, we discuss the basic properties of polynomial interpolation, including a discussion of its limitations. Section 3.2 then treats the important special case of extrapolation to the limit, and applies it to *numerical differentiation*, the problem of finding values of the derivative of a function for which one has only a routine computing function values. Section 3.3 treats piecewise polynomial interpolation in the simplest important case, that of *cubic splines*, and discusses their excellent approximation properties. Finally, Section 3.5 relates splines to another class of important interpolation functions, so-called *radial basis functions*.

3.1 Interpolation by Polynomials

The simplest class of interpolating functions are the polynomials.

3.1.1 Theorem. (Lagrange Interpolation Formulas). *For any pairwise distinct points $x_0, \ldots, x_n$, there is a unique polynomial p_n of degree $\leq n$ that interpolates $f(x)$ on $x_0, \ldots, x_n$. It can be represented explicitly as*

$$p_n(x) = \sum_{i=0:n} f(x_i) L_i(x), \tag{1.1}$$

where

$$L_i(x) := \prod_{\substack{j=0:n \\ j \neq i}} \frac{x - x_j}{x_i - x_j}. \tag{1.2}$$

p_n is called the *interpolation polynomial* to f at $x_0, \ldots, x_n$, and the $L_i(x)$ are referred to as *Lagrange polynomials*.

Proof. By definition, the L_i are polynomials of degree n with

$$L_i(x_j) = \begin{cases} 0 & \text{if } i \neq j, \\ 1 & \text{if } i = j. \end{cases}$$

From this it follows that (1.1) is a polynomial of degree $\leq n$ satisfying $p_n(x_j) = f(x_j)$ for $j = 0, \ldots, n$.

For the proof of uniqueness, we suppose that p is an arbitrary polynomial of degree $\leq n$ with $p(x_j) = f(x_j)$ for $j = 0, \ldots, n$. Then the polynomial $p_n - p$ is of degree $\leq n$ and has (at least) the $n + 1$ pairwise distinct zeros $x_0, \ldots, x_n$. Therefore, $p_n(x) - p(x)$ is divisible by the product $(x - x_0) \cdot \ldots \cdot (x - x_n)$. However, the degree of the divisor is $> n$; hence, this is possible only if $p_n(x) - p(x)$ is identically zero (i.e., if $p = p_n$). $\qquad\square$

Although this result solves the problem completely, there are many situations in which a different representation of the interpolation polynomial is more useful. Note that the interpolation polynomial is uniquely determined by the data, no matter in which form the polynomial is expressed. Hence one can pick the form of the interpolation polynomial according to other considerations, such as ease of use, computational cost, or numerical stability.

Linear and Quadratic Interpolation

To motivate the alternative approach we first look at linear and quadratic interpolation, the case of polynomials of degrees 1 and 2.

The linear interpolation problem is the case $n = 1$ and consists in finding for given $f(x_0)$ and $f(x_1)$ a linear polynomial p_1 with $p_1(x_0) = f(x_0)$ and $p_1(x_1) = f(x_1)$. The solution is well known from school: the straight line through the points (x_0, y_0), (x_1, y_1) is given by the equation

$$\frac{y - y_0}{x - x_0} = \frac{y_1 - y_0}{x_1 - x_0},$$

and, in the limiting case $x_1 \to x_0$, the line through x_0 with given slope y_0' by

$$\frac{y - y_0}{x - x_0} = y_0'.$$

The linear interpolation polynomial has therefore the representation

$$p_1(x) = f(x_0) + f[x_0, x_1](x - x_0), \tag{1.3}$$

where

$$f[x_0, x_1] := \begin{cases} \dfrac{f(x_1) - f(x_0)}{x_1 - x_0} & \text{if } x_1 \neq x_0, \\ f'(x_0) & \text{if } x_1 = x_0. \end{cases} \tag{1.4}$$

We call $f[x_0, x_1]$ a *first-order divided difference* of f. This notation for the slope, and the associated form (1.3) proves to be very useful. In particular, (1.3) is often a much more appropriate form to express linear polynomials numerically than using the representation $p(x) = mx + b$ in the power basis.

3.1.2 Example. (With optimal rounding, $B = 10$, $L = 5$). The straight line through the points $(6000, \frac{1}{3})$ and $(6001, -\frac{2}{3})$ may be represented, according to (1.3), by

$$p_1(x) \approx 0.33333 + \frac{0.33333 + 0.66667}{6000 - 6001}(x - 6000)$$
$$\approx 0.33333 - 1.0000(x - 6000). \tag{1.5}$$

In this form, the evaluation of $p_1(x)$ is stable and reproduces the function values at x_0 and x_1. However, if we convert (1.5) to standard form $p_1(x) = mx + b$, the resulting formula

$$\hat{p}_1(x) \approx -1.0000x + 6000.3$$

gives very inaccurate values at the interpolating points:

x	6000	6001
$p_1(x)$	0.33333	-0.66667
$\hat{p}_1(x)$	0.30000	-0.70000

From linear algebra, we are used to considering any two bases of a vector space as equivalent. However, for numerical purposes, the example shows significant differences. In particular, we conclude that expressing polynomials in the power basis $1, x, x^2, \ldots$ must be avoided to ensure numerical stability.

To find the quadratic interpolation polynomial p_2 (a parabola, or, in a limiting case, a straight line) to f at pairwise distinct points x_0, x_1, x_2, we modify the linear interpolation formula by a correction term that vanishes for $x = x_0$ and $x = x_1$:

$$p_2(x) := f(x_0) + f[x_0, x_1](x - x_0) + f[x_0, x_1, x_2](x - x_0)(x - x_1).$$

Because $x_2 \neq x_0, x_1$, the coefficient at our disposal, suggestively written as $f[x_0, x_1, x_2]$, can be determined from the interpolation requirement $p_2(x_2) = f(x_2)$. We find the *second-order divided difference*

$$f[x_0, x_1, x_2] := \frac{f[x_0, x_2] - f[x_0, x_1]}{x_2 - x_1} \quad \text{if } x_2 \neq x_1, x_0. \qquad (1.6)$$

3.1.3 Example. Let $f(0) = 1$, $f(1) = 3$, $f(3) = 2$. The parabola interpolating at 0, 1 and 3 is calculated from the divided differences

$$f(x_0) = 1, \quad f[x_0, x_1] = \frac{3-1}{1-0} = 2, \quad f[x_0, x_2] = \frac{2-1}{3-0} = \frac{1}{3}$$

and

$$f[x_0, x_1, x_2] = \frac{\frac{1}{3} - 2}{3 - 1} = -\frac{5}{6}$$

to

$$p_2(x) = 1 + 2x - \frac{5}{6}x(x - 1) = -\frac{5}{6}x^2 + \frac{17}{6}x + 1.$$

For stability reasons, the first of the two expressions for $p_2(x)$ is, in finite precision calculations, preferable to the second expression in the power basis.

Note that the divided difference notation (1.6), used with the variable x in place of x_2, allows us to write the linear interpolation formula (1.3) as an identity

$$f(x) = f(x_0) + f[x_0, x_1](x - x_0) + f[x_0, x_1, x](x - x_0)(x - x_1)$$

by adding the error term

$$f[x_0, x_1, x](x - x_0)(x - x_1).$$

Newton Interpolation Formulas

To obtain the cubic interpolating polynomial, we may correct the quadratic interpolation formula with a correction term that vanishes for $x = x_0, x_1, x_2$; by repetition of this process, we can find interpolating polynomials of any degree in a similar way. This yields the interpolation formulas of Newton. A generalization due to Hermite extends these formulas to the case where repetitions of some of the x_j are allowed; thus we do not assume that the x_j are pairwise distinct.

3.1.4 Theorem. (Newton Interpolation Formulas). *Let $D \subseteq \mathbb{R}$ be a closed interval, let $f : D \to \mathbb{R}$ be an $(n+1)$-times continuously differentiable function, and let $x_0, x_1, \ldots, x_n \in D$. Then:*

(i) There are uniquely determined continuous functions $f[x_0, \ldots, x_i, \cdot] : D \to \mathbb{R}$ $(i = 0, 1, \ldots, n)$ such that

$$f[x_0, \ldots, x_{i-1}, x] = f[x_0, \ldots, x_{i-1}, x_i] + f[x_0, \ldots, x_i, x](x - x_i) \quad (1.7)$$

for all $x \in D$.
(ii) For $i = 0, 1, \ldots, n$ the polynomial

$$p_i(x) := f[x_0] + f[x_0, x_1](x - x_0) + \cdots$$
$$+ f[x_0, \ldots, x_{i-1}, x_i](x - x_0) \cdots (x - x_{i-1}) \quad (1.8)$$

interpolates the function $f(x)$ at the points $x_0, \ldots, x_i \in D$.
(iii) We have

$$f(x) = p_i(x) + f[x_0, \ldots, x_i, x](x - x_0) \cdots (x - x_i). \quad (1.9)$$

Proof. To prove (1.7), it suffices to show that

$$g_i(x) := f[x_0, \ldots, x_{i-1}, x]$$

is $n-i$ times continuously differentiable in x and satisfies

$$g_i(x) = g_i(x_i) + (x - x_i)g_{i+1}(x). \quad (1.10)$$

We proceed by induction, starting with $g_0(x) = f[x] = f(x)$ for $i = 0$, where this follows from our discussion of the linear case. Hence suppose that (1.10) is true with $i-1$ in place of i. Then g_i is $n - i > 0$ times continuously differentiable and

$$g_i(x) = g_i(x_i) + (x - x_i) \int_0^1 g_i'(x_i + t(x - x_i))\, dt. \quad (1.11)$$

Thus, the function g_{i+1} defined by

$$g_{i+1}(x) = f[x_0, \dots, x_i, x] := \int_0^1 g_i'(x_i + t(x - x_i))\, dt$$

is still $n-i-1$ times continuously differentiable, and (1.10) holds for i. Equation (1.9) is also proved by induction; the case $i = 0$ was already treated. If we suppose that (1.9) holds with $i - 1$ in place of i, we can use (1.7) and the definition (3.1.8) to obtain

$$\begin{aligned}
f(x) &= p_{i-1}(x) + (f[x_0, \dots, x_{i-1}, x_i] + f[x_0, \dots, x_i, x](x - x_i)) \\
&\quad \times (x - x_0) \cdots (x - x_{i-1}) \\
&= p_i(x) + f[x_0, \dots, x_i, x](x - x_0) \cdots (x - x_i).
\end{aligned}$$

Thus (1.9) holds generally.

Finally, substitution of $x = x_j$ into (1.9) gives the interpolation property $f(x_j) = p_i(x_j)$ for $j = 0, \dots, i$. $\qquad \square$

Equation (1.8) is called the *Newton form* of the interpolation polynomial at $x_0, \dots, x_i$. To evaluate an interpolation polynomial in Newton form, it is advisable to use a Horner-like scheme. The recurrence

$$\begin{aligned}
v_n &:= f[x_0, \dots, x_n] \\
v_i &:= f[x_0, \dots, x_i] + (x - x_i)v_{i+1} \quad (i = n - 1, n - 2, \dots, 0)
\end{aligned}$$

gives rise to the expression

$$\begin{aligned}
v_i &= f[x_0, \dots, x_i] + f[x_0, \dots, x_{i+1}](x - x_i) + \cdots \\
&\quad + f[x_0, \dots, x_n](x - x_i) \cdots (x - x_{n-1})
\end{aligned}$$

from which $v_0 = p_n(x)$. A corresponding pseudo-MATLAB program is as follows.

3.1.5 Algorithm: Evaluation of the Newton Form

% Assume $d_i = f[x_0, \ldots, x_i]$

$v = d_n$;

for $i = n - 1 : -1 : 0$,

   $v = d_i + (x - x_i) * v$;

end;

% Now v is the interpolated approximation to $f(x)$

(Because MATLAB has no indices < 1, a true MATLAB code should interpolate instead $x_1, \ldots, x_n$ by a polynomial of degree $\leq n - 1$ and replace the zeros in this pseudo code with 1s.)

For the determination of the coefficients required in the Newton form, it is useful to have a general interpretation of the $f[x_0, \ldots, x_i]$ as higher order *divided differences*, a symmetry property for divided differences, and a formula for a limiting case. These generalize the fact that the slope of a line through two points of a curve is independent of the ordering of these points and becomes the tangent slope in the limit when these points merge.

3.1.6 Proposition.

(i) We have the divided difference representation

$$f[x_0, \ldots, x_i, x] = \frac{f[x_0, \ldots, x_{i-1}, x] - f[x_0, \ldots, x_{i-1}, x_i]}{x - x_i}$$

$$\text{for } x \neq x_i.$$

(ii) The derivative with respect to the last argument is

$$\frac{d}{dx} f[x_0, \ldots, x_{i-1}, x] = f[x_0, \ldots, x_{i-1}, x, x].$$

(iii) The divided differences are symmetric functions of their arguments, that is, for an arbitrary permutation π of the indices $0, 1, \ldots, i$, we have

$$f[x_0, x_1, \ldots, x_i] = f[x_{\pi 0}, x_{\pi 1}, \ldots, x_{\pi i}].$$

Proof. (i) follows directly from (1.7) and (ii) as limiting case from (i). To prove (iii), we note that the polynomial $p_i(x)$ that interpolates f at the points $x_0, \ldots, x_i$ has the highest coefficient $f[x_0, \ldots, x_i]$. A permutation π of the indices gives the same interpolation polynomial, but the representation (1.9) yields as highest coefficient $f[x_{\pi 0}, \ldots, x_{\pi i}]$ in place of $f[x_0, \ldots, x_i]$. Hence these must be the same. $\square$

3.1.7 Example. For the function f defined by

$$f(x) := \frac{1}{s - x}, \quad s \in \mathbb{C},$$

one can give closed formulas for all divided differences, namely

$$f[x_0, \ldots, x_i] = \frac{1}{(s - x_0)(s - x_1) \cdots (s - x_i)}. \tag{1.12}$$

Indeed, this holds for $i = 0$. If (1.12) is valid for some $i \geq 0$ then, for $x_{i+1} \neq x_i$,

$$
\begin{aligned}
f[x_0, \ldots, x_i, x_{i+1}] &= \frac{f[x_0, \ldots, x_i] - f[x_0, \ldots, x_{i-1}, x_{i+1}]}{x_i - x_{i+1}} \\[2mm]
&= \frac{\frac{1}{(s-x_0)\cdots(s-x_i)} - \frac{1}{(s-x_0)\cdots(s-x_{i-1})(s-x_{i+1})}}{x_i - x_{i+1}} \\[2mm]
&= \frac{(s - x_{i+1}) - (s - x_i)}{(s - x_0)\cdots(s - x_{i-1})(s - x_i)(s - x_{i+1})(x_i - x_{i+1})} \\[2mm]
&= \frac{1}{(s - x_0)\cdots(s - x_{i+1})},
\end{aligned}
$$

and by continuity, (1.12) holds in general.

Using the proposition, the coefficients of the Newton form can be calculated recursively from the function values $f[x_k] = f(x_k)$ if $x_0, \ldots, x_n$ are pairwise distinct: Once the values $f[x_0, \ldots, x_i]$ for $i < k$ are already known, then one obtains $f[x_0, \ldots, x_{k-1}, x_k]$ from

$$f[x_0, \ldots, x_i, x_k] = \frac{f[x_0, \ldots, x_{i-1}, x_k] - f[x_0, \ldots, x_i]}{x_k - x_i}, \quad i = 0, \ldots, k - 1.$$

To construct an interpolation polynomial of degree n in this way, one must use $O(n^2)$ operations; each evaluation in the Horner-like form then only takes $O(n)$ further operations. In practice, n is rarely larger than 5 or 6 and often only 2 or 3. The reason is that interpolation polynomials are not flexible enough to approximate typical functions with high accuracy, except over short intervals where a small degree is usually sufficient. Moreover, the more accurate spline interpolation (see Section 2.3) only needs $O(n)$ operations.

The Interpolation Error

To investigate the accuracy of polynomial interpolation we look at the interpolation errors

$$f(x) - p_i(x) = f[x_0, \ldots, x_i, x](x - x_0) \cdots (x - x_i). \qquad (1.13)$$

We first consider estimates for the divided difference term in terms of higher derivatives at some point in the interval hull $\square\{x_0, \ldots, x_i, x\}$.

3.1.8 Theorem. *Suppose that the function $f : D \subseteq \mathbb{R} \to \mathbb{R}$ is $n + 1$ times continuously differentiable.*

(i) If $x_0, \ldots, x_n \in D$, $0 \le i \le n$, then

$$f[x_0, \ldots, x_i] = \frac{f^{(i)}(\xi)}{i!} \quad \text{with } \xi \in \square\{x_0, \ldots, x_i\}. \qquad (1.14)$$

(ii) The error of the interpolation polynomial p_n to f at $x_0, \ldots, x_n \in D$ can be written as

$$f(x) - p_n(x) = \frac{f^{(n+1)}(\xi)}{(n + 1)!} q_n(x), \qquad (1.15)$$

where

$$q_n(x) := (x - x_0) \cdots (x - x_n) \qquad (1.16)$$

and $\xi \in \square\{x_0, \ldots, x_n, x\}$ depends on the interpolation points and on x.

Proof. We prove (i) by induction on n. For $n = 0$, (i) is obvious. We therefore assume that the assertion is true for some $n \ge 0$. If $f : D \to \mathbb{R}$ is $n + 2$ times continuously differentiable, then we can use the inductive hypothesis with $g(x) := f[x_0, x]$ in place of f and find, for all $i \le n$,

$$f[x_0, x_1, \ldots, x_{i+1}] = g[x_1, \ldots, x_{i+1}] = \frac{g^{(i)}(\xi)}{i!}$$

for some $\xi \in \square\{x_1, \ldots, x_{i+1}\}$.

Differentiating the integral representation

$$g(\xi) = f[x_0, \xi] = \int_0^1 f'(x_0 + t(\xi - x_0)) \, dt$$

from (1.11) i times with respect to ξ, and using the mean value theorem of integration it follows that

$$g^{(i)}(\xi) = \int_0^1 t^i f^{(i+1)}(x_0 + t(\xi - x_0))\, dt = f^{(i+1)}(x_0 + \tau(\xi - x_0)) \int_0^1 t^i dt$$

for some $\tau \in [0, 1]$. Therefore

$$f[x_0, \ldots, x_{i+1}] = \frac{g^{(i)}(\xi)}{i!} = \frac{f^{(i+1)}(\xi')}{i!} \int_0^1 t^i\, dt = \frac{f^{(i+1)}(\xi')}{(i+1)!}$$

with

$$\xi' := x_0 + \tau(\xi - x_0) \in \square\{x_0, \xi\} \subseteq \square\{x_0, \ldots, x_{i+1}\},$$

whence (i) is true in general.

Assertion (ii) follows from (i) and (1.9). $\qquad\square$

3.1.9 Corollary. *If $x, x_0, \ldots, x_n \in [a, b]$, then*

$$|f(x) - p_n(x)| \le \frac{\|f^{(n+1)}\|_\infty}{(n+1)!} |q(x)|. \tag{1.17}$$

This bounds the interpolation error as a product of a smoothness factor $\|f^{(n+1)}\|_\infty/(n+1)!$ depending on f only and a factor $|q(x)|$ depending on the interpolation points only.

Hermite Interpolation

For $x_0 = x_1 = \cdots = x_i = x$, $\square\{x_0, \ldots, x_i\} = \{x\}$, hence $\xi = x$ in relation (1.14), and we find

$$f[x, x, \ldots, x]\ (i+1 \text{ arguments}) = \frac{f^{(i)}(x)}{i!}. \tag{1.18}$$

One can use this property for the solution of the so-called *Hermite interpolation problem* to find a polynomial matching both the function values and one or more derivative values at the interpolation points. In the most important case we want, for pairwise distinct interpolation points $z_0, z_1, \ldots, z_m$, a polynomial p of degree $\le 2m + 1$ such that

$$p(z_j) = f(z_j) \quad \text{and} \quad p'(z_j) = f'(z_j) \quad \text{for } j = 0, \ldots, m. \tag{1.19}$$

To solve this problem one doubles each interpolation point, that is, one sets

$$x_{2j} := x_{2j+1} := z_j \quad (j = 0, \ldots, m),$$

and calculates the corresponding Newton interpolation polynomial $p = p_{2m+1}$. The error formula (1.13) now takes the form

$$f(x) - p(x) = f[z_0, z_0, \ldots, z_m, z_m, x](x - z_0)^2 \cdots (x - z_m)^2,$$

from which it easily follows that the interpolation requirement (1.19) is satisfied.

Similarly, in the general Hermite interpolation problem, an interpolation point z_j at which the function value and the values of the first k_j derivatives shall be matched has to be replaced by $k_j + 1$ identical interpolation points $x_{i_j} = \cdots = x_{i_j + k_j} = z_j$; then, again, the Newton interpolation polynomial gives the solution of minimal degree. A very special case of this is the situation where the function value and the first n derivative values shall be matched at a single point x_0; then (1.15), (1.18) and Theorem 3.1.4 give the representation

$$f(x) = f(x_0) + f'(x_0)(x - x_0) + \cdots + \frac{f^{(n)}(x_0)}{n!}(x - x_0)^n$$
$$+ \frac{f^{(n+1)}(\xi)}{(n+1)!}(x - x_0)^{n+1}$$

for some $\xi \in \square\{x, x_0\}$; and we see that the well-known Taylor formula with remainder term is a special case of Hermite interpolation.

In actual practice, the confluent interpolation points cause problems in the evaluation of the divided difference formulas because some of the denominators $x_i - x_k$ in the divided differences may become zero. However, by using the permutation symmetry, the calculation can be arranged so that everything works well. We simply compute the divided differences column by column in the following scheme.

$f(x_0)$				
	$f[x_0, x_1]$			
$f(x_1)$		$f[x_0, x_1, x_2]$		
	$f[x_1, x_2]$		$\cdots$	
$f(x_2)$		$f[x_1, x_2, x_3]$		
$\vdots$	$\vdots$	$\vdots$	$\cdots$	$f[x_0, x_1, \ldots, x_n]$
$f(x_{n-2})$		$f[x_{n-3}, x_{n-2}, x_{n-1}]$		
	$f[x_{n-2}, x_{n-1}]$		$\cdots$	
$f(x_{n-1})$		$f[x_{n-2}, x_{n-1}, x_n]$		
	$f[x_{n-1}, x_n]$			
$f(x_n)$				

The corresponding recursion for the $d_{ik} := f[x_i, x_{i+1}, \ldots, x_{k-1}, x_k]$ is

$$
d_{ik} = \begin{cases} \dfrac{d_{i+1,k} - d_{i,k-1}}{x_k - x_i} & \text{if } x_i \neq x_k, \\[2mm] \dfrac{f^{(k-i)}(x_i)}{(k-i)!} & \text{otherwise,} \end{cases}
$$

because by construction the case $x_i = x_k$ occurs only when $x_i = x_{i+1} = \cdots = x_k$, and then the $(k-i)$th derivative of f at x_i is known.

Convergence

Under suitable (but for applications often too restricted) conditions, it can be shown that when the number of interpolation points increases indefinitely, the corresponding sequence of interpolation polynomials converges to the function being interpolated.

In view of the close relation between Newton interpolation formulas and the Taylor expansion, we suppose that the function f has an absolutely convergent power series in the interval of interest. Then f can be viewed as a complex analytic function in an open region D of the complex plane containing that interval. (A comprehensive exposition of complex analysis from a computational point of view is in Henrici [43].)

We restrict the discussion to convex D because then the results obtained so far hold nearly without change (and with nearly identical proofs). We only need to replace interval hulls ($\square S$) by closed convex hulls (Conv S), and the statements (1.14), (1.15), and (1.17) by

$$
f[x_0, \ldots, x_i] \in \text{Conv}\left\{ \frac{f^{(i)}(\xi)}{i!} \,\middle|\, \xi \in \text{Conv}\{x_0, \ldots, x_i\} \right\},
$$

$$
|f(x) - p_n(x)| \leq \sup_{\xi \in \text{Conv}\{x_0, \ldots, x_i\}} \left| \frac{f^{(n+1)}(\xi)}{(n+1)!} \right| |q_n(x)|,
$$

and

$$
|f(x) - p_n(x)| \leq \frac{\left\| f^{(n+1)} \right\|_\infty}{(n+1)!} |q(x)|,
$$

where the ∞-norm is taken over a convex domain containing all arguments.

The complex point of view allows us to use tools from complex analysis that lead to a simple closed expression for divided differences. In the following,

$$
D[c; r] := \{ \xi \in \mathbb{C} \,|\, |\xi - c| \leq r \}
$$

denotes the closed complex disk with center $c \in \mathbb{C}$ and radius $r > 0$, ∂D its positively oriented boundary, and int D its interior.

3.1.10 Proposition. *Let f be a function analytic in the open set $D_0 \subseteq \mathbb{C}$. If the disk $D[c; r]$ is contained in D_0, then*

$$f[x_0, \ldots, x_n] = \frac{1}{2\pi i} \int_{\partial D} \frac{f(s)\, ds}{(s - x_0) \cdots (s - x_n)} \quad \text{for } x_0, \ldots, x_n \in \text{int } D.$$
$$(1.20)$$

Proof. For $n = 0$, the formula is just Cauchy's well-known integral formula

$$f(x) = \frac{1}{2\pi i} \int_{\partial K} \frac{f(s)}{s - x}\, ds \quad \text{for all } x \in \text{int } K.$$

Hence suppose that (1.20) is valid for $n - 1$ instead of n then, for $x_{n-1} \neq x_n$ (cf. Example 3.1.7),

$$\begin{aligned}
f[x_0, \ldots, x_n] &= \frac{f[x_0, \ldots, x_{n-1}] - f[x_0, \ldots, x_{n-2}, x_n]}{x_{n-1} - x_n} \\
&= \frac{1}{2\pi i} \int_{\partial K} \frac{\frac{f(s)}{(s-x_0)\cdots(s-x_{n-1})} - \frac{f(s)}{(s-x_0)\cdots(s-x_{n-2})(s-x_n)}}{x_{n-1} - x_n}\, ds \\
&= \frac{1}{2\pi i} \int_{\partial K} \frac{f(s)\, ds}{(s - x_0) \cdots (s - x_n)}.
\end{aligned}$$

This formula remains valid in the limit $x_n \to x_{n-1}$. So (1.20) holds for n and therefore in general. □

3.1.11 Corollary. *If $|f(\xi)| \leq M$ for all $\xi \in D[c; r]$ and $|x_j - c| \leq \rho < r$ for all $j = 0, \ldots, n$ then*

$$|f[x_0, \ldots, x_n]| \leq \frac{Mr}{(r - \rho)^{n+1}}.$$

Proof. By the previous proposition,

$$\begin{aligned}
|f[x_0, \ldots, x_n]| &= \frac{1}{|2\pi i|} \left| \int_{\partial K} \frac{f(s)\, ds}{(s - x_0) \cdots (s - x_n)} \right| \\
&\leq \frac{1}{2\pi} \int_{\partial K} \frac{|f(s)||ds|}{|s - x_0| \cdots |s - x_n|} \leq \frac{M}{2\pi(r - \rho)^{n+1}} \int_{\partial K} |ds| \\
&= \frac{M}{2\pi(r - \rho)^{n+1}} \cdot 2\pi r = \frac{Mr}{(r - \rho)^{n+1}}.
\end{aligned}$$

(Here, $\int |ds|$ denotes the unoriented line integral.) □

The fact that the number M is independent of n can be used to derive the desired statement about the convergence of interpolating polynomials as $n \to \infty$.

3.1.12 Theorem. *Let f be a function analytic in an open set $D \subseteq \mathbb{C}$ containing the disk $D[c; r]$. For an infinite sequence of complex numbers $x_j \in D[c; \rho]$ $(j = 0, 1, 2, \ldots)$, let p_n be the polynomial of degree $\leq n$ that interpolates the function at $x_0, x_1, \ldots, x_n$. If $r > 3\rho$, then for $x \in D[c; \rho]$, the sequence $p_n(x)$ converges uniformly to $f(x)$.*

Proof. If $x \in D[c; \rho]$ then $|x - x_j| \leq 2\rho$ for all $j = 0, 1, 2, \ldots$, so the interpolation error is bounded according to

$$|f(x) - p_n(x)| = |f[x_0, \ldots, x_n, x]||x - x_0| \cdots |x - x_n|$$
$$\leq \frac{Mr}{(r - \rho)^{n+2}}(2\rho)^{n+1} = \frac{Mr}{(r - \rho)} \left(\frac{2\rho}{r - \rho} \right)^{n+1}.$$

For $r > 3\rho$, $|\frac{2\rho}{r-\rho}| < 1$, so that $|f(x) - p_n(x)|$ is majorized by a sequence that does not depend on x and converges to zero. This proves uniform convergence on $D[c; \rho]$. $\square$

The hypothesis $r > 3\rho$ can be weakened a little by using more detailed considerations and more complex shapes in place of disks. However, a well-known example due to Runge shows that one cannot dispense with *some* condition of this sort.

3.1.13 Example. Interpolation of the function $f(x) := 1/(1 + x^2)$ at more and more equidistant interpolating points in the interval $[-5, 5]$ leads to divergence of the interpolating polynomials $p_n(x)$ for real x with $|x| \geq 3.64$, and convergence for $|x| \leq 3.63$. We illustrate this in Figure 3.1; for a proof, see, for example, Isaacson and Keller [46, Section 6.3.4].

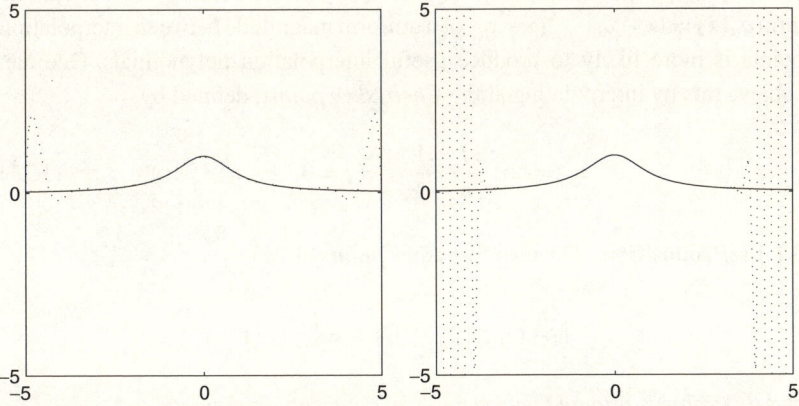

Figure 3.1. Equidistant polynomial interpolation in many points may be poor.

The hypothesis of the theorem is not satisfied because f has poles at $\pm i$, although f is analytic in a large open and convex region containing the interpolation interval $[-5, 5]$. For such functions, the derivatives $|f^{(n+1)}(\xi)|$ do not vary too much, and the error behavior (1.15) of the interpolation polynomial is mainly governed by the term $q_n(x) = (x - x_0) \cdots (x - x_n)$. The divergent behavior finds its explanation by observing that, when the interpolation points are equidistant and their number increases, this term develops huge spikes near the interval boundaries. This can be seen from the table

n	16	32	64	128	256
Q_n	$1.9 \cdot 10^1$	$3.1 \cdot 10^3$	$4.4 \cdot 10^7$	$4.6 \cdot 10^{16}$	$1.9 \cdot 10^{34}$

of quotients

$$Q_n = \max_{x \in [-5,5]} |q_n(x)| \Big/ \max_{x \in [-4,4]} |q_n(x)|.$$

Interpolation in Chebyshev Points

Although high-degree polynomial interpolation in equidistant points often leads to strong oscillations near the boundaries of the interpolation intervals destroying convergence, the situation is better when the interpolation points are more closely spaced there. Because interpolation on arbitrary intervals can be reduced to that on $[-1, 1]$ through a linear transformation of variables, we restrict the discussion to interpolation in the interval $[-1, 1]$, where the formulas are simplest.

The example just discussed suggests that a choice of the x_j that keeps the term $q_n(x) = (x - x_0) \cdots (x - x_n)$ of a uniform magnitude between interpolation points is more likely to produce useful interpolation polynomials. One can achieve this by interpolating at the *Chebyshev points*, defined by

$$x_j := \cos \frac{2j + 1}{2n + 2} \pi \quad (j = 0, \ldots, n). \tag{1.21}$$

3.1.14 Proposition. *For the Chebyshev points* (1.21),

$$|q_n(x)| \leq \frac{1}{2^n} \quad \text{for } x \in [-1, 1],$$

and this value is attained between any two interpolation points.

Proof. We introduce the *Chebyshev polynomials* T_n, defined recursively by

$$T_0(x) := 1, \quad T_1(x) := x, \quad T_2(x) = 2x^2 - 1,$$
$$T_{n+1}(x) := 2xT_n(x) - T_{n-1}(x) \quad (n = 1, 2, \ldots). \tag{1.22}$$

Clearly, the T_n are of degree n and, for $n > 0$, they have highest coefficient 2^{n-1}. Using the addition theorem $\cos(\alpha + \beta) + \cos(\alpha - \beta) = 2\cos\alpha\cos\beta$, a simple induction proof yields the relation

$$T_n(x) = \cos(n \arccos x) \quad \text{for all } n. \tag{1.23}$$

This shows that the polynomial $T_{n+1}(x)$ has all Chebyshev points as zeros, and hence $T_{n+1}(x) = 2^n q_n(x)$. Moreover, the extrema of $T_{n+1}(x)$ are ± 1, attained between consecutive zeros. This implies the assertion. □

For the interpolation error in interpolation at Chebyshev points, one can prove the following result.

3.1.15 Theorem. *The interpolation polynomial $p_n(x)$ interpolating an arbitrary s times continuously differentiable functions $f : [-1, 1] \to \mathbb{R}$ in the Chebyshev points* (1.21) *satisfies*

$$|f(x) - p_n(x)| = O\left(\frac{1}{n^s}\right).$$

Proof. See, for example, Conte and de Boor [12, Section 6.1]. □

This reference also shows that interpolation at the so-called *expanded Chebyshev points*, which, adapted to a general interval $[a, b]$, are given by

$$\frac{a + b}{2} + \frac{a - b}{2}\left(\cos\frac{2i + 1}{2n + 2}\pi\right)\Big/\left(\cos\frac{\pi}{2n + 2}\right)$$

is even slightly better.

3.2 Extrapolation and Numerical Differentiation

Extrapolation to the Limit

Extrapolation refers to the use of interpolation at points $x_0, \ldots, x_n$ for the approximation of a function f at a point $x \notin \square\{x_0, \ldots, x_n\}$. Because interpolation polynomials of low degree are generally rather inaccurate, and those of high degree are often bad already near the boundary of the interpolation interval and this behavior becomes worse outside the interval, extrapolation cannot be recommended in practice.

A very important exception is the case in which the interpolating points x_j $(j = 0, 1, 2, \ldots)$ form a sequence converging to zero, and the value of f at $x = 0$ is sought. The reason is that in this particular case, the usually offending term $q_n(x)$ behaves exceptionally well,

$$|q_n(0)| = |x_0 x_1 \cdots x_n|, \qquad (2.1)$$

and converges very rapidly to zero even when the x_j converge quite slowly.

3.2.1 Example. For the sequence $x_j := \frac{h}{2^j}$ $(j = 0, 1, 2, \ldots)$, one finds the following results:

n	1	2	4	8	16
$q_n(0)$	$0.5h^2$	$0.125h^3$	$9.8 \cdot 10^{-4} h^5$	$1.5 \cdot 10^{-11} h^9$	$1.1 \cdot 10^{-41} h^{17}$

Thus the extrapolation to the limit from values at a given sequence (usually $x_j = x^0/N_j$ for some slowly growing divergent sequence N_j) can be expected to give excellent results. It is a very valuable technique for getting function values at a point (usually $x = 0$) when the function becomes more and more difficult to evaluate as the argument approaches this point.

The Extrapolation Formulas of Neville

Because the value at the single point $x = 0$ is the only one of interest in extrapolation, the Newton interpolation polynomial is not the most common way to compute this value. Instead, one generally uses the extrapolation formulas of Neville, which give simultaneously the extrapolated values of many interpolation polynomials at a single argument.

3.2.2 Theorem. *The polynomials defined by*

$$p_{i0}(x) = f(x_i),$$

$$p_{ik}(x) := p_{i,k-1}(x) + (x_i - x) \frac{p_{i,k-1}(x) - p_{i-1,k-1}(x)}{x_{i-k} - x_i} \quad \text{for } k = 1, \ldots, i,$$
$$(2.2)$$

are the interpolation polynomials for f at $x_{i-k}, \ldots, x_i$.

Proof. Obviously $p_{ik}(x)$ is of degree $\leq k$. We show by induction on k the interpolation property

$$p_{ik}(x_l) = f(x_l) \quad \text{for } l = i - k, i - k + 1, \ldots, i. \qquad (2.3)$$

The assertion (2.3) is clearly true for $k = 0$. Hence suppose that (2.3) holds for $k - 1$ in place of k. In (2.2), the factor $(x - x_i)$ vanishes for $x = x_i$, and the fraction vanishes for $x = x_{i-1}, x_{i-2}, \ldots, x_{i-k+1}$; therefore,

$$p_{ik}(x_l) = p_{i,k-1}(x_l) = f(x_l) \quad \text{for } l = i, i - 1, \ldots, i - k + 1.$$

Moreover, for $x = x_{i-k}$, one obtains

$$\begin{aligned} p_{ik}(x_{i-k}) &= p_{i,k-1}(x_{i-k}) - (p_{i,k-1}(x_{i-k}) - p_{i-1,k-1}(x_{i-k})) \\ &= p_{i-1,k-1}(x_{i-k}) = f(x_{i-k}), \end{aligned}$$

so that (2.3) is valid for k and hence holds in general.

Thus $p_{ik}(x)$ is the uniquely determined polynomial of degree $\leq k$ which interpolates the function $f(x)$ at $x_{i-k}, \ldots, x_i$. □

The values $p_{ik}(0)$ are, for increasing k, successively better extrapolation approximations to $f(0)$, until a stage is reached where the limitations of polynomial interpolation (or rounding errors) increase the error again. In order to have a natural stopping criterion, one monitors the values $\delta_i := |p_{ii}(0) - p_{i,i-1}(0)|$, and stops the extrapolation if δ_i is no longer decreasing. Then one accepts $p_{ii}(0)$ as the best approximation for $f(0)$, and has the value of δ_i as a natural estimate for the error $f(0) - p_{ii}(0)$. (Of course, this is not a rigorous bound; the true error is unknown and might be larger.)

In the algorithmic formulation, we use only the x_i for $i > 0$ in accord with the lack of zero indices in MATLAB. We store $p_{ik}(0) - f(x_i)$ in p_{i+1-k}, overwriting old numbers no longer needed; the subtraction gives a slight improvement in final accuracy because the intermediate quantities are then smaller.

3.2.3 Algorithm: Neville Extrapolation to Zero

```
i = 1; p₁ = 0;
fₒₗₐ = f(x₁);
while 1,
    i = i + 1; pᵢ = 0;
    fₑₛₜ = f(xᵢ); df = fₑₛₜ − fₒₗₐ;
    for j = i − 1: −1:1,
        pⱼ = pⱼ₊₁ + (pⱼ₊₁ − pⱼ + df) ∗ xᵢ/(xⱼ − xᵢ);
    end,
    fₒₗₐ = fₑₛₜ; δₒₗₐ = δ;
    δ = abs(p₂ − p₁); if i > 2 & δ ≥ δₒₗₐ, break; end,
end,
fₑₛₜ = fₑₛₜ + p₁;
% best estimate |f(0) − fₑₛₜ| ≤ δ
```

For the frequent case where $x_i = x/q^i$ $(i = 1, 2, \ldots)$, the factor $x_i/(x_{i-k} - x_i)$ in the formula for p_{ik} becomes $1/(q^k - 1)$, and the algorithms takes the following form.

3.2.4 Algorithm: Neville Extrapolation to Zero for $x_i = x/q^i$

$i = 1$; $p_1 = 0$;
$f_{old} = f(x_1)$;
while 1,
    $i = i + 1$; $p_i = 0$; $Q = 1$;
    $f_{est} = f(x_i)$; $df = f_{est} - f_{old}$;
    for $j = i - 1: -1:1$,
        $Q = Q * q$;
        $p_j = p_{j+1} + (p_{j+1} - p_j + df)/(Q - 1)$;
    end,
    $f_{old} = f_{est}$; $\delta_{old} = \delta$;
    $\delta = \text{abs}(p_2 - p_1)$; if $i > 2$ & $\delta \geq \delta_{old}$, break; end,
end,
$f_{est} = f_{est} + p_1$;
% best estimate $|f(0) - f_{est}| \leq \delta$

We now demonstrate the power of the method with numerical differentiation; other important applications include numerical integration (see Section 4.4) and the solution of differential equations.

Numerical Differentiation

In practice, one often has functions f not given by arithmetical expressions and therefore not easily differentiable by automatic methods. However, derivatives are needed (or at least very useful) in many applications (e.g., to solve nonlinear algebraic or differential equations, or to find the extreme values of a function). In such cases, the common remedy is to resort to numerical differentiation. In the simplest case one approximates, for fixed x, the required value $f'(x)$ through a *forward difference quotient*

$$p(h) := f[x, x + h] = \frac{f(x + h) - f(x)}{h}, \quad h \neq 0$$

at a suitable value of h. By Taylor expansion, we find the error expansion

$$p(h) = f'(x) + \sum_{i = 1:s} \frac{f^{(i+1)}(x)}{(i + 1)!} h^i + O(h^{s+1}) \tag{2.4}$$

if f is $(s+2)$ times differentiable at x. If f is expensive to evaluate, a fixed $p(h)$ is used as approximation to $f'(x)$, with an error of $O(h)$. However, because for small h divided differences suffer from severe numerical instability due to cancellation, the accuracy achievable in finite precision arithmetic is rather low.

However, if one can afford to spend several function values for the computation of the derivative, higher accuracy can be achieved as follows. Because $p(h)$ is continuous at $h=0$, and (2.4) shows that it behaves locally like a polynomial, it is natural to calculate the derivative $f'(x) = p(0)$ by means of extrapolation: For a suitable sequence of numbers $h_i \neq 0$ converging to 0, one calculates the interpolation polynomial p_i using the values $p(h_j) = f[x, x + h_j]$ ($j = 0, \ldots, i$). Then the $p_i(0)$ can be expected to be increasingly accurate approximations to $f'(x)$, with an error of $O(h_0 \cdots h_n)$.

Another divided difference, the *central difference quotient*

$$
\begin{aligned}
f[x - h, x + h] &= \frac{f(x + h) - f(x - h)}{2h} \\
&= f'(x) + \sum_{i=1:s} \frac{f^{(2i+1)}(x)}{(2i + 1)!} h^{2i} + O(h^{2s+2}) \\
&=: p(h^2)
\end{aligned}
\tag{2.5}
$$

approximates the derivative $f'(x)$ with a smaller error of $O(h^2)$. Because in the asymptotic expansion only even powers of h occur, it is sensible to consider this expression as a function of h^2 instead of h. Because now h^2 plays the role of the previous h in the extrapolation method, the extrapolation error term is now $O(h_0^2 \cdots h_n^2)$. Thus h need not be chosen as small as for forward differences to reduce the truncation error to the same level. Hence central differences suffer from less cancellation and therefore lead to better results. The price to pay is that the calculation of the values $p(h_j^2) = f[x - h_j, \ x + h_j]$ for $j = 0, \ldots, n$ requires $2n + 2$ function evaluations $f(x \pm h_j)$, whereas for the calculation of $f[x, x + h_j]$ for $j = 0, \ldots, n$, only $n + 2$ function evaluations ($f(x + h_j)$ and $f(x)$) are necessary.

3.2.5 Example. We want to find the value of the derivative of the function given by $f(x) = \sin x$ at $x = 1$. Of course, the exact value is $f'(1) = \cos 1 = 0.540\,302\,305\,868\ldots$. The following table lists the central difference quotients $p(h_i^2) := f[x - h_i, x + h_i]$ for a number of values of $h_i > 0$. Moreover, the values $p_i(0)$ calculated by extrapolation are given for $i = 1, 2$. Correct digits are underlined. We used a pocket calculator with $B = 10$, $L = 12$, and optimal rounding, so that the working precision is $\varepsilon = \frac{1}{2} 10^{-11}$; but only 10 significant digits can be displayed.

i	h_i	$f[x - h_i, x + h_i]$	$p_i(0)$
0	0.04	0.540 158 236 9	
1	0.02	0.540 266 286 5	0.540 302 303 0
2	0.01	0.540 293 301 1	0.540 302 305 9
3	$0.5 \cdot 10^{-2}$	0.540 300 054 8	
4	10^{-4}	0.540 302 315 0	
5	$0.5 \cdot 10^{-4}$	0.540 302 280 0	
6	10^{-5}	0.540 302 150 0	
7	10^{-6}	0.540 304 500 0	
8	10^{-7}	0.540 295 000 0	

The accuracy of the central difference quotient first increases and attains its optimum for $h \approx 10^{-4}$ as h decreases. Smaller values than $h = 10^{-4}$ give less accurate results, because the calculation of $f[x - h, x + h]$ is hampered by cancellation. The increasing cancellation can be seen from the trailing zeros at small h.

The extrapolated value is correct to 10 places already for fairly large h_i; it is remarkable (but typical) that it is more accurate than the best attainable value for any of the central differences $f[x - h_i, x + h_i]$. To explain this behavior, we now turn to a stability analysis.

The Optimal Step Size

We want to estimate the order of magnitude of h that is optimal for the calculation of difference quotients and the resulting accuracy achievable for the approximation of $f'(x)$. In realistic circumstances, function values at arguments near x can be evaluated only with a relative accuracy of ε, say, so that

$$\tilde{f}(x \pm h) = f(x \pm h)(1 + O(\varepsilon)) \tag{2.6}$$

is calculated instead of $f(x \pm h)$. In the most favorable circumstances, ε is the machine precision; usually ε is larger. Inserting (2.6) into the definition (2.5) of the divided difference gives

$$\tilde{f}[x - h, x + h] = f[x - h, x + h] + O\left(|f(x)|\frac{\varepsilon}{h}\right).$$

Because $f'(x) = f[x - h, x + h] + O(|f'''(x)|h^2)$ by (2.5), we find that the total error

$$f'(x) - \tilde{f}[x - h, x + h] = O\left(|f(x)|\frac{\varepsilon}{h}\right) + O(|f'''(x)|h^2) \tag{2.7}$$

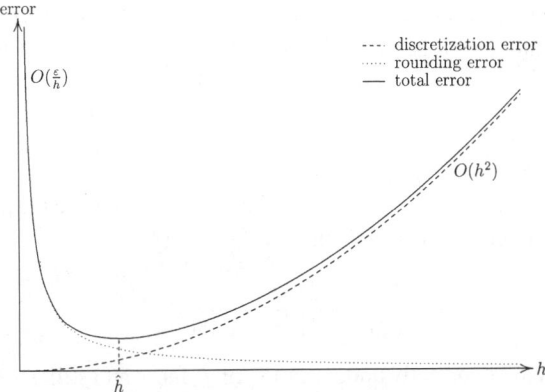

Figure 3.2. The numerical differentiation error in dependence on the step size h. The optimal step size $\hat{h}$ minimizing the total error is close to the point where discretization error and rounding error are equal.

consists of a contribution $O(\frac{\varepsilon}{h})$ due to errors in function evaluations and a contribution $O(h^2)$, the *discretization error* due to the finite difference approximation. The qualitative shape of the error curve is given in Figure 3.2.

By a similar argument one finds for a method that, in exact arithmetic, approximates $f'(x)$ with a discretization error of $O(h^s)$ a total error $\delta(h)$ (including errors in function evaluations) of

$$\delta(h) = O\left(|f(x)|\frac{\varepsilon}{h}\right) + O\left(|f^{(s+1)}(x)|h^s\right). \tag{2.8}$$

Limiting Accuracy

Under the natural assumption that the hidden constants in the Landau symbols do not change much for h in the relevant range, we can analyze the behavior of (2.8) by replacing it with

$$\delta(h) = a\frac{\varepsilon}{h} + bh^s$$

where a and b are positive real constants. By setting the derivative equal to zero, we find $h = (a\varepsilon/sb)^{1/(s+1)}$ as the value for which $\delta(h)$ is minimal, and the minimal value is $\delta_{min} = (s+1)bh^s = O(\varepsilon^{s/(s+1)})$. The optimal h is only a little smaller than the value of h where the discretization error term matches the function evaluation error term, which is $h = (a\varepsilon/b)^{1/(s+1)}$ and gives a total error of the same magnitude.

If we take into account the dependence of a and b on f, we find that the optimal magnitude of h is

$$h_{opt} = O\left(\left| \frac{f(x)}{f^{(s+1)}(x)} \right|^{1/(s+1)} \varepsilon^{1/(s+1)} \right). \tag{2.9}$$

Because in practice, $f^{(s+1)}(x)$ is not available, one cannot use the optimal value h_{opt} of h; a simpler heuristic substitute is

$$h = \left| \frac{f(x)}{f[x + h_0, x - h_0]} \right| \varepsilon^{1/(s+1)}.$$

This value behaves correctly under scaling of f and x and still gives total errors of the optimal order $O(\varepsilon^{s/(s+1)})$, but now with a suboptimal hidden factor.

In particular, for the approximation of $f'(x)$ by the central difference quotient $f[x - h, x + h]$, cf. (2.5), we have $s = 2$ in (2.8), and the optimal magnitude of h is $O(\varepsilon^{1/3})$, achieving the minimal total error of magnitude $O(\varepsilon^{2/3})$. This is corroborated in the above example, where $\varepsilon = \frac{1}{2} 10^{-11}$ is the machine precision.

Extrapolation of $p(h^2)$ using three values $f[x - h_0, x + h_0]$, $f[x - h_1, x + h_1]$, and $f[x - h_2, x + h_2]$ with $h_i = h/2^i$ gives a discretization error of $O(h^6)$, hence $O(\varepsilon^{1/7})$ as optimal magnitude for h and nearly full accuracy $O(\varepsilon^{6/7})$ as optimal magnitude for the minimal total error. Again, this is consistent with the example.

Note that if we use for the calculation the forward difference quotient $f[x, x + h]$, cf. (2.4) we only have $s = 1$ in (2.8), and the optimal choice $h = O(\varepsilon^{1/2})$ only gives the much inferior accuracy $O(\varepsilon^{1/2})$.

Higher Derivatives

For the approximation of higher derivatives one has to use higher order divided differences. The experience with the first derivative suggests to use also central differences with arguments symmetric around the point where the derivative is to be approximated; then again odd powers of h cancel, and we can use extrapolation for $h^2 \to 0$.

To approximate second derivatives $f''(x)$ one can use

$$p(h^2) := 2f[x - h, x, x + h] = \frac{f(x + h) - 2f(x) + f(x - h)}{h^2} \quad \text{if } h \neq 0;$$

then $f''(x) = p(0)$, and the asymptotic expansion

$$p(h^2) = f''(x) + \sum_{i=1:s} \frac{2f^{(2i+2)}(x)}{(2i + 2)!} h^{2i} + O(h^{2s+2})$$

holds for sufficiently often differentiable f. Because of the division by h^2, the total error now behaves like $O(\frac{\varepsilon}{h^2}) + O(h^2)$, giving an optimal error of $O(\varepsilon^{1/2})$ for h of order $O(\varepsilon^{1/4})$. With quadratic extrapolation, the error looks like $O(\frac{\varepsilon}{h^2}) + O(h^6)$, giving an improved optimal error of $O(\varepsilon^{3/4})$ for h of order $O(\varepsilon^{1/8})$. Note that the step size h must be chosen now much larger, reflecting the fact that otherwise cancellation is much more severe.

To approximate third derivatives $f'''(x)$ one uses similarly

$$p(h^2) := 6f[x - 2h, x - h, x, x + h, x + 2h]$$
$$= \frac{\frac{1}{2}(f(x + 2h) - f(x - 2h)) - (f(x + h) - f(x - h))}{h^3} \quad \text{if } h \neq 0;$$

then $f'''(x) = p(0)$ and $p(h^2) = f'''(x) + O(h^2)$. The optimal h is now of order $O(\varepsilon^{1/5})$, giving an error of order $O(\varepsilon^{2/5})$. Again, extrapolation improves on this.

3.3 Cubic Splines

As shown, polynomial interpolation has good approximation properties on narrow intervals but may be poor on wide intervals. This suggests the use of piecewise polynomial functions to keep the advantages and overcome the problems associated with polynomial interpolation.

Piecewise Linear Interpolation

To set the stage, we first look at the simple case of piecewise linear interpolation.

3.3.1 Definition.

(i) A *grid* on $[a, b]$ is a set $\Delta = \{x_1, \ldots, x_n\}$ satisfying

$$a = x_1 < x_2 < \cdots < x_{n-1} < x_n = b. \tag{3.1}$$

$x_1, \ldots, x_n$ are called the *nodes* of Δ, and $h := \max\{|x_{j+1} - x_j| \mid j = 1, \ldots, n - 1\}$ is called the *mesh size* of Δ. A grid $\Delta = \{x_1, \ldots, x_n\}$ is called *equispaced* if

$$x_i = a + (i - 1)h \quad \text{for } i = 1, \ldots, n;$$

the mesh size is then $h = (b - a)/(n - 1)$.

(ii) A function $p : [a, b] \to \mathbb{R}$ is called *piecewise linear* over the grid Δ if it is continuous and agrees on each interval $[x_i, x_{i+1}]$ with a linear polynomial.

(iii) On the space of continuous, real-valued functions f defined on the interval $[a, b]$, we define the norms

$$\|f\|_\infty := \sup\{|f(x)| \mid x \in [a, b]\}$$

and

$$\|f\|_2 = \sqrt{\int_a^b f(x)^2 \, dx}.$$

For piecewise linear interpolation, it is most natural to choose the grid Δ as the set of interpolation points; then the interpolation condition

$$S(x_j) = f(x_j) \quad \text{for } j = 1, \ldots, n \tag{3.2}$$

automatically ensures continuity in $[a, b]$, and we find the unique interpolant

$$S(x) = f(x_j) + f[x_j, x_{j+1}](x - x_j) \quad \text{for all } x \in [x_j, x_{j+1}].$$

We see immediately that an arbitrarily accurate approximation is possible when the data points are sufficiently closely spaced (i.e., if the mesh size h is sufficiently small). If, in addition, we assume sufficient smoothness of f, then we can bound the achieved accuracy as follows.

3.3.2 Theorem. *Let $S(x)$ be a piecewise linear interpolating function on the grid $\Delta = \{x_1, \ldots, x_n\}$ with mesh size h over $[a, b]$. If the function $f(x)$ to be interpolated is twice continuously differentiable, then*

$$|f(x) - S(x)| \le \frac{h^2}{8} \|f''\|_\infty \quad \text{for all } x \in [a, b].$$

Proof. For $x \in [x_j, x_{j+1}]$, the relation

$$f(x) - S(x) = f[x_j, x_{j+1}, x](x - x_j)(x - x_{j+1})$$

implies the bound

$$|f(x) - S(x)| \le \frac{1}{2} \|f''\|_\infty |(x - x_j)(x - x_{j+1})| \le \frac{1}{2} \|f''\|_\infty \frac{(x_{j+1} - x_j)^2}{4}$$

$$\le \frac{h^2}{8} \|f''\|_\infty. \qquad \square$$

In order to increase the accuracy by three decimal places, it is necessary to decrease h^2 by a factor of 1000, which requires the introduction of at least

$\sqrt{1000} \approx 32$ subintervals between any two already existing grid points. This means 32 times as much work; cubic polynomial interpolation would already be of order $O(h^4)$, reducing the work factor to a more reasonable factor of $\sqrt[4]{1000} \approx 5.6$.

Thus, in many problems, piecewise linear interpolation is either too inaccurate or too slow to give a satisfying accuracy. However, this can be remedied by using piecewise polynomials of higher degree with sufficient smoothness.

Splines

We now add smoothness requirements that define a class of piecewise polynomials with excellent approximation properties.

A *spline* of order k over a grid Δ is a $k - 2$ times continuously differentiable function $S : [a, b] \rightarrow \mathbb{R}$ such that the $(k - 2)$nd derivative $S^{(k-2)}$ is piecewise linear (over Δ). A spline S of order k is piecewise a polynomial of degree at most $k - 1$; indeed, S must agree between two adjacent nodes with a polynomial of degree at most k because the $(k - 2)$nd derivative is linear. In particular, splines of order 4 are piecewise cubic polynomials; they are called *cubic splines*.

Splines have excellent approximation properties. Because cubic splines are satisfactory for many interpolation problems, we restrict to these and refer for higher order splines to de Boor [15]. (For splines in several variables, see de Boor [16].)

Important examples of splines are the so-called B-splines. A *cubic basis spline*, short a *B-spline* over the grid $\Delta = \{\bar{x}_1, \ldots, \bar{x}_n\}$, is a cubic spline $S_l(x)$ defined over the extended grid $\bar{x}_{-2} < \bar{x}_{-1} < \cdots < \bar{x}_{n+3}$ with the property that (for some $l = 0, 1, \ldots, n + 1$) $S_l(\bar{x}_l) > 0$ and $S_l(x) = 0$ for $x \notin (\bar{x}_{l-2}, \bar{x}_{l+2})$. Cubic basis splines exist on every grid and are determined up to a constant factor by the (extended) grid Δ and the index l; see Section 3.5.

3.3.3 Example. For the equispaced case, one easily checks that the functions defined by

$$B_k(x) := B\left(\frac{x}{h} - k\right),$$

where

$$B(x) := \begin{cases} \frac{1}{4}(2 - |x|)^3 - (1 - |x|)^3 & \text{for } |x| \leq 1, \\ \frac{1}{4}(2 - |x|)^3 & \text{for } 1 < |x| < 2, \\ 0 & \text{for } |x| \geq 2 \end{cases} \tag{3.3}$$

are B-splines on the equispaced grid with $x_l = x_0 + lh$ (cf. Figure 3.3).

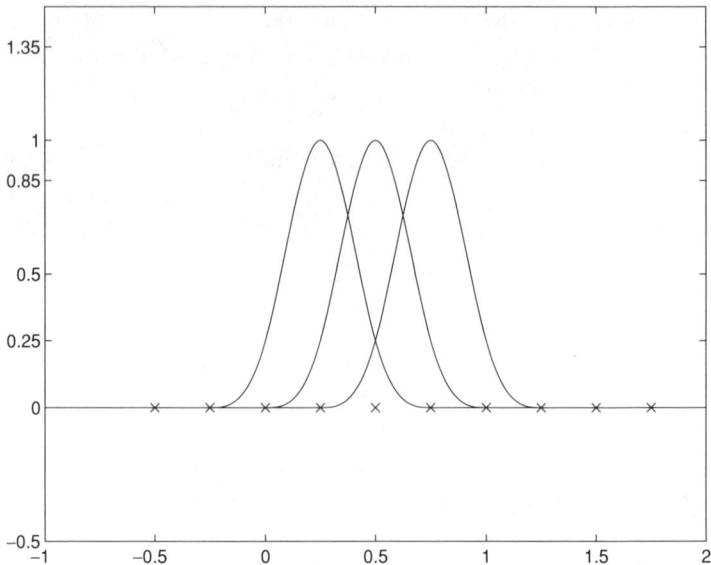

Figure 3.3. Cubic B-splines for an equally spaced grid, $h = \frac{1}{4}$.

For interpolation by cubic splines, it is again most natural to choose the grid Δ as the set of interpolation points. The interpolation condition (3.2) again ensures continuity in $[a, b]$, but differentiability conditions must be enforced by suitable constraints. We represent S in $[x_j, x_{j+1}]$ as a cubic Hermite interpolation polynomial. Using the abbreviations

$$\beta_j = f[x_j, x_{j+1}], \quad \gamma_j = f[x_j, x_{j+1}, x_j], \quad \delta_j = f[x_j, x_{j+1}, x_j, x_{j+1}],$$
(3.4)

we find

$$S(x) = f(x_j) + \beta_j(x - x_j) + \gamma_j(x - x_j)(x - x_{j+1})$$
$$+ \delta_j(x - x_j)^2(x - x_{j+1}) \quad \text{for all } x \in [x_j, x_{j+1}].$$
(3.5)

Because the derivatives of S at x_j and x_{j+1} are not specified, the constants γ_j and δ_j are still undetermined and at our disposal, and must be determined by imposing the condition that a cubic spline is twice continuously differentiable at the nodes (elsewhere, this is the case automatically). The calculation of the corresponding coefficients γ_j and δ_j is very cheap but a little involved because, as we show, a tridiagonal system of linear equations must be solved.

3.3.4 Theorem. *Let* $\Delta = \{x_1, \ldots, x_n\}$ *be a grid over* $[a, b]$ *with spacings*

$$h_i = x_{i+1} - x_i \quad (i = 1, \ldots, n - 1).$$

The function S given by (3.5) for $x \in [x_j, x_{j+1}]$ *is a cubic spline, interpolating f on the grid, precisely when there are numbers* m_j $(j = 1, \ldots, n)$ *so that*

$$\gamma_j = 2m_j + m_{j+1}, \quad \delta_j = \frac{m_{j+1} - m_j}{h_j} \quad (j = 1, \ldots, n - 1), \tag{3.6}$$

$$(1 - q_j)m_{j-1} + 2m_j + q_j m_{j+1} = f[x_{j-1}, x_j, x_{j+1}] \quad (j = 2, \ldots, n - 1), \tag{3.7}$$

where

$$q_j = \frac{h_j}{h_{j-1} + h_j} \in [0, 1] \quad (j = 2, \ldots, n - 1).$$

Proof. Because $S(x)$ is piecewise cubic, $S''(x)$ is piecewise linear, and if we knew $S''(x)$, then we would know $S(x)$. Because, however, $S''(x)$ is still unknown, we define

$$m_j = \frac{1}{6} S''(x_j) \quad \text{for } j = 1, \ldots, n.$$

Formula (3.5) gives the following derivatives for $x \in [x_j, x_{j+1}]$:

$$S'(x) = \beta_j + \gamma_j(2x - x_j - x_{j+1})$$
$$+ \delta_j(2(x - x_j)(x - x_{j+1}) + (x - x_j)^2),$$
$$S''(x) = 2\gamma_j + 2\delta_j(3x - 2x_j - x_{j+1}).$$

Thus we obtain the following continuity condition on S'':

$$6m_j = 2\gamma_j - 2\delta_j h_j \quad \text{for } x \to x_j + 0 \quad \text{in } S''(x),$$
$$6m_{j+1} = 2\gamma_j + 4\delta_j h_j \quad \text{for } x \to x_{j+1} - 0 \text{ in } S''(x);$$

but this is equivalent to (3.6).

To get (3.7), we derive continuity conditions on S' as $x \to x_j + 0$ and $x \to x_{j+1} - 0$:

$$S'(x_j) = \beta_j - \gamma_j h_j = \beta_j - (2m_j + m_{j+1})h_j,$$
$$S'(x_{j+1}) = \beta_j + \gamma_j h_j + \delta_j h_j^2 = \beta_j + (m_j + 2m_{j+1})h_j.$$

We can therefore formulate the continuity condition on S' as

$$\beta_j - (2m_j + m_{j+1})h_j = \beta_{j-1} + (m_{j-1} + 2m_j)h_{j-1};$$

this is equivalent to

$$f[x_{j-1}, x_j, x_{j+1}] = \frac{f[x_j, x_{j+1}] - f[x_{j-1}, x_j]}{x_{j+1} - x_{j-1}} = \frac{\beta_j - \beta_{j-1}}{h_{j-1} + h_j}$$

$$= \frac{(2m_j + m_{j+1})h_j + (m_{j-1} + 2m_j)h_{j-1}}{h_{j-1} + h_j}$$

for $j = 2, \ldots, n-1$, and this simplifies to (3.7).   □

Because $f[x_{j-1}, x_j, x_{j+1}]$ can be computed from the known function values, the previous equation gives $n - 2$ defining equations for the n unknowns m_j. We can pose various additional conditions to obtain the two missing equations. We now examine four different possibilities.

CASE 1: A simple requirement is the *free node* condition for x_2 and x_{n-1}. In this case, we demand that $S(x)$ be given by the same cubic polynomial in each of the two neighboring sets of intervals $[x_1, x_2]$, $[x_2, x_3]$ and $[x_{n-2}, x_{n-1}]$, $[x_{n-1}, x_n]$; thus x_2 and x_{n-1} are, in a certain sense, "artificial" nodes of the spline. As a consequence, S''' cannot have jumps at x_2 or x_{n-1}. Since S''' shouldn't have a jump at x_2, we must have $\delta_1 = \delta_2$, so

$$\frac{m_2 - m_1}{x_2 - x_1} = \frac{m_3 - m_2}{x_3 - x_2}.$$

This leads to the condition

$$m_3 = \frac{(x_3 - x_1)m_2 - (x_3 - x_2)m_1}{x_2 - x_1} = \frac{(h_1 + h_2)m_2 - h_2 m_1}{h_1}.$$

Plugging this into (3.7) for $j = 2$ gives

$$h_1(h_1 + h_2)f[x_1, x_2, x_3] = h_1^2 m_1 + 2h_1(h_1 + h_2)m_2 + h_1 h_2 m_3$$
$$= (h_1^2 - h_2^2)m_1 + (2h_1 + h_2)(h_1 + h_2)m_2$$
$$= (h_1 - h_2)(h_1 + h_2)m_1$$
$$\quad + (2h_1 + h_2)(h_1 + h_2)m_2,$$

so

$$\left(1 - \frac{h_2}{h_1}\right)m_1 + \left(2 + \frac{h_2}{h_1}\right)m_2 = f[x_1, x_2, x_3]. \qquad (3.8)$$

Similarly, the free node condition at x_{n-1} gives the determining equation

$$\left(2 + \frac{h_{n-2}}{h_{n-1}}\right) m_{n-1} + \left(1 - \frac{h_{n-2}}{h_{n-1}}\right) m_n = f[x_{n-2}, x_{n-1}, x_n]. \quad (3.9)$$

From (3.7) and these two additional equations, we obtain the tridiagonal linear system $Am = d$ with

$$m = \begin{pmatrix} m_1 \\ \vdots \\ m_n \end{pmatrix}, \quad d = \begin{pmatrix} f[x_1, x_2, x_3] \\ f[x_1, x_2, x_3] \\ f[x_2, x_3, x_4] \\ \vdots \\ f[x_{n-2}, x_{n-1}, x_n] \\ f[x_{n-2}, x_{n-1}, x_n] \end{pmatrix},$$

$$A_{ii} = \begin{cases} 1 - h_2/h_1 & \text{for } i = 1, \\ 1 - h_{n-2}/h_{n-1} & \text{for } i = n, \\ 2 & \text{otherwise,} \end{cases}$$

$$A_{i,i+1} = \begin{cases} 2 + h_2/h_1 & \text{for } i = 1, \\ q_j & \text{otherwise,} \end{cases}$$

$$A_{i,i-1} = \begin{cases} 2 + h_{n-2}/h_{n-1} & \text{for } i = n, \\ 1 - q_j & \text{otherwise,} \end{cases}$$

$$A_{ik} = 0 \quad \text{if } |k - i| > 1.$$

The computational cost to obtain the m_i and, therefore, the spline coefficients is only of order $O(n)$. If we have an equispaced grid, then we get the following matrix:

$$A = \begin{bmatrix} 0 & 3 & & & & & 0 \\ \frac{1}{2} & 2 & \frac{1}{2} & & & & \\ & \frac{1}{2} & 2 & \frac{1}{2} & & & \\ & & \ddots & \ddots & \ddots & & \\ & & & \frac{1}{2} & 2 & \frac{1}{2} & \\ & & & & \frac{1}{2} & 2 & \frac{1}{2} \\ 0 & & & & & 3 & 0 \end{bmatrix}$$

In this important special case, the solution of the linear system $Am = b$ simplifies a little; m_2 and m_{n-1} may be determined directly, then m_3 to m_{n-2} may be determined by Gaussian elimination, and m_1 and m_n

may be determined by substitution. Because of diagonal dominance, pivoting is unnecessary.

CASE 2: Another possibility to determine two additional equations for the m_j is to choose two additional points, $x_0 = x_1 - h_0$ and $x_{n+1} = x_n + h_n$, near a and b and demand that S interpolate the function f at these points. For x_0, we obtain the additional condition

$$
\begin{aligned}
f[x_0, x_1, x_2] &= \gamma_1 + \delta_1(x_0 - x_1') \\
&= \frac{3}{2}(m_1 + m_2) + \frac{m_2 - m_1}{h_1}\left(x_0 - \frac{x_1 + x_2}{2}\right),
\end{aligned}
$$

so

$$
f[x_0, x_1, x_2] = \frac{(2h_1 + h_0)m_1 + (h_1 - h_0)m_2}{h_1}.
$$

Similarly, we obtain as additional condition for x_{n+1}:

$$
f[x_{n-1}, x_n, x_{n+1}] = \frac{(2h_{n-1} + h_n)m_n + (h_{n-1} - h_n)m_{n-1}}{h_{n-1}}.
$$

This leads to a tridiagonal linear system analogous to that in Case 1 but with another right side d and other boundary values for A. This linear system can also be solved without pivoting, provided that

$$
\frac{x_1 - x_0}{h_1}, \frac{x_{n+1} - x_n}{h_{n-1}} \in \left[-\frac{1}{4}, \frac{5}{4}\right], \tag{3.10}
$$

because then

$$
\left\| I - \frac{1}{2}A \right\|_\infty \le \frac{3}{4} \tag{3.11}
$$

and A is an H-matrix.

CASE 3: If f is *periodic* with period in $[a, b]$, we can construct S to be a *periodic spline* with $x_0 = x_{n-1}$, $x_{n+1} = x_2$, and $m_1 = m_n$. In this case, we obtain a linear system for $m_1, \ldots, m_{n-1}$ from (3.7), whose matrix

A has the following form:

$$
\begin{pmatrix}
\times & \times & 0 & \cdots & 0 & \times \\
\times & \times & \times & & 0 & 0 \\
0 & \times & \times & & & 0 \\
\vdots & & & \ddots & & \vdots \\
0 & 0 & & & \ddots & \times \\
\times & 0 & 0 & \cdots & \times & \times
\end{pmatrix}
$$

Here, $\left\| I - \frac{1}{2}A \right\|_\infty \le \frac{1}{2}$, so that we may again apply Gaussian elimination without pivoting; however, because of the additional corner elements outside the band, nonzero elements are produced in the last rows and columns of the factorization. Thus the expense increases a little, but remains of order $O(n)$.

CASE 4: If the derivative of f at the end points x_1 and x_n is known (or can reliably be estimated), we can set $x_0 = x_1$ and $x_{n+1} = x_n$; to obtain m_1 and m_n, we then compute the divided differences $f[x_1, x_1, x_2]$ and $f[x_{n-1}, x_n, x_n]$ from these derivative values. This corresponds to the requirement that $S'(x) = f'(x)$ for $x \in \{x_1, x_n\}$. The spline so obtained is called a *complete spline interpolant*.

In order to give a bound for the error of approximation $\|f - S\|_\infty$, we first prove the following.

3.3.5 Lemma. *Suppose f is twice continuously differentiable in $[a, b]$. Then, for every complete spline interpolant for f on $[a, b]$,*

$$\|S''\|_\infty \le 3\|f''\|_\infty.$$

Proof. For a complete spline interpolant, $\|I - \frac{1}{2}A\| \le \frac{1}{2}$; therefore, Proposition 2.4.2 implies

$$\|A^{-1}\|_\infty \le 1.$$

Because $f[x, x', x''] = \frac{1}{2}f''(\xi)$ for some $\xi \in \square\{x, x', x''\}$, we have $\|d\|_\infty \le \frac{1}{2}\|f''\|_\infty$. Because S'' is piecewise linear, we have

$$
\begin{aligned}
\|S''\|_\infty &= \max_{i=1:n} S''(x_i) = 6 \max_{i=1:n} |m_i| = 6\|m\|_\infty \\
&= 6\|A^{-1}d\|_\infty \le 6\|A^{-1}\|_\infty \|d\|_\infty \le 6 \cdot 1 \cdot \frac{1}{2} \cdot \|f''\|_\infty.
\end{aligned}
$$

$\square$

The following error bound for the complete spline interpolant results, as well as similar bounds for the errors in derivatives.

3.3.6 Theorem. *Suppose f is four times continuously differentiable in $[a, b]$. Then the bounds*

$$\|f - S\|_\infty \le \frac{h^4}{16}\|f^{(4)}\|_\infty,$$

$$\|f' - S'\|_\infty \le \frac{h^3}{2}\|f^{(4)}\|_\infty,$$

$$\|f'' - S''\|_\infty \le \frac{h^2}{2}\|f^{(4)}\|_\infty,$$

hold for every complete spline interpolant with mesh size h.

Proof. For the proof, we utilize the cubic spline $\bar{S}$ with

$$\bar{S}''(x_j) = f''(x_j) \quad \text{for } j = 1, \dots, n,$$
$$\bar{S}'(x_j) = f'(x_j) \quad \text{for } j = 1 \text{ and } j = n,$$

where the existence of $\bar{S}$ follows by integrating the piecewise linear function through $f''(x_j)$ twice. Lemma 3.3.5 implies

$$\|f'' - S''\|_\infty \le \|f'' - \bar{S}''\|_\infty + \|(\bar{S} - S)''\|_\infty$$
$$\le \|f'' - \bar{S}''\|_\infty + 3\|(\bar{S} - f)''\|_\infty = 4\|f'' - \bar{S}''\|_\infty.$$

Because $\bar{S}''$ is the piecewise linear interpolant to f'' at the x_i, Theorem 3.3.2 implies the bounds

$$\|f'' - \bar{S}''\|_\infty \le \frac{h^2}{8}\|(f'')''\|_\infty,$$

so

$$\|f'' - S''\|_\infty \le \frac{h^2}{2}\|f^{(4)}\|_\infty. \tag{3.12}$$

To obtain the corresponding inequalities for f and f', we set $e := f - S$. Because $e(x_j) = 0$, the relationship

$$e(x) = (x - x_j)(x - x_{j+1})e[x_j, x_{j+1}, x]$$

then follows for $x \in [x_j, x_{j+1}]$, $j = 1, \dots, n - 1$. Formula (3.12) and

$|(x - x_j)(x - x_{j+1})| \le h^2/4$ then imply

$$\|f - S\|_\infty = \|e\|_\infty \le \frac{h^2}{4} \cdot \frac{1}{2}\|e''\|_\infty \le \frac{h^4}{16}\|f^{(4)}\|_\infty.$$

For the derivative of e, we obtain

$$\begin{aligned} e'(x) &= (e[x, x_{j+1}] - e[x_j, x_{j+1}]) - (e[x, x_{j+1}] - e[x, x]) \\ &= (x - x_j)e[x, x_j, x_{j+1}] - (x_{j+1} - x)e[x, x, x_{j+1}], \end{aligned}$$

so (3.12) implies

$$\|f' - S'\|_\infty = \|e'\|_\infty \le h \cdot \frac{1}{2}\|e''\|_\infty + h \cdot \frac{1}{2}\|e''\|_\infty = h\|e''\|_\infty \le \frac{h^3}{2}\|f^{(4)}\|_\infty.$$

$\square$

3.3.7 Remarks.

(i) Comparison with Hermite interpolation shows that the approximation error has the same order $(O(h^4))$. In addition (and this can also be proved for Hermite interpolation), the first and second derivatives are also approximated with orders $O(h^3)$ and $O(h^2)$. (The present simple proof does not yield optimal coefficients in the error bounds; see de Boor [15] for best estimates.)

(ii) For equispaced grids (i.e., $x_{i+1} - x_i = h$ for each i), it is possible to show that the first derivative is also approximated with order $O(h^4)$ *at the grid points*. Computation of the complete spline interpolant for a function f is therefore also useful for numerical differentiation of f.

(iii) Theorem 3.3.6 holds without change for periodic splines in place of complete splines. It also holds for splines defined by the free node condition and splines defined through additional interpolation points, when the error expressions are multiplied by the (usually small) constant factor $\frac{1}{4}(1 + 3\|A^{-1}\|_\infty)$.

An Optimality Property of Cubic Splines

We now prove an important optimality theorem for complete spline interpolants.

3.3.8 Theorem. *Suppose f is twice continuously differentiable in $[a, b]$, and suppose $S(x)$ is a complete spline interpolant for f on the grid $a = x_1 < \cdots <$*

$x_n = b$. Then

$$\|S''\|_2^2 = \|f''\|_2^2 - \|f'' - S''\|_2^2 \leq \|f''\|_2^2.$$

Proof. To prove the statement, we compute the expression

$$
\begin{aligned}
\varepsilon &:= \|f''\|_2^2 - \|f'' - S''\|_2^2 - \|S''\|_2^2 \\
&= \int_a^b \left((f''(x))^2 - (f''(x) - S''(x))^2 - S''(x)^2 \right) dx \\
&= 2 \int_a^b (f''(x) - S''(x)) S''(x)\, dx \\
&= 2 \sum_{i=2:n} \int_{x_{i-1}}^{x_i} \left(f''(x) - S''(x) \right) S''(x)\, dx.
\end{aligned}
$$

Because $S''(x)$ is differentiable when $x \in (x_{i-1}, x_i)$, we obtain

$$\varepsilon = 2 \sum_{i=2:n} \left((f'(x) - S'(x)) S''(x) \Big|_{x_{i-1}}^{x_i} - \int_{x_{i-1}}^{x_i} (f'(x) - S'(x)) S'''(x)\, dx \right)$$

by integration by parts. Because the expression $(f'(x) - S'(x))S''(x)$ is continuous on all of $[a, b]$, and because $S'''(x)$ exists and is constant in the interval $[x_{i-1}, x_i]$, we furthermore obtain

$$\varepsilon = 2 \left[(f'(x) - S'(x)) S''(x) \Big|_a^b - \sum_{i=2:n} (f(x) - S(x)) S'''(x) \Big|_{x_{i-1}}^{x_i} \right].$$

Because f' and S' agree by assumption on the boundary, the first expression vanishes. The second expression also vanishes because f and S agree at the nodes x_j $(j = 1, \ldots, n)$. Therefore, $\varepsilon = 0$, and the assertion follows. $\qquad\square$

3.3.9 Remarks.

(i) The optimality theorem states that the 2-norm of the second derivative of the complete cubic spline interpolant is minimal over all twice continuously differentiable functions that take on specified values at $x_1, \ldots, x_n$ and whose derivatives take on specified values at x_1 and x_n. This optimality property gave rise to the name "spline," borrowed from the name for a bendable metal strip used by engineers in drafting. If such a physical spline were forced to go through the data points, then the resulting curve would be characterized by a minimum stress energy that is approximated by $\int \|f''\|_2^2$.

(ii) The 2-norm of the second derivative is an (approximate) measure for the total curvature of the function. This is because the curvature of a function $f(x)$ is

$$K(x) = \frac{f''(x)}{\sqrt{1 + f'(x)^2}}.$$

Therefore, if the derivative of the function is small or almost constant, then the curvature is approximately equal or proportional to the second derivative of the function, respectively. Thus, we can approximately interpret the theorem in the following geometric way: the complete spline interpolant has a total curvature that is at most as large as that of the original function. The theorem furthermore says that the complete spline interpolant is approximately optimal over all interpolating functions in the sense of minimal total curvature and therefore is more graphically esthetic.

(iii) The same proof produces analogous optimality for periodic splines and the so-called *natural splines,* which are characterized by the condition $S''(x_1) = S''(x_n) = 0$. The latter are of course appropriate only for the approximation of functions f with $f''(x_1) = f''(x_n) = 0$.

3.4 Approximation by Splines

In practice, we frequently know function values only approximately because they come from experimental measurements or expensive simulations. Because of their minimal curvature property, splines are ideal for fitting noisy data (x_i, y_i) $(i = 1, \ldots, m)$ lying approximately on a smooth curve $y = f(x)$ of unknown parametric form.

To find a suitable functional dependence, one represents $f(x)$ as a cubic spline $S(x)$. If we interpolated noisy data with a node at each value we would obtain a very unsmooth curve reflecting the noise introduced by the inaccuracies of the function values. Therefore, one approximates the function by a spline with a few nodes only.

Approximation by Interpolation

A simple and useful method is the following: We choose a few of the data points as nodes and consider the spline interpolating f at these points. We then compute the error interval

$$\mathbf{e} = \Box \{S(x) - f(x) \mid x \in M\}$$

corresponding to the set M of data points that were ignored. If the maximum error is then larger than a previously specified accuracy bound, then we choose as additional nodes those two data points at which the maximum positive and negative error occurred, and interpolate again. We then compute the maximal error again, and so on, until the approximation is sufficiently accurate. The computational expense for this technique is $O(nm)$, where m is the total number of data points and n is the maximum number of data points that are chosen to be nodes.

A problem with all approximation methods is the treatment of *outliers*, that is, particularly erroneous data points that are better ignored because they should not make any contribution to the interpolating function. We can eliminate isolated outliers for linearly ordered data points $x_1 < x_2 < \cdots < x_m$ as follows. We write $\varepsilon_l = S(x_l) - f(x_l)$ and, for $i < m$,

$$e_i = \begin{cases} \varepsilon_i & \text{if } |\varepsilon_i| \leq |\varepsilon_{i+1}| \\ \varepsilon_{i+1} & \text{otherwise.} \end{cases}$$

We then take $\mathbf{e} = \square\{e_i \mid i = 1, \ldots, m-1\}$ as error interval in the previous method.

Approximation by Least Squares

A more sophisticated method that better reflects the stochastic nature of the noise uses the method of least squares. Here, one picks as approximating function the spline that minimizes the expression

$$\sum_{i=1}^{m} w_i (y_i - S(x_i))^2, \tag{4.1}$$

where the w_i are appropriately chosen *weights*.

Regarding the choice of weights, we note that function values should make a contribution to the approximating function in proportion to how precise they are; thus the weights depend, among other things, on the accuracy of the function values. In the following, we assume that the function values are all equally precise; then a simple choice would be to set each w_i to 1. However, the distance between two adjacent data points should be reflected in the weights, too, because individual data points should provide a smaller contribution in regions where they are close together than in regions where they are far apart. A reasonable choice is, for example,

$$w_i := \begin{cases} (x_2 - x_1)/2 & \text{for } i = 1, \\ (x_{i+1} - x_{i-1})/2 & \text{for } i = 2, \ldots m-1, \\ (x_m - x_{m-1})/2 & \text{for } i = m, \end{cases}$$

because the expression (4.1) is then approximately equal to the integral $\int (f(x) - S(x))^2 dx = \| f - S \|_2^2$ (see the trapezoidal rule in Section 4.3). With this choice of weights, we approximately minimize the 2-norm of the error function.

The choice of the number of nodes is more delicate because using too few nodes gives a poor approximation, whereas using too many produces wiggly curves fitting not only the information in the data but also their noise. A suitable way is to proceed stepwise, essentially as explained in the approximation by interpolation.

Minimization of (4.1) proceeds easiest when we represent the spline $S(x)$ as a linear combination of B-splines. The fact that these functions have small compact support with little overlap ensures that the matrix of the associated least squares problem is sparse and well conditioned.

In the following, we allow the nodes $\bar{x}_1 < \bar{x}_2 < \cdots < \bar{x}_m$ of the spline to differ from the data points x_i. Usually, much fewer nodes are used than data points are available.

To minimize (4.1), we assume that S is of the form

$$S(x) := \sum_{l=1}^{n} z_l S_l(x),$$

where the z_l are parameters to be determined and where the S_l are B-splines over a fixed grid Δ. Thus, we seek the minimum of

$$h(z) = \sum_{i=1}^{m} w_i (y_i - S(x_i))^2 = \sum_{i=1}^{m} F_i(z)^2 = \| F(z) \|_2^2,$$

with

$$F_i(z) = \sqrt{w_i}(y_i - S(x_i)) = \sqrt{w_i} \left(y_i - \sum_{l=1}^{n} z_l S_l(x_i) \right).$$

Written in matrix form, the previous equation is $F(z) = D(y - Bz)$, where

$$D = \text{Diag}(\sqrt{w_1}, \ldots, \sqrt{w_n}) \in \mathbb{R}^{m \times m},$$

$$B = \begin{pmatrix} S_1(x_1) & \cdots & S_n(x_1) \\ \vdots & & \vdots \\ S_1(x_m) & \cdots & S_n(x_m) \end{pmatrix} \in \mathbb{R}^{m \times n},$$

$$y = \begin{pmatrix} y_1 \\ \vdots \\ y_m \end{pmatrix} \in \mathbb{R}^m, \quad z = \begin{pmatrix} z_1 \\ \vdots \\ z_n \end{pmatrix} \in \mathbb{R}^m.$$

We therefore need to solve the linear least squares problem

$$\|Dy - DBz\|_2 = \min! \tag{4.2}$$

Here, selection of the basis splines leads to a matrix B with a simple, sparse form, in that there are at most four nonzero elements in each row of B. For example, when $n = 7$, this leads to the following form, when there are precisely $1, 4, 2, 1, 2, 1$ data points in the intervals $[\bar{x}_i, \bar{x}_{i+1}]$, $i = 1, 2, 3, 4, 5, 6$, respectively:

$$
B = \begin{pmatrix}
\times & \times & \times &   &   &   &   & 0 \\
\times & \times & \times & \times &   &   &   &   \\
\times & \times & \times & \times &   &   &   &   \\
\times & \times & \times & \times &   &   &   &   \\
\times & \times & \times & \times &   &   &   &   \\
       & \times & \times & \times & \times &   &   &   \\
       & \times & \times & \times & \times &   &   &   \\
       &   & \times & \times & \times & \times &   &   \\
       &   &   & \times & \times & \times & \times &   \\
       &   &   & \times & \times & \times & \times &   \\
0   &   &   &   & \times & \times & \times &
\end{pmatrix}
$$

To solve problem (4.2), we may use the normal equations

$$(DB)^T DBz = (DB)^T Dy \tag{4.3}$$

(Theorem 2.2.7) because the matrix for this special problem is generally well-conditioned. Formula (4.3) is equivalent to

$$(B^T W B)z = B^T W y, \tag{4.4}$$

where $W = \mathrm{Diag}(w_1, \ldots, w_n)$. Because the element

$$(B^T W B)_{jk} = \sum_i w_i S_j(x_i) S_k(x_i)$$

vanishes whenever S_j and S_k do not overlap, the matrix $B^T W B$ is a 7-band matrix. Moreover, $B^T W B$ is positive definite, so that a solution of (4.4) via a Cholesky factorization succeeds without problems.

Because B contains a total of at most $4m$ nonzero elements (and therefore on average $4m/n$ nonzero elements per column) and (by symmetry) only four bands must be computed, the computation of the left side of the normal equations requires at most $4n \cdot 4m/n \cdot 3 = O(m)$ operations. Similarly, the formation of the right side also takes $O(m)$ operations. Because the solution of the 7-band system takes $O(n)$ operations and $n < m$, a total of $O(m)$ operations are required to solve (4.4).

A comparison with approximation by interpolation shows that a factor of $O(n)$ is saved. Nonetheless, the least squares method is faster only beyond a certain n because the constant associated with the expression for the cost is now larger than before.

The least squares method can also be used for interpolation; then, in contrast to the case already considered, the data points y_i and the nodes $\bar{x}_i$ do not need to correspond. According to de Boor [15],

$$\bar{x}_i := \begin{cases} x_i & \text{for } i = 1, m \\ (x_{i-1} + x_i + x_{i+1})/3 & \text{otherwise} \end{cases}$$

is a nearly optimal choice of nodes.

Parametric Splines

A case that often occurs in practice (e.g., when modeling forms such as airplane wings) is that in which we seek a smooth curve that connects the pairs (x_i, y_i) and (x_{i+1}, y_{i+1}). Here, the x_i $(i = 1, \ldots, m)$ are in general *not* necessarily distinct. We cannot approximate such curves by a function f with $f(x_i) \approx y_i$. To solve this problem, we introduce an artificial parameter s and view the pairs (x, y) as a function of s. This function can then be approximated by two splines. We obtain good results when we take the following arclength approximation for s:

$$s_1 := 0, \quad s_{i+1} := s_i + \sqrt{(x_{i+1} - x_i)^2 + (y_{i+1} - y_i)^2}.$$

Then, s_i is the length of the piecewise linear curve which connects (x_1, y_1) with (x_i, y_i) through the points $(x_2, y_2), \ldots, (x_{i-1}, y_{i-1})$. If we now specify splines S_x and S_y corresponding to a grid on $[0, s_m]$ with $S_x(s_i) \approx x_i$ and $S_y(s_i) \approx y_i$ $(i = 1, \ldots, m)$, then the curve $(x, y) = \big(S_x(s), S_y(s)\big)$ is a smooth curve which approximates the given point set.

3.4.1 Remarks.

(i) The constructed curve is invariant under orthogonal transformations (rotations and reflections).

(ii) The parametric curve can be easily graphed, since we may simply graph $x = S_x(s)$ and $y = S_y(s)$.

(iii) If we are looking for a closed curve then $x_{m+1} = x_1$ and $y_{m+1} = y_1$, and we can approximate with periodic splines.

3.5 Radial Basis Functions

Radial basis functions are real-valued functions defined on $\mathbb{R}^d$ that are radially symmetric with respect to some center x_i, and hence have the form $\varphi(\|x - x_i\|_2)$ for a suitable univariate function $\varphi : \mathbb{R}_+ \to \mathbb{R}$. Most analyzed and used are radial basis functions with

$$\varphi(r) = \sqrt{r^2 + c^2} \quad \text{(multiquadric)}$$
$$\varphi(r) = r^2 \log r^2 \quad \text{(thin plate spline)}$$
$$\varphi(r) = e^{-c^2 r^2} \quad \text{(Gaussian)}$$

Strong interpolation and approximation theorems are known for the interpolation and approximation by linear combinations of radial basis functions; they are based on the fact that good radial basis functions *grow* with $\|x\|$, but their higher order divided differences decay (see, e.g., Light [56]). The multivariate analysis is quite involved; here, we only relate univariate radial basis functions to splines, to motivate that we may indeed expect good approximation properties.

3.5.1 Theorem. *The space of all piecewise linear functions with possible breakpoints (nodes) at an infinite grid $\ldots < x_i < x_{i+1} < \ldots$ agrees with the space of all linear combinations of the functions $\Phi_i (i = 1, \ldots, n)$ defined by $\Phi_i(x) := |x - x_i|$.*

Proof. All piecewise linear functions over the grid $\ldots < x_i < x_{i+1} < \ldots$ have a representation as a linear combination

$$S(x) = \sum_i S(x_i) B_i(x)$$

of the linear B-splines

$$B_i(x) = \begin{cases} (x - x_{i-1})/(x_i - x_{i-1}) & \text{if } x_{i-1} \leq x \leq x_i, \\ (x - x_{i+1})/(x_i - x_{i+1}) & \text{if } x_i \leq x \leq x_{i+1}, \\ 0 & \text{otherwise.} \end{cases}$$

Now, whereas the functions

$$\Phi_i(x) := |x - x_i|$$

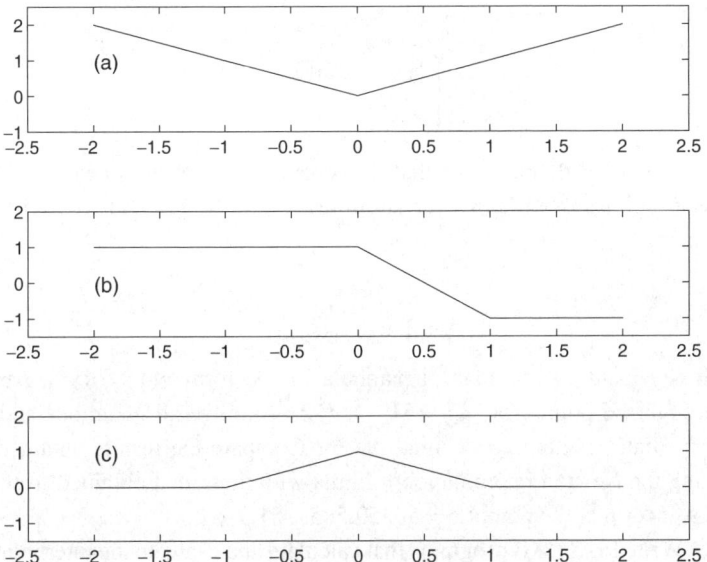

Figure 3.4. Linear B-splines as linear combinations of simple radial basis functions (a) Φ_i, (b) $\Phi_{i+1,i}$, (c) B_i.

are unbounded and wedge-shaped, the differenced functions

$$\Phi_{i+1,i} := \frac{\Phi_{i+1} - \Phi_i}{x_{i+1} - x_i}$$

are constant outside $[x_i, x_{i+1}]$, and the twice differenced functions

$$B_i(x) = \frac{\Phi_{i+1,i} - \Phi_{i,i-1}}{2}$$

are the linear B-splines over the grid. In particular, $B_i(x)$ and therefore all piecewise linear functions are linear combinations of $\Phi_i(x)$. Conversely, because each $\Phi_i(x)$ is piecewise linear, all their linear combinations are piecewise linear. Thus approximation by piecewise linear functions is equivalent to approximation by linear combinations of $|x - x_i|$. □

More generally, higher order splines can be written as linear combinations of functions of the form

$$\Phi_i(x) = \varphi(x - x_i)$$

centered at x_i where for order s splines,

$$\varphi(x) = \begin{cases} |x|^s & \text{if } s \text{ is odd,} \\ x|x|^{s-1} & \text{if } s \text{ is even.} \end{cases} \tag{5.1}$$

Indeed, it is not difficult to see that, in generalization of the linear case, one obtains B-splines of order s over an arbitrary grid by $(s + 1)$-fold repeated differencing.

3.6 Exercises

1. Fit (by hand calculations) a parabola of the form $g(x) = ax^2 + bx + c$ through the points $(50, \frac{1}{3})$, $(51, \frac{2}{3})$, $(52, \frac{11}{3})$. Give all quantities and intermediate results to 5 decimal places. Compare the results obtained by using the Newton interpolation formula with those that obtained from the power form at the points $x = 50, 50.5, 51, 51.5, 52$.

2. (a) Write MATLAB programs that calculate and evaluate the interpolation polynomial for a MATLAB function f at n pairwise distinct interpolation points $x_1, \ldots, x_n$.

   (b) Interpolate the functions $f_1(x) := e^{x/10}$ and $f_2(x) := 1/(1 + x^2)$ at the equidistant interpolation points $x_i := -5 + 10i/n$, $i = 0, \ldots, n$.

   For f_1 and f_2 and $n = 1, 4, 8, 16$, plot f and p_n and calculate an estimate of the maximum error

   $$\max_{x \in I} |p_n(x) - f(x)|$$

   by evaluating the interpolation polynomial p_n at 101 equidistant points in the intervals (i) $I = [-1, 1]$ and (ii) $I = [-5, 5]$.

   (c) For both functions and both intervals, determine the value of n where the estimated error becomes least.

3. Let the function $f : \mathbb{R} \to \mathbb{R}$ be sufficiently differentiable.

   (a) Show that divided differences satisfy

   $$\frac{d}{dx_v} f[x_0, x_1, \ldots, x_n] = f[x_0, x_1, \ldots, x_{v-1}, x_v, x_v, x_{v+1}, \ldots, x_n]$$

   for $v = 0, \ldots, n$.

   (b) Find and prove an analogous expression for $\frac{d^k}{dx_v^k} f[x_0, x_1, \ldots, x_n]$.

4. Show that the Lagrange polynomials can be written as

   $$L_j(x) = q[x, x_j]/q[x_j, x_j],$$

   where $q(x) = (x - x_0) \cdots (x - x_n)$.

5. Given $n + 1$ points $x_0, \ldots, x_n$ and an analytical function $f : \mathbb{R} \to \mathbb{R}$, the matrices $S_n(f)$, $A \in \mathbb{R}^{(n+1)\times(n+1)}$ are defined by $S_n(f) := S$ with

$$
S_{ik} := \begin{cases} f[x_i, \ldots, x_k] & \text{for } i \leq k, \\ 0 & \text{for } i > k \end{cases} \qquad (i, k = 0, \ldots, n)
$$

and

$$
A := \begin{pmatrix} x_0 & 1 & & & 0 \\ & \ddots & & \ddots & \\ & & & x_{n-1} & 1 \\ 0 & & & & x_n \end{pmatrix}.
$$

Prove for $\alpha, x_0, \ldots, x_n \in \mathbb{R}$, and arbitrary n times differentiable functions f, g:

(a) $S_n(\alpha) = \alpha I$, $S_n(f \pm g) = S_n(f) \pm S_n(g)$, $S_n(x \cdot f) = A S_n(f)$.

(b) If p is a polynomial then $S_n(p) = p(A)$.

(c) $S_n(f \cdot g) = S_n(f) \cdot S_n(g)$.

(d) If $h := fg$ then

$$
h[x_i, \ldots, x_k] = \sum_{j=i:k} f[x_i, \ldots, x_j] g[x_j, \ldots, x_k].
$$

(e) Specialize (d) to get a formula for the kth derivative $(fg)^{(k)}$.

Hint: Use induction to get (b) from (a); deduce (c) from (b) and interpolation.

6. Let $d_i = f[x_0, \ldots, x_i]$ and

$$
A = \begin{pmatrix} x_0 & & & & 0 & -d_0 \\ 1 & x_1 & & & & -d_1 \\ & \ddots & \ddots & & & \vdots \\ & & 1 & x_{n-1} & & -d_{n-2} \\ 0 & & & 1 & & d_n x_n - d_{n-1} \end{pmatrix},
$$

$$
D = \begin{pmatrix} 1 & & & 0 \\ & \ddots & & \\ & & 1 & \\ 0 & & & d_n \end{pmatrix}.
$$

Show that $p_n(x) = \det(Dx - A)$ is the Newton interpolation polynomial for f at $x_0, \ldots, x_n$.

7. Let $(x_0, y_0, y_0') := (-1, -3, 19)$, $(x_1, y_1, y_1') := (0, 0, -1)$, and $(x_2, y_2, y_2') := (1, 3, 19)$. Calculate the Hermite interpolation polynomial $p(x)$ of degree 5 with $p(x_\nu) = y_\nu$ and $p'(x_\nu) = y_\nu'$, $\nu = 0, 1, 2$.

8. Let $p_n(x)$ be the polynomial of degree $\leq n$ that interpolates the sufficiently smooth function $f(x)$ at the pairwise distinct points $x_0, \ldots, x_n$.

   (a) Show that for $k \leq n$ the kth derivative of the error function $f(x) - p_n(x)$ has at least $n - k + 1$ zeros in the smallest interval $[a, b]$ containing all x_ν.

   (b) Use (a) to prove for arbitrary $k \in \mathbb{N}$ the existence of some $\xi \in [a, b]$ such that

$$f[x_0, x_1, \ldots, x_k] = \frac{f^{(k)}(\xi)}{k!}. \qquad (6.1)$$

   (c) Show that the relation (6.1) holds even when certain interpolation points coincide.

   (d) Show that, for the Hermite interpolation polynomial $\bar{p}_n(x)$ that coincides with the values of the function f and its first derivative at the points $x_0, \ldots, x_n$, the remainder formula

$$f(x) - \bar{p}_n(x) = \frac{f^{(2n+2)}(\xi)}{(2n + 2)!} q_n^2(x)$$

with $\xi \in \square\{x_0, \ldots, x_n, x\}$ and q_n given by $q_n(x) := (x - x_0) \cdots (x - x_n)$ holds for arbitrary $x \in \mathbb{R}$.

9. Let D be a closed disk with fixed center $c \in \mathbb{C}$ and radius r. Let $x_\nu \in D$, $\nu = 0, \ldots, k$ be given points, and let

$$\rho := \max_{\nu = 0, \ldots, k} |x_\nu - c|.$$

Then for every analytic function $f : D \subseteq \mathbb{C} \to \mathbb{C}$,

$$|f[x_0, x_1, \ldots, x_k]| \leq \frac{r}{(r - \rho)^{k+1}} \sup_{\xi \in D} |f(\xi)| =: S(r).$$

Now, let $f(x) := e^x$, $x \in \mathbb{C}$.

   (a) Give an explicit expression for $S(r)$.

   (b) For which value $\bar{r}$ of r is $S(r)$ minimal (with $c, x_0, x_1, \ldots, x_k \in \mathbb{C}$ fixed)?

   (c) How big is the overestimation factor

$$q_k^{\text{over}} := \frac{S(\bar{r})}{|f[x_0, x_1, \ldots, x_k]|}$$

for $x_0 = x_1 = \cdots = x_k = c$ and $k = 1, 2, 4, 8, 16, 32$?

10. The Chebyshev points

$$x_j := \cos \frac{2j+1}{2(n+1)} \pi \quad (j=0,\dots,n)$$

minimize the expression

$$\sup_{\xi \in [-1,1]} |q_n(\xi)|,$$

where $q_n(x) := (x - x_0) \cdots (x - x_n)$.

Show this for $n = 1$, and express the points x_0, x_1 in terms of square roots.

11. Write a MATLAB program that calculates the derivative $f'(x)$ of a given differentiable function f at the point $x \in \mathbb{R}$. For this purpose use the central difference quotient

$$f[x - h_i, x + h_i] := \frac{f(x + h_i) - f(x - h_i)}{2h_i}$$

for $h_i := 2^{-i} h_0$, $i = 0, 1, 2, \dots$, where $h_0 > 0$ is given. Let p_n be the interpolation polynomial of degree $\leq n$ satisfying

$$p_n(h_i^2) = f[x - h_i, x + h_i], \quad i = 0, \dots, n.$$

Choose the degree n for which

$$|p_{n+1}(0) - p_n(0)| \geq |p_n(0) - p_{n-1}(0)|$$

holds for the first time. Then $\frac{1}{2}(p_n(0) + p_{n-1}(0))$ is an estimate of $f'(x)$ and $\frac{1}{2}|p_n(0) - p_{n-1}(0)|$ is a bound for the error.

Determine in this way an approximation and an estimated error for the derivative of $f(x) := e^x$ at $x = 1$ using in succession, the starting values $h_0 := 1, 0.1, 0.01, 0.001$, and compare the results with the true error.

12. Suppose, for the linear function $f(x) := a + bx$, with $a, b \neq 0$, the first derivative $f'(0) = b$ is estimated from

$$\delta_h = \frac{f(h) - f(-h)}{2h}$$

in binary floating point arithmetic with mantissa length L and correct rounding. Let a and b be given binary floating point numbers and let h be a power of 2, so that multiplication with h and division by $2h$ can be performed exactly. Give a bound for the relative error $|(\tilde{\delta}_h - f'(0))/f'(0)|$ of the value $\tilde{\delta}_h$ calculated for δ_h. How does this bound behave as $h \to 0$?

13. A lot of high quality public domain software is freely available on the World Wide Web (WWW). The NETLIB site at

$$\texttt{http://www.netlib.org/}$$

contains a rich collection of mathematical software, papers, and databases.

(a) Search the NETLIB repository for the key word "interpolation" and explore some of the links provided.

(b) Get a suitable program for spline interpolation, and try it out on some simple examples. (To get the program work in a MATLAB environment, you may need to learn something about how to make mex-files that allow you to access FORTRAN or C programs from within MATLAB.)

14. Let S_f be the spline function corresponding to the grid $x_1 < x_2 < \cdots < x_n$ that interpolates f at the points $x_0, x_1, \ldots, x_{n+1}$. Show that the mapping $f \rightarrow S_f$ is linear, that is, that

$$S_{f+g} = S_f + S_g \quad \text{and} \quad S_{\alpha f} = \alpha S_f.$$

15. For an extended grid $x_{-2} < x_{-1} < x_0 < \cdots < x_n < x_{n+1} < x_{n+2} < x_{n+3}$, define functions B_k $(k = 1, \ldots, n)$ by

$$B_k(x)$$

$$:= \begin{cases} \alpha_{k,k-2}(x - x_{k-2})^3 & \text{for } x_{k-2} < x \le x_{k-1}, \\ \alpha_{k,k-2}(x - x_{k-2})^3 + \alpha_{k,k-1}(x - x_{k-1})^3 & \text{for } x_{k-1} < x \le x_k, \\ \alpha_{k,k+2}(x_{k+2} - x)^3 + \alpha_{k,k+1}(x_{k+1} - x)^3 & \text{for } x_k < x \le x_{k+1}, \\ \alpha_{k,k+2}(x_{k+2} - x)^3 & \text{for } x_{k+1} < x < x_{k+2}, \\ 0 & \text{otherwise}, \end{cases}$$

where

$$\alpha_{k,j} = \frac{x_{k+2} - x_{k-2}}{\displaystyle\prod_{\substack{|i-k|\le 2 \\ i\ne k}} (x_i - x_j)}.$$

(a) Show that these formulas define cubic splines, the *B-splines* over the grid.

(b) Extend the grids $(-5, -2, -1, 0, 1, 2, 5)$ and $(-3, -2, -1, 0, 1, 2, 3)$ by three arbitrary points on both sides and plot the nine resulting B-splines (as defined previously) and their sum on the range of the extended grids, using the plotting features of MATLAB. The figure should consist of two subplots, one for each grid, and be self-explaining; so do not forget the labeling of the axes and a plot title.

Hint: The MATLAB functions `subplot`, `title`, `text`, `xlabel`, and `ylabel` may be useful. Try also `set(gca,'xlim',[-6,6])`, and experiment with `set` and `get`.

(c) The plot suggests a conjecture; can you formulate and prove it?

16. For given grid points $x_1 < x_2 < \cdots < x_n$ and function values $f(x_j)$, let $S(x)$ be the cubic spline with

$$S(x) = \alpha_j + \beta_j(x - x_j) + \gamma_j(x - x_j)(x - x_{j+1})$$
$$+ \delta_j(x - x_j)^2(x - x_{j+1}) \quad \text{for } x \in [x_j, x_{j+1}],$$

satisfying $S(x_j) = f(x_j)$, $j = 1, \ldots, n$ and the free node condition at x_2 and x_{n-1}.

(a) Write a MATLAB subroutine that computes the coefficients $\alpha_j, \beta_j, \gamma_j$, and δ_j $(j = 1, \ldots, n - 1)$ from the data points by solving the linear tridiagonal system for the m_j (cf. Theorem 3.3.5).

(b) Write a MATLAB program that evaluates the spline function $S(x)$ at an arbitrary point x. The program should only do three multiplications per evaluation.
Hint: Imitate Horner's method.

(c) Does your program work correctly when you interpolate the function

$$f(x) := \cosh x, \quad x \in [-2, 2],$$

at the equidistant points $-2 = x_1 < \cdots < x_{11} = 2$?

17. Let S be a cubic spline with $m_j := S''(x_j)$ for $j = 1, \ldots, n$. S is called a *natural spline*, provided $m_1 = m_n = 0$.

(a) Show that the natural spline that interpolates a function g minimizes g'' for all twice continuously differentiable interpolants f.

(b) Natural splines are not very appropriate for approximating functions because near the boundary, the error of the approximation is typically $O(h^2)$ instead of the expected $O(h^3)$. To demonstrate this, determine the natural spline interpolant for the parabola $f(x) = x^2$ over the grid $\Delta = \{-h, 0, h\}$ and determine the maximum error in the interval $[-h, h]$. (However, natural splines *are* the right choice if it is known that the function is C^2 and linear on both sides outside the interpolation interval.)

18. *Quadratic splines.* Suppose that the function values $f(x_i), i = 1, \ldots, n$, are known at the equidistant points $x_i = a + (i-1)h$, where $h = (b-a)/(n-1)$. Show that there exists a unique quadratic (i.e. order 2) spline function $S(x)$ over the grid $x_0' < \cdots < x_n'$, where $x_i' = a + (i - \frac{1}{2})h$, $i = 0, \ldots, n$ that interpolates f at $x_1, \ldots, x_n$.

Hint: Formulate a system of equations for the unknown coefficients of S in each subinterval $[x'_{i-1}, x'_i]$.

19. *Dilation equations.* Dilation equations relate splines at two different scales. This provides the starting point for the multiresolution analysis of sound or images by means of so-called wavelets. (For details, see, e.g., Daubeches [13], De Vore and Lucier [21].)

    (a) Show that

$$
B(x) = \begin{cases}
1 - \frac{1}{3}x^2 & \text{if } |x| \le 1, \\
\frac{1}{6}(3 - |x|)^2 & \text{if } 1 < |x| < 3, \\
0 & \text{if } |x| \ge 3
\end{cases}
$$

    is a quadratic (B-)spline.

    (b) Prove that $B(x)$ satisfies for all $x \in \mathbb{R}$ the *dilation equation*

$$
B(x) = \frac{1}{4}B(2x - 3) + \frac{3}{4}B(2x - 1) + \frac{3}{4}B(2x + 1) + \frac{1}{4}B(2x + 3).
$$

    (c) Derive a similar dilation equation for the cubic B-spline in (3.3).

20. (a) Show, for φ as in (5.1), that every linear combination

$$
S(x) = \sum_{i = 1:n} \alpha_k \varphi(x - x_k) \tag{6.2}
$$

    is $k - 1$ times differentiable, and $S^{(k-1)}(x)$ is piecewise linear with nodes at $x_1, \ldots, x_n$.

    (b) Write the cubic B-spline in (3.3) as a linear combination (6.2). Check by plotting the graph for both expressions in MATLAB.

    Hint: First match the second derivatives, then determine the free parameters to force compact support.

4

Numerical Integration

This chapter treats numerical methods for the approximation of one-dimensional definite integrals of the form $\int_a^b f(x)\,dx$ or $\int_a^b \omega(x)f(x)\,dx$. After a discussion of general accuracy and convergence results, we consider the highly accurate Gaussian quadrature rules, most suitable for smooth functions, possibly with known endpoint singularities. Based on the trapezoidal rule, we then derive adaptive step size methods for the integration of difficult integrands.

The final two sections show how the methods for integration can be extended to multistep methods for solving initial value problems for systems of coupled ordinary differential equations. However, the latter is a vast subject, and we barely scratch the surface.

A thorough treatment of all aspects of numerical integration of univariate functions is in Davis and Rabinowitz [14]. The integration of functions of several variables is treated in Stroud [91], Engels [25], and Krommer and Ueberhuber [54]. The solution of ordinary differential equations is covered thoroughly by Hairer et al. [35, 36].

4.1 The Accuracy of Quadrature Formulas

In this section, we look at general approximation properties of formulas that use a finite number of values of a function f to approximate a definite integral of the form

$$I(f) := \int_a^b f(x)\,dx. \tag{1.1}$$

Because the integral (1.1) is linear in f, it is desirable that the approximating formulas have the same property. We therefore only consider formulas of the following form.

179

4.1.1 Definition. A formula of the form

$$Q(f) := \sum_{j=0:N} \alpha_j f(x_j) \tag{1.2}$$

is called an *(N + 1)-point quadrature* rule with the *nodes* $x_0, \ldots, x_N$ and corresponding *weights* $\alpha_0, \ldots, \alpha_N$.

The nodes usually lie in the interval of integration $[a, b]$ because the integrand is possibly undefined outside of $[a, b]$; for example, when $f(x) = \sqrt{(x - a)(b - x)}$.

In order for the formulas to remain numerically stable for large N, we require that all weights be nonnegative; there is then no cancellation when we evaluate $Q(f)$ for functions of constant sign. Indeed, the maximum amplification factor for the roundoff error is given by $\sum |\alpha_j| / \sum \alpha_j$ for function values of approximately constant magnitude, and this quotient is 1 precisely whenever the α_j are all nonnegative.

In order to obtain good approximation properties, we further require that the quadrature rule integrates exactly all f in a particular class, where this class must be meaningfully chosen. In the simplest case, the interval of integration is finite and the integrand $f(x)$ is continuous in $[a, b]$. In a formally only slightly more general case, the integrand can be divided into the product of a simple but possibly singular factor $\omega(x)$, the so-called *weight function*, and a continuous factor $f(x)$. In this case, instead of (1.1), the integral has the form

$$I_\omega(f) = \int_a^b \omega(x) f(x) \, dx. \tag{1.3}$$

Numerical evaluation of (1.3) is then still done approximately with a formula of the form (1.2), where the weights α_j are now dependent on the weight function $\omega(x)$.

4.1.2 Examples. Apart from the most important trivial weight function $\omega(x) = 1$, frequently used weight functions are as follows:

(i) $\omega(x) = x^\alpha$ with $\alpha > -1$ for $[a, b] = [0, 1]$,
(ii) $\omega(x) := 1/\sqrt{1 - x^2}$ for $[a, b] = [-1, 1]$,
(iii) $\omega(x) := e^{-x}$ for $[a, b] = [0, \infty]$,
(iv) $\omega(x) := e^{-x^2}$ for $[a, b] = [-\infty, \infty]$.

The last two examples show how an appropriate choice of the weight function makes numerical integration over an infinite interval possible with only a finite number of function values.

Normalized Quadrature Formulas

In general, it is often useful to use a linear transformation to transform the interval of integration to the simplest possible interval. We demonstrate this for various weight functions, namely:

CASE 1: The trivial weight function $\omega(x) = 1$: One usually normalizes to $[a, b] = [-1, 1]$. If

$$\int_{-1}^{1} f(x) \, dx = \sum \alpha_i f(x_i)$$

holds for polynomials of degree $\leq n$, then the substitution $t := c + xh$ results in the formula

$$\int_{c-h}^{c+h} f(t) \, dt = h \sum \alpha_i f(c + x_i h)$$

for polynomials f of degree $\leq n$.

CASE 2: The algebraic weight $\omega(x) = x^\alpha$ ($\alpha > -1$): One usually normalizes to $[a, b] = [0, 1]$. If

$$\int_{0}^{1} x^\alpha f(x) \, dx = \sum \alpha_i f(x_i)$$

holds for polynomials of degree $\leq n$, then the substitution $t := xh$ results in the formula

$$\int_{0}^{h} t^\alpha f(t) \, dt = h^{\alpha+1} \sum \alpha_i f(x_i h)$$

for polynomials f of degree $\leq n$.

CASE 3: The exponential weight $\omega(x) = e^{-\lambda x}$ ($\lambda > 0$): One usually normalizes to $[a, b] = [0, \infty]$, $\lambda = 1$. If

$$\int_{0}^{\infty} e^{-x} f(x) \, dx = \sum \alpha_i f(x_i)$$

holds for polynomials of degree $\leq n$, then the substitution $t := c + \lambda^{-1} x$ results in the formula

$$\int_{c}^{\infty} e^{-\lambda t} f(t) \, dt = \lambda^{-1} e^{-\lambda c} \sum \alpha_i f(c + \lambda^{-1} x_i)$$

for polynomials f of degree $\leq n$.

Approximation Order

Because continuous functions can be approximated arbitrarily closely by polynomials, it is reasonable to require that the quadrature formula (1.2) reproduces the integral (1.3) exactly for low-degree polynomials.

4.1.3 Definition. A quadrature rule

$$Q(f) := \sum_j \alpha_j f(x_j) \tag{1.4}$$

has *approximation order* (or simply *order*) $n + 1$ with respect to the weight function $\omega(x)$ if

$$\int_a^b \omega(x) f(x)\, dx = Q(f) \quad \text{for all polynomials } f \text{ of degree } \leq n. \tag{1.5}$$

In particular, when f is constant we have the following.

4.1.4 Corollary. *A quadrature rule (1.4) has a positive approximation order precisely when it satisfies the* consistency *condition*

$$\sum_j \alpha_j = \int_a^b \omega(x)\, dx = I(1). \tag{1.6}$$

We first consider the accuracy of quadrature rules, and investigate when we can ensure the convergence of a sequence of quadrature rules to the integral (1.1). The following two theorems give information in an important case.

4.1.5 Theorem. *Suppose Q is a quadrature rule with nodes in $[a, b]$ and nonnegative weights. If Q has the order $n+1$ with respect to the weight function $\omega(x)$, then the bound*

$$\left| \int_a^b \omega(x) f(x)\, dx - Q(f) \right| \leq \left(\int_a^b |\omega(x)| + \sum_{j=0:n} |\alpha_j| \right) \| p_n - f \|_\infty$$

holds for every continuous function $f : [a, b] \to \mathbb{C}$ and all polynomials p_n of degree $\leq n$. In particular, for nonnegative weight functions,

$$\left| \int_a^b \omega(x) f(x)\, dx - Q(f) \right| \leq 2 I(1) \| p_n - f \|_\infty. \tag{1.7}$$

Proof. By assumption Q has order $n + 1$ so, for all polynomials p_n of degree $\leq n$ we have

$$\int_a^b \omega(x) p_n(x)\,dx = Q(p_n).$$

It follows that

$$\int_a^b \omega(x) f(x)\,dx = \int_a^b \omega(x)(f(x) - p_n(x))\,dx + Q(p_n)$$

$$= \int_a^b \omega(x)(f(x) - p_n(x))\,dx + \sum_{j=0:n} \alpha_j p_n(x_j).$$

With this we have

$$\left| \int_a^b \omega(x) f(x)\,dx - Q(f) \right|$$

$$= \left| \int_a^b \omega(x)\,(f(x) - p_n(x))\,dx + \sum_{j=0:n} \alpha_j(p_n(x_j) - f(x_j)) \right|$$

$$\leq \int_a^b |\omega(x)| \|f - p_n\|_\infty\,dx + \sum_{j=0:n} |\alpha_j| \|f - p_n\|_\infty$$

$$= \left(\int_a^b |\omega(x)| + \sum_{j=0:n} |\alpha_j| \right) \|f - p_n\|_\infty.$$

In particular, if $\omega(x)$ and all α_j are nonnegative, then (1.6) implies

$$\int_a^b |\omega(x)| + \sum_{j=0:n} |\alpha_j| = \int_a^b \omega(x) + \sum_{j=0:n} \alpha_j = 2I(1). \qquad \square$$

4.1.6 Theorem. *Let Q_l ($l = 1, 2, \ldots$) be a sequence of quadrature rules with nodes in the bounded interval $[a, b]$, nonnegative weights, having order $n_l + 1$ with respect to some nonnegative weight function $\omega(x)$. If $n_l \to \infty$ as $l \to \infty$, then*

$$\int_a^b \omega(x) f(x)\,dx = \lim_{l \to \infty} Q_l(f)$$

for every continuous function $f : [a, b] \to \mathbb{C}$.

Proof. By the Weierstrass approximation theorem, there is a sequence of polynomials $p_k(x)$ ($k = 0, 1, 2, \ldots$) of degree k with

$$\lim_{k \to \infty} \|p_k - f\|_\infty = 0.$$

For sufficiently large l we have $n_l \geq k$, so that Theorem 4.1.5 may be applied to each of these polynomials p_k; we thus obtain

$$\lim_{l \to \infty} \left| \int_a^b \omega(x) f(x) \, dx - Q_l(f) \right| = 0.$$

$\square$

4.1.7 Remarks.

 (i) Theorems of approximation theory imply that actually

$$\| f - p_k \|_\infty = O(k^{-m}) \quad \text{as } k \to \infty$$

for m-times continuously differentiable functions; the approximation error (1.7) therefore diminishes at least as rapidly.

(ii) Instead of requiring nonnegative weights it suffices to assume that the sums $\sum |\alpha_i|$ remain bounded.

For the most important case, that of constant weight $\omega(x) = 1$, we now prove some statements about the accuracy of transformed quadrature rules over small intervals.

4.1.8 Proposition. *If the relationship*

$$\int_{c-h}^{c+h} f(t) \, dt = h \sum \alpha_i f(c + x_i h) + O(h^{n+2}) \quad (h \to 0) \qquad (1.8)$$

holds for all $(n + 1)$-times continuously differentiable functions f in $[c - h, c + h]$, then the corresponding quadrature rule

$$Q(f) = \sum \alpha_i f(x_i) \qquad (1.9)$$

for $(n + 1)$-times continuously differentiable functions f in $[-1, 1]$ has order $n + 1$ with respect to $\omega(x) = 1$.

Proof. Suppose $k \leq n$ and $f(t) := (t - c)^k$. Then (1.8) holds, and we obtain

$$\frac{h^{k+1} - (-h)^{k+1}}{k+1} = h \sum \alpha_i (x_i h)^k + O(h^{n+2}).$$

This implies

$$\int_{-1}^{1} t^k \, dt = \frac{1 - (-1)^{k+1}}{k+1} = \sum \alpha_i x_i^k + O(h^{n+1-k}).$$

Because $n + 1 - k > 0$, we obtain the relationship

$$\int_{-1}^{1} t^k \, dt = Q(x^k)$$

in the limit $h \to 0$. The normalized quadrature rule (1.9) is therefore exact for all x^k with $k \leq n$. It is thus also exact for linear combinations of x^k ($k \leq n$), and therefore for all polynomials of degree $\leq n$, that is, it has order $n + 1$. □

In practice we are interested in the converse of the previous proposition because we start with a normalized quadrature rule and we desire a bound for the error of integration over the original interval. We obtain this with the following theorem.

4.1.9 Theorem. *Suppose* $Q(f) = \sum \alpha_i f(x_i)$ *is a normalized quadrature rule over the interval* $[-1, 1]$ *with nonnegative weights* α_i *and with order* $n + 1$ *with respect to* $\omega(x) = 1$. *Then the error bound*

$$\left| \int_{c-h}^{c+h} f(t) \, dt - h \sum \alpha_i f(c + x_i h) \right| \leq \frac{16}{(n+1)!} \left(\frac{h}{2} \right)^{n+2} \left\| f^{(n+1)} \right\|_{\infty}$$

holds for every $N + 1$ *times continuously differentiable function* f *on* $[c - h, c + h]$, *where the maximum norm is over* $[c - h, c + h]$.

Proof. Let $p_n(x)$ denote the polynomial of degree $\leq n$ that interpolates $g(x) := f(c + xh)$ at the Chebychev points. Theorem 3.1.8 and Proposition 3.1.14 then imply

$$\| g - p_n \|_{\infty} \leq \frac{\left\| g^{(n+1)} \right\|_{\infty}}{2^n (n+1)!}$$

over the interval $[-1, 1]$. This and application of Theorem 4.1.5 to $Q(g)$ give

$$\left| \int_{c-h}^{c+h} f(t) \, dt - h \sum \alpha_i f(c + x_i h) \right| = h \left| \int_{-1}^{1} g(x) \, dx - Q(g) \right|$$

$$\leq 4h \cdot \| g - p_n \|_{\infty} \leq \frac{4h}{2^n} \frac{\left\| g^{(n+1)} \right\|_{\infty}}{(n+1)!}$$

$$= \frac{16}{(n+1)!} \left(\frac{h}{2} \right)^{n+2} \left\| f^{(n+1)} \right\|_{\infty}. □$$

If we wish to compute the error bounds explicitly, we can bound $\| f^{(n+1)} \|_{\infty}$ with interval arithmetic by computing $|f^{(n+1)}([c - h, c + h])|$, provided an arithmetic expression for $f^{(n+1)}$ is available or $f^{(n+1)}$ is computed recursively via automatic differentiation from an arithmetic expression for f. However, the error of integration can already be bounded in terms of the nth derivative; in particular we have the following.

4.1.10 Theorem. *If the quadrature rule* $Q(f) = \sum \alpha_i f(x_i)$, *normalized on* $[-1, 1]$, *has nonnegative weights* α_j *and order* $n + 1$ *with respect to* $\omega(x) = 1$, *then*

$$\left| \int_{c-h}^{c+h} f(t)\, dt - h \sum \alpha_i f(c + x_i h) \right|$$
$$\leq \frac{16}{n!} \left(\frac{h}{2} \right)^{n+1} \mathrm{rad}\left(f^{(n)}([c-h, c+h]) \right).$$

Proof. Let $p_{n-1}(x)$ denote the polynomial of degree $n - 1$, which interpolates $g(x) := f(c + xh)$ at the Chebychev points, and let $q(x) := 2^{1-n} T_n(x)$ be the normalized Chebychev polynomial of degree n. Then $|q(x)| \leq 2^{1-n}$ for $x \in [-1, 1]$. With a suitable constant γ, we now define

$$p_n(x) := p_{n-1}(x) + \gamma q(x).$$

We then have

$$g(x) - p_{n-1}(x) = \frac{g^{(n)}(\xi)}{n!} q(x) = \frac{h^n}{n!} f^{(n)}(c + \xi h) q(x)$$

for some $\xi \in [-1, 1]$, so

$$|g(x) - p_n(x)| = \left| \frac{h^n}{n!} f^{(n)}(c + \xi h) - \gamma \right| |q(x)|$$
$$\leq \left| \frac{h^n}{n!} f^{(n)}([c-h, c+h]) - \gamma \right| \cdot 2^{1-n}.$$

We now choose γ in such a way that the expression on the right side is minimal. This is the case when $\gamma = \frac{h^n}{n!} \mathrm{mid}\, f^{(n)}([c-h, c+h])$. Then we get

$$\|g - p_n\|_\infty \leq \frac{2}{n!} \left(\frac{h}{2} \right)^n \mathrm{rad}\left(f^{(n)}([c-h, c+h]) \right).$$

As in the proof of Proposition 4.1.9, we finally obtain

$$\left| \int_{c-h}^{c+h} f(t)\, dt - h \sum \alpha_i f(c + x_i h) \right|$$

$$\leq \frac{16}{n!} \left(\frac{h}{2} \right)^{n+1} \operatorname{rad} f^{(n)}([c - h, c + h]). \qquad \square$$

4.1.11 Remarks.

(i) The bound for the error is still of order $O(h^{n+2})$ because the radius of the enclosure for the nth derivative is always $O(h)$.

(ii) Under the assumptions of the theorem, the transformed quadrature rule for the interval $[c - h, c + h]$ also has order $n + 1$ because the nth derivative of a polynomial of degree n is constant and the radius of a constant function is zero, whence the error is zero.

(iii) We can avoid recursive differentiation if we use the Cauchy integral theorem to bound the higher order derivatives. Indeed, if f is analytic in the complex disk $D[c; r]$ with radius $r > h$ centered at c, and if $|f(\xi)| \leq M$ for $\xi \in D[c; r]$, then Corollary 3.1.11 implies the bound

$$\frac{1}{(n + 1)!} \left\| f^{(n+1)} \right\|_\infty \leq \frac{Mr}{(r - h)^{n+2}},$$

and the bound in Proposition 4.1.9 becomes $16Mrq^{n+2}$, with $q = h/(2r - 2h)$. For sharper bounds along similar lines, see Eiermann [24] and Petras [80].

In order to apply these theorems in practice, we must know how to determine the $2n + 2$ parameters α_j and x_j in the quadrature rule (1.4) in order to obtain a predetermined approximation accuracy. In the next section, we integrate interpolating polynomials to determine, for arbitrary weights, so-called interpolatory quadrature rules, among them the Gaussian quadrature rules, which have particularly high order.

4.2 Gaussian Quadrature Formulas

When we require that all polynomials of degree $\leq n$ are integrated exactly, we obtain a class of $(n + 1)$-point quadrature formulas that are sufficiently accurate for many purposes. In this vein, we have the following.

4.2.1 Definition. An $(n + 1)$-point quadrature formula $Q(f)$ is an *interpolatory quadrature formula* for the weight function $\omega(x)$ provided

$$Q(f) = \int_a^b \omega(x) f(x) \, dx$$

for all polynomials f of degree $\leq n$, that is, when $Q(f)$ has order $n + 1$ with respect to $\omega(x)$.

The following theorem gives information concerning existence and uniqueness of interpolatory quadrature formulas.

4.2.2 Theorem. *To each weight $\omega(x)$ and $n + 1$ arbitrarily preassigned pairwise distinct nodes $x_0, \ldots, x_n$, there is exactly one quadrature formula with order $n + 1$, namely*

$$Q(f) = \sum_{j=0:n} \alpha_j f(x_j)$$

with

$$\alpha_j := \int_a^b \omega(x) L_j(x) \, dx,$$

with the Lagrange polynomials

$$L_j(x) = \prod_{i \neq j} \frac{x - x_i}{x_j - x_i}.$$

Proof. Suppose $f(x)$ is a polynomial of degree $\leq n$. Then Lagrange's interpolation formula (see Theorem 3.1.1)

$$f(x) = \sum_{j=0:n} L_j(x) f(x_j)$$

implies, with

$$\int_a^b \omega(x) f(x) \, dx = \sum_{j=0:n} \int_a^b \omega(x) L_j(x) f(x_j) \, dx$$

the desired representation.   □

The interpolatory quadrature formulas corresponding to the constant weight function $\omega(x) = 1$ and equidistant nodes $x_j = a + j h_n$ $(j = 0, \ldots, n)$ are called the *Newton-Cotes quadrature rules* of designed order $n + 1$. If n is even, the true order is $n + 2$. Indeed, a symmetry argument shows that $(2x - a - b)^{n+1}$ (and hence any polynomial of degree $n + 1$) is integrated exactly to zero when n is even.

For $n = 1$, we obtain the *simple trapezoidal rule*

$$T_1(f) = \frac{b - a}{2}(f(a) + f(b))$$

of order $n + 1 = 2$, for $n = 2$ the *simple Simpson rule* (also called *Kepler's barrel rule*)

$$S_1(f) = \frac{b - a}{6}\left(f(a) + 4f\left(\frac{a + b}{2}\right) + f(b)\right)$$

of order $n + 2 = 4$, and for $n = 4$ the *simple Milne rule*

$$M_1(f) = \frac{b - a}{90}\left(7f(a) + 32f\left(\frac{3a + b}{4}\right) + 12f\left(\frac{a + b}{2}\right)\right.$$
$$\left. + 32f\left(\frac{a + 3b}{4}\right) + 7f(b)\right) \tag{2.1}$$

of order $n + 2 = 6$. If $b - a = O(h)$, Theorem 4.1.9 implies that the error is $O(h^3)$ for the simple trapezoidal rule, $O(h^5)$ for the simple Simpson rule, and $O(h^7)$ for the simple Milne rule.

In the section on adaptive integration, we meet these rules again as the first members of infinite sequences $T_N(f)$, $S_N(f)$, and $M_N(f)$ of low order quadrature rules.

Unfortunately, the Newton-Cotes formulas have negative weights α_j when n is large, more precisely for $n = 8$ and $n \geq 10$. For example, $\sum |\alpha_j| / \sum \alpha_j \approx 10^{11}$ for $n = 40$, which leads to a tremendous amplification of the roundoff error introduced during computation of the $f(x_j)$. We must therefore choose the nodes carefully if we want to maintain nonnegative weights.

As already shown in Section 3.1, many functions can be interpolated better when the nodes are chosen closer together near the boundary, rather than equidistant. We also use this principle here because in general the better we interpolate a function, the better its integral is approximated. As an example, we treat normalized quadrature formulas in with respect to $\omega(x) = 1$ on the interval $[-1, 1]$. Here, an advantageous distribution of the nodes is given by the roots of the Chebychev polynomial $T_n(x) = \cos(n \cdot \arccos(x))$ (see Section 3.1). However, the extreme values of the Chebychev polynomial, that is,

$$x_j = \cos(\pi j/n) \quad \text{for } j = 0, \ldots, n$$

have proved in practice to be more appropriate nodes for integration. The corresponding interpolatory quadrature formulas are called *Clenshaw-Curtis formulas*. We give without proof explicit formulas for their weights:

$$
\alpha_j =
\begin{cases}
\dfrac{2}{n}\left(1 - 2 \displaystyle\sum_{k=1:\lfloor n/2 \rfloor} \dfrac{\cos(2\pi jk/n)}{4k^2 - 1}\right) & \text{for } j = 1, \ldots, n-1, \\[4mm]
\dfrac{1}{n}\left(1 - 2 \displaystyle\sum_{k=1:\lfloor n/2 \rfloor} \dfrac{1}{4k^2 - 1}\right) = \dfrac{1}{n(2\lfloor n/2 \rfloor + 1)} & \text{for } j = 0, n.
\end{cases}
$$

Because $\alpha_j \geq \alpha_n$, all α_j are positive for this choice of nodes, so the quadrature formula with these weights α_j is numerically stable. In practice, in order to avoid too many additional cosine evaluations, one tabulates the x_j and α_j for various values of n (e.g., for $n = 1, 2, 4, 8, 16, \ldots$).

Until now we have specified the quadrature formulas in such a way that the x_j were somehow fixed beforehand and the weights α_j were then computed. We have thus not made use of the fact that we can also choose the nodes in order to get a higher order with the same number of nodes. In order to find optimal nodes, we prove the following proposition, which points out the conditions under which an order higher than $n + 1$ can be achieved.

4.2.3 Proposition. *An interpolatory quadrature formula with nodes $x_0, \ldots, x_n$ has order $n + 1 + k$ precisely whenever the polynomial*

$$q(x) := (x - x_0) \cdots (x - x_n)$$

obeys the relationship

$$\int_a^b \omega(x)q(x)x^i dx = 0 \quad \text{for } i = 0, \ldots, k - 1. \tag{2.2}$$

Proof. Every polynomial f of degree $n + k$ has the form

$$f(x) = p_n(x) + q(x)r(x) \tag{2.3}$$

for some polynomial p_n of degree $\leq n$ (the interpolating polynomial at the nodes $x_0, \ldots, x_n$) and some polynomial $r(x)$ of degree $\leq k - 1$. Thus $Q(f)$ has order of approximation $n + 1 + k$ precisely whenever the relationship

$$\int_a^b \omega(x) f(x)\,dx = Q(f) = \sum_{i=0:n} \alpha_i f(x_i),$$

or equivalently

$$\int_a^b \omega(x) f(x)\,dx = \sum_{i=0:n} \alpha_i p_n(x_i) = \int_a^b \omega(x) p_n(x)\,dx$$

holds for all f of the form (2.3).

If we consider the difference between the left and right sides, we see that this is the case precisely when

$$\int_a^b \omega(x) q(x) r(x)\,dx = 0$$

for all polynomials $r(x)$ of degree $\leq k - 1$, that is, whenever (2.2) holds. □

Equation (2.2) places k conditions on the $n + 1$ nodes. We may therefore hope to satisfy these conditions with $k = n + 1$ with an appropriate choice of nodes. The order of approximation $2n + 2$ so obtained would thus then be twice as large as before. As we shall now show, for nonnegative weights this is indeed possible, in a unique way.

We need some additional notation for this. We introduce the abbreviation

$$\langle f, g \rangle := \int_a^b \omega(x) f(x) g(x)\,dx \tag{2.4}$$

for these frequently occurring integrals. Whenever the weight function $\omega(x)$ is continuous in (a, b) and the conditions

$$\int_a^b \omega(x)\,dx = \mu_0 < \infty, \quad \omega(x) > 0 \quad \text{for } x \in (a, b) \tag{2.5}$$

hold, then $\langle \cdot\,, \cdot \rangle$ is a positive definite scalar product in the space of real continuous bounded functions in the open interval (a, b).

4.2.4 Definition. A sequence of polynomials $p_i(x)$ $(i = 0, 1, 2, \ldots)$ of degree i is called a *system of orthogonal polynomials* over $[a, b]$ with respect to the weight function $\omega(x)$ if the condition

$$\langle p_i, p_j \rangle = 0 \quad \text{for } i = 0, 1, \ldots, j - 1$$

is satisfied.

To every weight function $\omega(x)$ satisfying (2.5) that is continuous in (a, b) there corresponds a system of orthogonal polynomials that is unique to within normalization. Such a system may be determined, for example, from the sequence $1, x, \ldots, x^n$ with Gram-Schmidt orthogonalization.

Based on the following theorem, we can obtain quadrature formulas with optimal order from such systems of orthogonal polynomials.

4.2.5 Theorem. *Suppose $\omega(x)$ is a weight function which is continuous in (a, b) and which satisfies (2.5). Then for each $n \geq 0$ there is exactly one quadrature formula*

$$G_n(f) := \sum_{j=0:n} \alpha_j f(x_j)$$

with approximation order $2n + 2$. Its weights

$$\alpha_k = \int_a^b L_k(x)^2 \omega(x) \, dx \tag{2.6}$$

are nonnegative and its nodes x_j all lie in the interval $]a, b[$. The nodes are the roots of the $(n + 1)$st orthogonal polynomial $p_{n+1}(x)$ with respect to the weight function $\omega(x)$. In particular, the relationship

$$\lim_{n \to \infty} G_n(f) = \int_a^b \omega(x) f(x) \, dx \tag{2.7}$$

holds for all continuous functions f in $[a, b]$.

The uniquely determined quadrature formula $G_n(f)$ is called the *$(n+1)$-point Gaussian quadrature rule* for the weight function $\omega(x)$ in $[a, b]$.

Proof.

(i) Suppose G_n satisfies

$$\int_a^b \omega(x) f(x) \, dx = G_n(f) = \sum_{i=0:n} \alpha_i f(x_i) \tag{2.8}$$

for every polynomial f of degree $\leq 2n + 1$. By Proposition 4.2.3, $\langle q, x^i \rangle = 0$ then holds for $i = 0, \ldots, n$. Thus $q(x)$ is a polynomial of degree $n + 1$ that is orthogonal to all polynomials of degree $\leq n$, so $q(x)$ is proportional to

$p_{n+1}(x)$, that is, the roots of $p_{n+1}(x)$ correspond to those of $q(x)$ and hence to the nodes. By Theorem 4.2.2, the quadrature formula is thus unique.

We now determine the sign of the weights and the location of the nodes. For this, we again use the Lagrange polynomial $L_k(x)$, and choose the two special polynomials

$$f_1(x) := L_k(x)^2, \quad f_2(x) := xL_k(x)^2$$

of degree $\leq 2n + 1$ for f in (2.8). Because $L_k(x_j) = \delta_{kj}$, we obtain

$$\alpha_k = G_n(f_1) = \int_a^b L_k(x)^2 \omega(x)\,dx \geq 0$$

and

$$x_k = \frac{1}{\alpha_k} G_n(f_2) = \frac{\int_a^b x L_k(x)^2 \omega(x)\,dx}{\int_a^b L_k(x)^2 \omega(x)\,dx} \in \,]a, b[.$$

(ii) Conversely, if the x_i are the roots of $p_{n+1}(x)$, the weights are given by (2.6), and f is a polynomial of degree $\leq 2n + 1$, then

$$f(x) - \sum_{j=0:n} L_j(x)^2 f(x_j) = g(x) p_{n+1}(x)$$

for some polynomial g of degree $\leq n$, since the left side has each x_j as a root. Hence

$$\int_a^b f(x)\,dx - G_n(f) = \int_a^b \left(f(x) - \sum_{j=0:n} L_j(x)^2 f(x_j) \right) dx$$

$$= \int_a^b g(x) p_{n+1}(x)\,dx = 0$$

by (2.2). Thus G_n has indeed the approximation order $2n + 2$.
Statement (2.7) now follows from Theorem 4.1.6. □

Gaussian quadrature rules often give excellent results. However, despite the good order of approximation, they are not universally applicable. To apply them, it is essential (unlike for formulas with equidistant weights) to have a function given by an expression (or a program) because the nodes x_i are in general irrational.

Tabulated nodes and weights for Gaussian quadrature rules for many different weight functions can be found, for example, in Abramowitz and Stegun [1].

4.2.6 Example. Suppose $\omega(x) = 1$ and $[a, b] = [-1, 1]$. The corresponding polynomials are (up to a constant normalization factor) the *Legendre polynomials*. Orthogonalization of $1, x, \ldots, x^n$ gives the sequence

$$p_0(x) = 1, \quad p_1(x) = x, \quad p_2(x) = x^2 - \frac{1}{3}, \quad p_3(x) = x^3 - \frac{3}{5}x,$$

and, continuing recursively for arbitrary j,

$$p_{j+1}(x) = xp_j(x) - \frac{j^2}{4j^2 - 1}p_{j-1}(x). \tag{2.9}$$

For $n \leq 2$, we obtain the following nodes and weights. For $n = 0$,

$$x_0 = 0, \quad \alpha_0 = 2,$$

for $n = 1$,

$$x_0 = \sqrt{1/3}, \quad \alpha_0 = 1,$$
$$x_1 = \sqrt{1/3}, \quad \alpha_1 = 1,$$

and for $n = 2$,

$$x_0 = -\sqrt{3/5}, \quad \alpha_0 = 5/9,$$
$$x_1 = 0, \quad \alpha_1 = 8/9,$$
$$x_2 = \sqrt{3/5}, \quad \alpha_2 = 5/9.$$

If we substitute $z := c + hx$ (and add error terms from [1]), we obtain for $n = 1$ the formula

$$\int_{c-h}^{c+h} f(z)\, dz = h \cdot \left(f\left(c - \frac{h}{\sqrt{3}}\right) + f\left(c + \frac{h}{\sqrt{3}}\right) \right) + \frac{h^5}{135} f^{(4)}(\xi),$$

which requires one function evaluation less than Simpson's rule, but has an error term of the same order $O(h^5)$. For $n = 2$, we obtain the formula

$$\int_{c-h}^{c+h} f(z)\, dz = \frac{h}{9}(5f(c - h\sqrt{3/5}) + 8f(c) + 5f(c + h\sqrt{3/5}))$$
$$+ \frac{h^7}{3150} f^{(6)}(\xi),$$

with the same number of function evaluations as Simpson's rule, but an error term of a higher order $O(h^7)$.

A 3-term recurrence relation like (2.9) is typical for orthogonal polynomials:

4.2.7 Proposition. *Let $p_k(x)$ be polynomials of degree $k = 0, 1, \ldots$ with highest coefficient 1, orthogonal with respect to the weight function $\omega(x)$. Then, with $p_{-1}(x) = 0$, $p_0(x) = 1$, we have*

$$p_j(x) = (x - a_j)p_{j-1}(x) - b_j p_{j-2}(x) \quad \text{for } j \geq 1, \tag{2.10}$$

with the constants

$$a_j = \int \omega(x)xp_{j-1}^2(x)\,dx \Big/ \int \omega(x)p_{j-1}^2(x)\,dx \quad (j \geq 1),$$

$$b_j = \int \omega(x)p_{j-1}^2(x)\,dx \Big/ \int \omega(x)p_{j-2}^2(x)\,dx \quad (j \geq 2).$$

Proof. We fix j and expand the polynomial $xp_{j-1}(x)$ of degree j as a linear combination

$$xp_{j-1}(x) = \sum_{l=0:j} \alpha_l p_l(x).$$

Taking inner products with $p_k(x)$ and using the orthogonality, we find $\langle xp_{j-1}, p_k \rangle = \alpha_k \langle p_k, p_k \rangle$, hence

$$\alpha_k = \langle xp_{j-1}, p_k \rangle / \langle p_k, p_k \rangle.$$

By comparing the highest coefficients, we find $\alpha_j = 1$, hence

$$\langle xp_{j-1}, p_j \rangle / \langle p_j, p_j \rangle = 1. \tag{2.11}$$

By definition of the inner product, $\langle xp_{j-1}, p_k \rangle = \langle p_{j-1}, xp_k \rangle = 0$ for $k < j - 2$ since p_{j-1} is orthogonal to all polynomials of degree $< j - 1$. Thus $\alpha_k = 0$ for $k < j - 2$, and we find

$$xp_{j-1}(x) = p_j(x) + \alpha_{j-1}p_{j-1}(x) + \alpha_{j-2}p_{j-2}(x).$$

This gives (2.10) with

$$a_j = \alpha_{j-1} = \langle xp_{j-1}, p_{j-1} \rangle / \langle p_{j-1}, p_{j-1} \rangle,$$
$$b_j = \alpha_{j-2} = \langle xp_{j-1}, p_{j-2} \rangle / \langle p_{j-2}, p_{j-2} \rangle$$
$$= \langle p_{j-1}, p_{j-1} \rangle / \langle p_{j-2}, p_{j-2} \rangle$$

because $\langle xp_{j-1}, p_{j-2} \rangle = \langle xp_{j-2}, p_{j-1} \rangle = \langle p_{j-1}, p_{j-1} \rangle$ by (2.11). $\qquad\square$

If the coefficients of the three-term recurrence relation are known, one can compute the nodes and weights for the corresponding Gaussian quadrature rules in a numerically stable way by solving a tridiagonal eigenvalue problem (cf. Exercise 13). (This is preferable to forming the p_j explicitly and finding their zeros, which is much less stable.)

4.3 The Trapezoidal Rule

Gaussian rules have error terms that depend on high derivatives of f, and hence are not suitable for the integration of functions with little smoothness. However, even nondifferentiable continuous functions may always be integrated by an approximation with Riemann sums. In this section, we refine this approach.

We assume that we know the values $f(x_j)$ ($j = 0, \ldots, N$) of a function $f(x)$ on a grid $a = x_0 < x_1 < \cdots < x_N = b$. When we review the analytical definition of the integral in terms of Riemann sums, we see that it is possible to approximate the integral by summing the areas of rectangles with sides of lengths $x_i - x_{i-1}$ and $f(x_{i-1})$ or $f(x_i)$ (Figure 4.1).

We also immediately see that we obtain a better approximation if we replace the rectangles with trapezoids obtained by joining a point $(x_{i-1}, f(x_{i-1}))$ with the next point $(x_i, f(x_i))$; that is, we replace $f(x)$ by a piecewise linear function through the data points $(x_i, f(x_i))$ and compute the area underneath this piecewise linear function in terms of a sum of trapezoidal areas. Because the area of a trapezoid is given by the product of its base and average height, we obtain the approximation

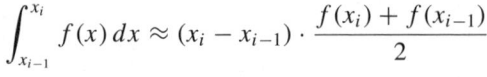

$$\int_{x_{i-1}}^{x_i} f(x)\,dx \approx (x_i - x_{i-1}) \cdot \frac{f(x_i) + f(x_{i-1})}{2}$$

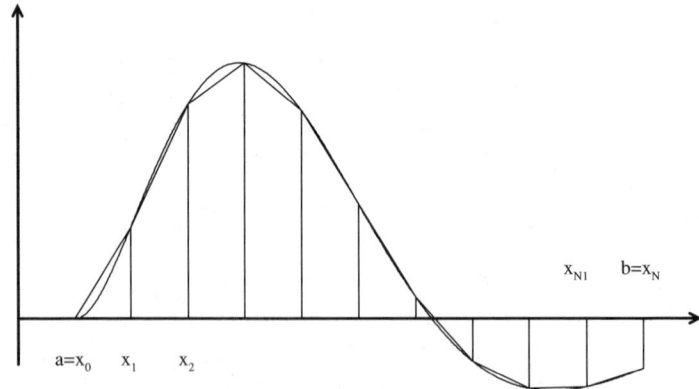

Figure 4.1. The integral as a sum of trapezoidal areas.

for the integral of f over the subinterval $[x_{i-1}, x_i]$. By summing over all subintervals, we obtain the following approximation for the integral over the entire interval $[a, b]$:

$$\int_a^b f(x)\,dx = \sum_{i=1:N} \int_{x_{i-1}}^{x_i} f(x)\,dx \approx \sum_{i=1:N} (x_i - x_{i-1}) \frac{f(x_i) + f(x_{i-1})}{2}.$$

After combining terms with the same index i, we obtain the *trapezoidal rule* over the grid $a = x_0 < x_1 < \cdots < x_N = b$:

$$\begin{aligned}
T(f) :&= \frac{x_1 - x_0}{2} f(x_0) + \sum_{i=1:N-1} \frac{x_{i+1} - x_{i-1}}{2} f(x_i) \\
&+ \frac{x_N - x_{N-1}}{2} f(x_N).
\end{aligned} \tag{3.1}$$

For equispaced nodes $x_i = a + i \cdot h_N$ obtained by subdividing the interval $[a, b]$ into N equal parts of length $h_N := (b - a)/N$, the trapezoidal rule simplifies to the *equidistant trapezoidal rule*

$$T_N(f) := h_N \sum_{i=1:N-1} f(x_i) + \frac{h_N}{2}(f(a) + f(b)) \tag{3.2}$$

with

$$h_N = (b - a)/N, \quad x_i = a + i h_N.$$

4.3.1 Remark. Usually, the nodes in an equidistant formula are not stored but are computed while summing $T_N(f)$. However, whenever N is somewhat larger, the nodes are not computed accurately with the formula $x_i = x_{i-1} + h$ because rounding errors accumulate. For this reason, it is strongly recommended that the slightly more expensive formula $x_i = a + ih$ be used. (In most cases, the dominant cost is that for getting the function values, anyway.) Roundoff errors can be further reduced if we sum the terms defining $T_N(f)$ in higher precision; in most cases, the additional expense required for this is negligible compared with the cost for function evaluations.

Clearly, the order of approximation of the trapezoidal rule is only $n + 1 = 2$, so Theorem 4.1.6 does not apply. Nevertheless, convergence to the integral follows because continuous functions may be approximated arbitrarily closely by piecewise linear functions. Indeed, for twice continuously differentiable functions, an error bound can be derived, using the following lemma.

4.3.2 Lemma. *If f is twice continuously differentiable in $[x_{i-1}, x_i]$, then there is a $\xi_i \in [x_{i-1}, x_i]$ such that*

$$\int_{x_{i-1}}^{x_i} f(x)\,dx = (x_i - x_{i-1})\frac{f(x_i) + f(x_{i-1})}{2} - \frac{(x_i - x_{i-1})^3}{12} f''(\xi_i).$$

Proof. We use divided differences. For $x \in [x_{i-1}, x_i]$, we have

$$\begin{aligned}
\int_{x_{i-1}}^{x_i} f(x)\,dx &= \int_{x_{i-1}}^{x_i} (f(x_{i-1}) + f[x_{i-1}, x_i](x - x_{i-1}) \\
&\quad + f[x_{i-1}, x_i, x](x - x_{i-1})(x - x_i))\,dx \\
&= f(x_{i-1})(x_i - x_{i-1}) + f[x_{i-1}, x_i]\frac{(x_i - x_{i-1})^2}{2} \\
&\quad + \int_{x_{i-1}}^{x_i} f[x_{i-1}, x_i, x](x - x_{i-1})(x - x_i)\,dx.
\end{aligned}$$

The mean value theorem for integrals states that

$$\int_a^b f(x)g(x)\,dx = f(\xi)\int_a^b g(x)\,dx \quad \text{for some } \xi \in [a, b],$$

provided f and g are continuous in $[a, b]$ and g has a constant sign on $[a, b]$. Applying this to the last integral with $g(x) := (x - x_{i-1})^2 - (x_i - x_{i-1})(x - x_{i-1}) = (x - x_{i-1})(x - x_i) \leq 0$, we obtain, after slight modification,

$$\begin{aligned}
\int_{x_{i-1}}^{x_i} f(x)\,dx &= (x_i - x_{i-1})\frac{f(x_i) + f(x_{i-1})}{2} + f[x_{i-1}, x_i, \xi_i'] \\
&\quad \times \int_{x_{i-1}}^{x_i} ((x - x_{i-1})^2 - (x_i - x_{i-1})(x - x_{i-1}))\,dx \\
&= (x_i - x_{i-1})\frac{f(x_i) + f(x_{i-1})}{2} - \frac{f''(\xi_i)}{2} \cdot \frac{(x_i - x_{i-1})^3}{6}
\end{aligned}$$

for suitable $\xi_i', \xi_i \in [a, b]$. $\square$

The following theorem for the trapezoidal rule results.

4.3.3 Theorem. *For the trapezoidal rule on a grid with mesh size h, the bound*

$$\left| \int_a^b f(x)\,dx - T(f) \right| \leq \frac{h^2}{12}(b - a)\|f''\|_\infty$$

holds for all twice continuously differentiable functions f defined on $[a, b]$.

Proof. From Lemma 4.3.2, the error is

$$\left| \int_a^b f(x)\,dx - T(f) \right| = \left| -\sum_{i=1:N} \frac{(x_i - x_{i-1})^3}{12} f''(\xi_i) \right|$$

$$\leq \sum_{i=1:N} \frac{(x_i - x_{i-1})h^2}{12} \|f''\|_\infty = \frac{h^2}{12}(b-a)\|f''\|_\infty.$$

$\square$

4.3.4 Remark. The formula in Theorem 4.3.3 indeed gives the correct order $O(h^2)$, but is otherwise fairly rough because it does not take account of variations in f'' in the interval $[a, b]$. As we see from the first expression in the proof, we may allow large steps $x_i - x_{i-1}$ everywhere where $|f''|$ is small; however, in order to maintain a small error, we must choose closely spaced nodes at those places where $|f''|$ is large. This already points to adaptive methods, which shall be considered in the next section.

In order to increase the accuracy with which the trapezoidal rule approximates an integral by a factor of 100, h must be decreased to $h/10$. This requires approximately 10 times as many nodes and therefore approximately 10 times as many function evaluations. That is too expensive in practice. Fortunately, significant speed-ups are possible if the nodes are equidistant.

The key is the observation that, for equispaced grids, we can replace the norm bound in Theorem 4.3.3 by an asymptotic expansion. This fact is very important because it allows the use of extrapolation techniques to improve the accuracy (see Section 4.4).

By Theorem 4.3.3, $T_N(f)$, given by the equidistant trapezoidal formula (3.2), has an error approximately proportional to h_N^2. More precisely, we show here that it behaves like a polynomial in h_N^2, provided f has sufficiently many derivatives. To this end, we express the individual trapezoidal areas $\frac{h}{2}(f(x_i) + f(x_{i-1}))$ in integral form. Because we may use a linear transformation to get $x_{i-1} = -1, x_i = 1, h = 2$, we first prove the following.

4.3.5 Lemma. *There exist numbers γ_{2j} with $\gamma_0 = 1$ and polynomials $p_j(x)$ of degree j such that the relationship*

$$g(1) + g(-1) = \int_{-1}^1 \left(\sum_{j=0:\lfloor m/2 \rfloor} \gamma_{2j} g^{(2j)}(x) + p_{m+1}(x) g^{(m+1)}(x) \right) dx \quad (3.3)$$

holds for $m \geq 0$ and all $(m + 1)$-times continuously differentiable functions $g : [-1, 1] \to \mathbb{C}$.

Proof. We define

$$I_m(g) := \int_{-1}^{1} \left(\sum_{j=0:m-1} \gamma_j g^{(j)}(x) + p_m(x) g^{(m)}(x) \right) dx$$

for suitable constants γ_j and polynomials $p_m(x)$, and determine these such that the conditions

$$I_m(g) = g(1) + g(-1) \quad \text{for } m \geq 1, \tag{3.4}$$

$$\gamma_m = 0 \quad \text{for odd } m \geq 1 \tag{3.5}$$

follow, which imply the assertion. By choosing

$$\gamma_0 := 1, \quad p_1(x) := x,$$

we find

$$I_1(g) = \int_{-1}^{1} (g(x) + x g'(x)) \, dx$$

$$= \int_{-1}^{1} (x g(x))' \, dx = x g(x) \big|_{-1}^{1} = g(1) + g(-1),$$

and (3.4) holds for $m = 1$. Also, with

$$\gamma_j := \frac{1}{2} \int_{-1}^{1} p_j(x) \, dx, \quad p_{j+1}(x) := \int_{-1}^{x} (\gamma_j - p_j(z)) \, dz \quad \text{for } j \geq 1,$$

we find

$$p'_{j+1}(x) = \gamma_j - p_j(x), \quad p_{j+1}(-1) = p_{j+1}(1) = 0,$$

so that, for $(m+1)$-times continuously differentiable functions g,

$$I_{m+1}(g) - I_m(g) = \int_{-1}^{1} \left((\gamma_m - p_m(x)) g^{(m)}(x) + p_{m+1}(x) g^{(m+1)}(x) \right) dx$$

$$= \int_{-1}^{1} \left(p'_{m+1}(x) g^{(m)}(x) + p_{m+1}(x) g^{(m)'}(x) \right) dx$$

$$= \int_{-1}^{1} \left(p_{m+1}(x) g^{(m)}(x) \right)' dx = p_{m+1}(x) g^{(m)}(x) \big|_{-1}^{1}$$

$$= 0.$$

Thus, (3.4) is valid for all $m \geq 1$.

In order to prove (3.5), we assume that (3.5) is valid for odd $m < 2k$; this is certainly the case when $k = 0$. We then plug the function $g(x) := x^m$ for $m = 2k + 1$ into (3.4) and obtain

$$I_m(g) = 1 + (-1)^m.$$

However, the integral can be computed directly:

$$I_m(g) = \int_{-1}^{1} \left(\sum_{j=0:m-1} \gamma_j \frac{m!}{(m-j)!} x^{m-j} + p_m(x)m! \right) dx$$

$$= m! \left(\sum_{j=0:m-1} \gamma_j \frac{1 - (-1)^{m+1-j}}{(m+1-j)!} + 2\gamma_m \right).$$

Because m is odd, the even terms in the sum vanish; but the odd terms also vanish because of the induction hypothesis. Therefore, $m! \cdot 2\gamma_m = 0$, so that (3.5) holds in general. $\qquad\square$

4.3.6 Remark. We may use the same argument as for (3.5) to obtain

$$\sum_{i=0:k} \frac{\gamma_{2i}}{(2k - 2i + 1)!} = \frac{1}{(2k)!} \quad \text{for } k \geq 0$$

for $m = 2k$; from this, the γ_{2i} (related to the so-called *Bernoulli numbers*) may be computed recursively. However, we do not need their numerical values.

From (3.3), we deduce the following representation of the error term of the equidistant trapezoidal rule.

4.3.7 Theorem. (Euler-MacLaurin Summation Formula). *If f is an $(m + 1)$-times continuously differentiable function, then*

$$T_N(f) = \int_a^b f(x)\, dx + \sum_{j=1:\lfloor m/2 \rfloor} c_j h_N^{2j} + r_m(f), \tag{3.6}$$

where the constants, independent of N, are

$$c_j = 2^{-2j} \gamma_{2j} \left(f^{(2j-1)}(b) - f^{(2j-1)}(a) \right) \quad (j = 1, \ldots, \lfloor m/2 \rfloor),$$

and where the remainder term $r_m(f)$ is bounded by

$$|r_m(f)| \leq C_m h_N^{m+1}(b - a) \left\| f^{(m+1)} \right\|_\infty,$$

for some constant C_m depending only on m. In particular, $r_m(f) = 0$ for polynomials f of degree $\leq m$.

Proof. In order to use Lemma 4.3.5, we substitute

$$x_i(t) := a + \left(i + \frac{t-1}{2}\right)h, \quad h := h_N$$

and set $g_i(t) := f(x_i(t))$. We then have $x_{i-1} = x_i(-1)$, $x_i = x_i(1)$, so the lemma implies

$$\frac{h}{2}(f(x_i) + f(x_{i-1})) - \int_{x_{i-1}}^{x_i} f(x)\,dx$$

$$= \frac{h}{2}(g_i(1) + g_i(-1)) - \frac{h}{2}\int_{-1}^1 g_i(t)\,dt$$

$$= \frac{h}{2}\int_{-1}^1 \left(\sum_{j=1:\lfloor m/2\rfloor} \gamma_{2j}g_i^{(2j)}(t) + p_{m+1}(t)g_i^{(m+1)}(t)\right)dt$$

$$= \frac{h}{2}\left(\sum_{j=1:\lfloor m/2\rfloor} \gamma_{2j}\left(g_i^{(2j-1)}(1) - g_i^{(2j-1)}(-1)\right)\right.$$

$$\left. + \int_{-1}^1 p_{m+1}(t)g_i^{(m+1)}(t)\,dt\right)$$

$$= \sum_{j=1:\lfloor m/2\rfloor} \left(\frac{h}{2}\right)^{2j} \gamma_{2j}\left(f^{(2j-1)}(x_i) - f^{(2j-1)}(x_{i-1})\right) + \frac{h}{2}\tilde{r}_m(g_i),$$

(3.7)

where the remainder term

$$\tilde{r}_m(g_i) := \int_{-1}^1 p_{m+1}(t)g_i^{(m+1)}(t)\,dt$$

$$= \left(\frac{h}{2}\right)^{m+1}\int_{-1}^1 p_{m+1}(t)f^{(m+1)}(x_i(t))\,dt$$

is bounded by

$$|\tilde{r}_m(g_i)| \leq 2C_m h^{m+1}\left\|f^{(m+1)}\right\|_\infty$$

with constant

$$C_m = \frac{1}{2^{m+2}}\int_{-1}^1 |p_{m+1}(t)|\,dt.$$

By summing (3.7) over $i = 1, \ldots, N$ we obtain

$$T_N(f) - \int_a^b f(x)\,dx$$

$$= \sum_{j=1:\lfloor m/2 \rfloor} \left(\frac{h}{2}\right)^{2j} \gamma_{2j}\left(f^{(2j-1)}(b) - f^{(2j-1)}(a)\right) + r_m(f),$$

where the remainder term

$$r_m(f) = \frac{h}{2} \sum_{i=1:N} \tilde{r}_m(g_i)$$

is bounded by

$$|r_m(f)| \le \frac{h}{2} \sum_{j=1:N} 2C_m h^{m+1} \left\| f^{(m+1)} \right\|_\infty = C_m h^{m+1}(b-a) \left\| f^{(m+1)} \right\|_\infty. \qquad \square$$

The Euler-MacLaurin summation formula now allows us to combine the trapezoidal rule with extrapolation procedures and thus introduce very effective quadrature rules.

4.4 Adaptive Integration

If we compute a sequence of values $T_N(f)$ for $N = N_0, 2N_0, 4N_0, \ldots$, then we can obtain better quadrature rules by taking appropriate linear combinations. To see this, we consider

$$T_N(f) = \frac{h_N}{2}\left(f(a) + f(b) + 2 \sum_{i=1:N-1} f(a + ih_N)\right),$$

$$T_{2N}(f) = \frac{h_N}{4}\left(f(a) + f(b) + 2 \sum_{i=1:N-1} f(a + ih_N)\right.$$

$$\left. + 2 \sum_{i=1:N} f\left(a + \left(i - \frac{1}{2}\right)h_N\right)\right)$$

$$= \frac{1}{2}\left(T_N(f) + h_N \sum_{i=1:N} f\left(a + \left(i - \frac{1}{2}\right)h_N\right)\right)$$

Hence, if we know $T_N(f)$, then to compute $T_{2N}(f)$ it is sufficient to compute the expression

$$R_N(f) := h_N \sum_{i=1:N} f\left(a + \left(i - \frac{1}{2}\right)h_N\right) \qquad (4.1)$$

with N additional function evaluations, and we thus obtain

$$T_{2N}(f) = \frac{1}{2}\left(T_N(f) + R_N(f)\right). \tag{4.2}$$

The expression $R_N(f)$ is called the *rectangle rule*. The linear combination

$$
\begin{aligned}
S_N(f) &:= \frac{4T_{2N}(f) - T_N(f)}{3} = \frac{T_N(f) + 2R_N(f)}{3} \\
&= \frac{h_N}{6}\left(f(a) + f(b) + 2 \sum_{i=1:N-1} f(a + ih_N) \right. \\
&\qquad \left. + 4 \sum_{i=1:N} f\left(a + \left(i - \frac{1}{2}\right)h_N\right) \right)
\end{aligned}
\tag{4.3}
$$

is called the *(composite) Simpson rule* because it generalizes the simple Simpson rule, which is obtained as $S_1(f)$. Because

$$I(f) := \int_a^b f(x)\,dx = \lim_{N\to\infty} T_N(f),$$

we also have

$$\lim_{N\to\infty} R_N(f) = \lim_{N\to\infty} S_N(f) = I(f).$$

However, although $T_N(f)$ and $R_N(f)$ approximate the integral $I(f)$ with an error of order $O(h_N^2)$, the error in Simpson's rule $S_N(f)$ has the more advantageous order $O(h_N^4)$. This is due to the fact that the linear combination (4.3) exactly eliminates the term $c_1 h_N^2$ in the Euler-MacLaurin summation formula (Theorem 4.3.7). We give another proof that also provides the following error bound.

4.4.1 Theorem. *Suppose f is a four-times continuously differentiable function on $[a, b]$. Then the error in Simpson's rule $S_N(f)$ is bounded by*

$$\left| \int_a^b f(x)\,dx - S_N(f) \right| \le \frac{h_N^4}{2880}(b - a)\left\| f^{(4)} \right\|_\infty.$$

Proof. We proceed similarly to the derivation of the error bound for the trapezoidal rule, and first compute a bound for the error on a subinterval $[x_{i-1}, x_i]$. Because the equation $S_N(f) = I(f)$ holds for $f(x) = x^i$ ($i = 0, 1, 2, 3$), Simpson's rule is exact for all polynomials of degree ≤ 3. By Theorem 3.1.4, we find for the simple Simpson rule $S(f)$ applied to $[x_{i-1}, x_i]$,

$$\int_{x_{i-1}}^{x_i} f(x)\, dx - S(f)$$

$$= \int_{x_{i-1}}^{x_i} f[y_0, y_1, y_1, y_2, y](y - y_0)(y - y_1)^2(y - y_2)\, dy,$$

where $y_0 := x_{i-1}$, $y_1 := (x_{i-1} + x_i)/2$, and $y_2 := x_i$. Because $(y - y_0)(y - y_1)^2(y - y_2) \leq 0$ for $y \in [x_{i-1}, x_i]$, we may apply the mean value theorem for integrals to obtain

$$\int_{x_{i-1}}^{x_i} f(x)\, dx - S(f)$$

$$= f[y_0, y_1, y_1, y_2, \xi'] \int_{x_{i-1}}^{x_i} (y - y_0)(y - y_1)^2(y - y_2)\, dy$$

$$= \left(\frac{x_{i+1} - x_i}{2}\right)^5 f[y_0, y_1, y_1, y_2, \xi'] \int_{-1}^{1} t^2(t^2 - 1)\, dt$$

$$= -\frac{4}{15}\left(\frac{h_N}{2}\right)^5 f[y_0, y_1, y_1, y_2, \xi']$$

$$= -\frac{1}{90}\left(\frac{h_N}{2}\right)^5 f^{(4)}(\xi)$$

for suitable $\xi', \xi \in [x_{i-1}, x_i]$. By summation over all subintervals $[x_{i-1}, x_i]$, $i = 1, \ldots, N$, we then obtain

$$\left|\int_a^b f(x)\, dx - S_N(f)\right| \leq N \left(\frac{h_N}{2}\right)^5 \frac{\|f^{(4)}\|_\infty}{90} = \frac{h_N^4}{2880}(b - a)\|f^{(4)}\|_\infty. \quad \square$$

The linear combination

$$M_N := \frac{16 S_{2N} - S_N}{15} \tag{4.4}$$

is called the *(composite) Milne rule* because it generalizes the simple Milne rule, which is obtained as $M_1(f)$. The Milne rule similarly eliminates the factors c_1 and c_2 in the Euler-MacLaurin summation formula (Theorem 4.3.7), and thus has an error of order $O(h_N^6)$. We may continue this elimination procedure; with a new notation we obtain the following *Romberg triangle*.

$$
\begin{aligned}
&T_{0,0} = T_N \\
&T_{1,0} = T_{2N} \quad T_{1,1} = S_N \\
&T_{2,0} = T_{4N} \quad T_{2,1} = S_{2N} \quad T_{2,2} = M_N \\
&T_{3,0} = T_{8N} \quad T_{3,1} = S_{4N} \quad T_{3,2} = M_{2N} \quad T_{3,3} \quad \cdots \\
&\quad \cdots \qquad\qquad \cdots \qquad\qquad \cdots \qquad\qquad \cdots
\end{aligned}
$$

We now formulate this notation precisely and bound the error.

4.4.2 Theorem. *Suppose f is $(2m + 2)$-times continuously differentiable on $[a, b]$. If we set*

$$
T_{0,0} := T_N(f),
$$

$$
T_{i,0} := T_{2^i N}(f) = \frac{1}{2}(T_{i-1,0} + R_{2^{i-1}N}(f)) \quad (i = 1, 2, \ldots)
$$

and

$$
T_{i,k} := T_{i,k-1} + \frac{T_{i,k-1} - T_{i-1,k-1}}{4^k - 1} \quad \text{for } k = 1, \ldots, i, \tag{4.5}
$$

then $T_{i,k}(f) = T_{i,k}$ has order $2k + 2$ with respect to $\omega(x) = 1$, and

$$
\int_a^b f(x)\, dx = T_{i,k}(f) + O\left(h_{2^i N}^{2k+2}\right) \tag{4.6}
$$

holds for each $i \geq 0$ and $k \leq m$.

Proof. We show by induction on k that constants $e_j^{(k)}$ ($j > k$) exist such that

$$
T_{i,k} = \int_a^b f(x)\, dx + \sum_{j=k+1:m} e_j^{(k)} h_{2^i N}^{2j} + O\left(h_{2^i N}^{2m+2}\right).
$$

By Theorem 4.3.7, this is valid for $k = 0$. It follows for $k > 0$ from

$$
\begin{aligned}
T_{i,k} &= T_{i,k-1} + \frac{T_{i,k-1} - T_{i-1,k-1}}{4^k - 1} \\
&= \frac{4^k T_{i,k-1} - T_{i-1,k-1}}{4^k - 1} \\
&= \int_a^b f(x)\, dx + \sum_{j=k+1:m} e_j^{(k-1)} \frac{4^k - 2^{2j}}{4^k - 1} h_{2^i N}^{2j} + O\left(h_{2^i N}^{2m+2}\right)
\end{aligned}
$$

with

$$e_j^{(k)} := e_j^{(k-1)} \frac{4^k - 2^{2j}}{4^k - 1} \quad (j > k).$$

Formula (4.6) follows. Proposition 4.1.8 then shows that the order of $T_{i,k}$ is $2k + 2$. □

4.4.3 Remarks.

(i) To compute $T_{n,n}$, an exponential number 2^n of function evaluations are required; the resulting error is of order $O(h^{2n+2})$. For errors of the same order, Gaussian quadrature rules only require $n + 1$ function evaluations. Thus the method is useful only for reasonably small values of n.

(ii) The Romberg triangle is equivalent to the Neville extrapolation formulas (Algorithm 3.2.4 for $q = 2$) applied to the extrapolation of $(x_l, f_l) = (h_N^2, T_N(f))$, $N = 2^l$ to $x = 0$. The equivalence is explained by the Euler-MacLaurin summation formula (Theorem 4.3.7): $T_N(f)$ is approximately a polynomial p_n of degree n in h_N^2, and the value of this polynomial p_n at zero then approximates the integral.

(iii) One can use Neville extrapolation also to compute an approximation to the integral $\int f(x)\,dx$ from $T_{N_l}(x)$ for arbitrary sequences of N_l in place of $N_l = 2^l$. The *Bulirsch sequence* $N_l = 1, 2, 3, 4, 6, \ldots, 2^i, 3 \cdot 2^{i-1}, \ldots$ is a recommended alternative choice that makes good use of old function values and often needs considerably fewer function values to achieve the same accuracy.

A further improvement is possible by using rational extrapolation formulas (see, e.g., Stoer and Bulirsch [90]) instead of Neville's formula because the former often converge more rapidly.

(iv) It can be shown that the weights of $T_{i,k}$ are all positive and the bound

$$\left| \int_a^b f(x)\,dx - T_{i,k} \right| \le \gamma_{2k+2} h_{2^i N}^{2n+2} \left\| f^{(2n+2)} \right\|_\infty \quad (n = \min(k, m))$$

holds. Thus, for $k \le m$, the kth column of the Romberg triangle converges to $I(f)$ as $i \to \infty$ with order $h_{2^i N}^{2k+2}$. By Theorem 4.1.6, the $T_{i,k}$ even converge to $I(f)$ as $i + k \to \infty$ for arbitrary continuous functions, and indeed, the better that the functions can be approximated by polynomials, the better the convergence.

In practice, we wish to be able to estimate the error approximately without computation of higher order derivatives. For this purpose, the sizes of the last correction $\varepsilon_{i,k} := |T_{i,k-1} - T_{i,k}|$ may be chosen as an error estimate for $T_{i,k}$ in the Romberg triangle. This is reasonable if we can guarantee that, under appropriate assumptions, the error $|I(f) - T_{i,k}|$ is actually bounded by ε_{ik}. By Theorem 4.4.2, we have $|I(f) - T_{i,k}| = O(h^{2k+2})$ with $h = 2^{-i}(b - a)$ and $\varepsilon_{ik} = O(h^{2k})$ because $|I(f) - T_{i,k-1}| = O(h^{2k})$. We thus see that the relation $|I(f) - T_{i,k}| \le \varepsilon_{ik}$ is usually true for sufficiently large i.

However, two arguments show that we cannot unconditionally trust this error estimate. When roundoff errors result in $T_{i,k} = T_{i,k-1}$, we have $\varepsilon_{ik} = 0$ even though the error is almost certainly larger. Moreover, the previous argument holds only asymptotically as $i \to \infty$. However, we really need an error estimate for small i. The previously mentioned bound then does not necessarily hold, as we see from the following example of a rapidly decreasing function.

4.4.4 Example. If we integrate $I := \int_0^1 (1 + \frac{1}{2} e^{-25x}) \, dx$ exactly, we obtain $I = 1.02000\dots$. With numerical integration with the Romberg triangle (with $N = 1$), a desired relative error smaller than 0.5%, and computation with four significant decimal places, we obtain:

$$
\begin{aligned}
&T_{0,0} = 1.250 \\
&T_{1,0} = 1.125 \quad T_{1,1} = 1.083 \\
&T_{2,0} = 1.063 \quad T_{2,1} = 1.042 \quad T_{2,2} = 1.039,
\end{aligned}
$$

where $\varepsilon_{22} = |T_{2,2} - T_{2,1}| = 0.003$, so $I \approx 1.039 \pm 0.003$. However, comparison with the exact value gives $|I - T_{2,2}| \approx 0.019$; the actual error is therefore more than six times larger than the estimated error.

Thus, we need to choose a reasonable security factor, say 10, and estimate

$$I(f) \approx T_{i,k} \pm 10|T_{i,k-1} - T_{i,k}|. \tag{4.7}$$

Adaptive Integration by Step Size Control

The Romberg procedure converges only slowly whenever the function to be integrated has problems in the sense that higher derivatives do not exist or are large at least one point (small m in Theorem 4.4.2). We may ameliorate this with a more appropriate subdivision in which we partition the original interval into subintervals, selecting smaller subintervals where the function has such problems. However, in practice we do not usually know these points beforehand; we thus need a *step size control*, that is, a procedure that, during the computation subdivides recursively those intervals that contribute too large an error in the integration.

We obtain such a procedure if we work with a small part of the Romberg triangle, $T_{i,k}$ with $k \le i \le n$ (and $N = 1$), to compute $T_{n,n}$ ($n = 2$ or 3). If the resulting approximation is not accurate enough, we then halve the original interval and compute the subintegrals separately. The $T_{i,k}$ with $k \le i \le n-1$ for the resulting subintervals can be computed easily from the already computed function values, and the only additional function values needed to compute the $T_{n,k}$ are those involved in $R_N(f)$, $N = 2^{n-1}$. This can be used in the implementation to increase the effectiveness.

To avoid infinite loops, we may prescribe a lower bound $h_{\min}$ for the length of the subintervals. If we fall below this length, we accept $T_{n,n}$ as the value of the subintegral and we issue a warning that the desired precision has not been attained. To avoid exceeding a prescribed error bound Δ for the entire integral $I[a, b] = \int_a^b f(x)\,dx$, it is appropriate to require that every subintegral $I[\underline{x}, \bar{x}] = \int_{\underline{x}}^{\bar{x}} f(x)\,dx$ has an error of at most $\Delta(\bar{x} - \underline{x})/(b - a)$; that is, we assign to each subintegral that portion of the error corresponding to the ratio of the length of the subinterval to the total length. However, this *uniform* error strategy has certain disadvantages.

This is because in practice, many functions to be integrated behave poorly primarily on the boundary; for example, when the derivative does not exist there (as with $\sqrt{x}\cos x$ for $x = 0$). The subintegrals on the boundary are then especially inaccurate, and it is advisable to allow a larger error for these in order to avoid having to subdivide too often. This can be achieved when we assign the maximum error per interval with a *boundary dominant* error strategy. For example, we may proceed as follows: If the maximum allowed error in the interval $[\underline{x}, \bar{x}]$ is Δ_0, we allow errors Δ_1 and Δ_2 for the intervals $[\underline{x}, \check{x}]$ and $[\check{x}, \bar{x}]$ produced from the subdivision, where

$$\Delta_1 = \tfrac{2}{3}\Delta_0, \quad \Delta_2 = \tfrac{1}{3}\Delta_0 \quad \text{if } \underline{x} = a, \bar{x} < b,$$

$$\Delta_1 = \tfrac{1}{3}\Delta_0, \quad \Delta_2 = \tfrac{2}{3}\Delta_0 \quad \text{if } a < \underline{x}, \bar{x} = b,$$

$$\Delta_1 = \tfrac{1}{2}\Delta_0, \quad \Delta_2 = \tfrac{1}{2}\Delta_0 \quad \text{otherwise.}$$

If we use the portion of the Romberg triangle to $T_{n,n}$, $1 + 2^n s$ function evaluations are required for s accepted subintervals. If we process the intervals from left to right, then due to the splitting procedure, up to $\log((b - a)/h_{\min})$ incompletely processed intervals may occur, and for each of them, 2^{n-1} already available function values, the upper interval bound, and the maximum error need to be stored (cf. Figure 4.2). Thus, $(2^{n-1} + 2)\log((b - a)/h_{\min})$ storage locations are necessary to store information that is not immediately needed. More sophisticated integration routines use variable order quadrature formulas

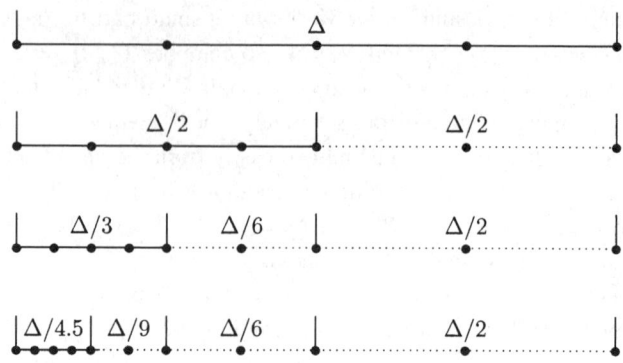

Figure 4.2. The boundary dominant error strategy, $n = 2$.

(e.g., a variable portion of the Romberg triangle), and determine the optimal order adaptively by an order control similar to that discussed in Section 4.5.

The boundary dominant error strategy can also be applied whenever the integrand possesses singularities (or singular derivatives) at known points in the interior. In this case, we subdivide the original interval *a priori* according to the singularities, then we treat each subinterval separately.

4.5 Solving Ordinary Differential Equations

In this section, we treat (for reason of space quite superficially) numerical approximation methods for the solution of initial value problems for systems of coupled ordinary differential equations. To simplify the presentation, we limit ourselves to autonomous problems of the form

$$y_i'(t) = F_i(y_1(t), \dots, y_n(t)), \quad y_i(t_0) = y_i^0 \quad (i = 1, \dots, n), \qquad (5.1)$$

or, in short vector notation,

$$y' = F(y), \quad y(t_0) = y^0.$$

However, this does not significantly limit applicability of the results. In particular, all methods for solving (5.1) can be generalized in a straightforward way to nonautonomous systems

$$y' = F(t, y), \quad y(t_0) = y^0$$

by making the system autonomous with the additional function $y_0 = t$ and the additional differential equation (and initial condition)

$$y_0' = 1, \quad y_0(t_0) = t_0.$$

(In practice, this transformation is not explicitly applied, but the method is transformed to correspond to it.) Similarly, higher order differential equations can be reduced to the standard form (5.1); for example, the problem

$$x'' = G(t, x), \quad x(t_0) = x^0, \quad x'(t_0) = v^0$$

is equivalent to (5.1) with

$$y = \begin{pmatrix} x \\ x' \\ t \end{pmatrix}, \quad y^0 = \begin{pmatrix} x^0 \\ v^0 \\ t_0 \end{pmatrix}, \quad F\begin{pmatrix} x \\ x' \\ t \end{pmatrix} = \begin{pmatrix} x' \\ G(t, x) \\ 1 \end{pmatrix}.$$

We pose the problem more precisely as follows. Suppose a continuous function $F : D \subseteq \mathbb{R}^n \to \mathbb{R}^n$, a point $y_0 \in \text{int}(D)$, and a real interval $[a, b]$ are given. We seek a continuously differentiable function $y : [a, b] \to \mathbb{R}^n$ with $y'(t) = F(y(t))$, $y(a) = y^0$. Each such function is called a *solution* to the initial value problem $y' = F(y)$, $y(a) = y^0$ in the interval $[a, b]$.

Peano's theorem provides rather general conditions for the existence of a solution. It asserts that (5.1) has a solution in $[a, a + h]$ for sufficiently small h and that this solution can be continued until it reaches the boundary of D (i.e., until either $t \to \infty$ or $\|y(t)\| \to \infty$, or $y(t)$ approaches the boundary of D as t increases).

Because most differential equations cannot be solved explicitly, numerical approximation methods are essential. Peano's continuation property is the basis for the development of such methods: From knowledge of the previously computed solution, we compute a new solution point a short distance $h > 0$ away, and we repeat such *integration steps* until the upper limit of the interval $[a, b]$ is reached.

One traditionally divides numerical methods for the initial value problem (5.1) into two classes: *one-step methods* and *multistep (or multivalue) methods*. One-step methods (also called *Runge-Kutta methods*) are memoryless and only make use of the most recently computed solution point to compute a new solution point, whereas multistep methods retain some memory of the history by storing, using, and updating a matrix containing old and new information.

The successful numerical solution of initial value problems requires a thorough knowledge of the propagation of rounding error and discretization error. In this chapter, we give only an elementary introduction to multistep methods (which are closely related to interpolation and integration techniques), and treat the numerical issues by means of examples rather than analysis. For an in-depth treatment of the whole subject, we refer the reader to Hairer et al. [35, 36].

Euler's Method

The prototypical one-step method is the *Euler's method*. In it, we first replace y' by a divided difference to obtain

$$\frac{y(t_1) - y(t_0)}{t_1 - t_0} \approx y'(t_0) = F(y^0)$$

or, with $h_i := t_1 - t_0$,

$$t_1 = t_0 + h_i, \quad y^1 = y^0 + h_i F(y^0) \approx y(t_1).$$

Continuing in the same way with variable step size h_i, we obtain the formulas

$$t_{i+1} = t_i + h_i, \quad y^{i+1} = y^i + h_i F(y^i)$$

for $i = 0, \ldots, N - 1$, and we expect to have $y(t_i) \approx y^i$ for $i = 1, \ldots, N$. To define an approximating function $\tilde{y}(t)$ for the solution of the initial value problem (5.1), we then simply interpolate these $N + 1$ points; for example, with methods from Chapter 3.

One can show that, for constant step sizes $h_i = h$, Euler's method has a global error of order $O(h)$ on intervals of length $O(1)$, and thus converges to the solution as $h \to 0$. However, as the following example shows, sometimes very small step sizes are needed to get an acceptable solution.

4.5.1 Example. Application of Euler's method with constant step size $h_i = h$ to the problem

$$y' = \lambda y, \quad y(0) = 1$$

leads to the approximation

$$y^i = a_h^{\lambda t_i} \quad \text{with } a_h = (1 + h\lambda)^{\frac{1}{h\lambda}},$$

instead of the exact solution

$$y(t_i) = e^{\lambda t_i}.$$

The method does converge because $a_h \to e$ as $h \to 0$. The following table shows what happens for larger step sizes.

$h\lambda$	.01	.1	1	10	100	1000
a	2.71	2.59	2.00	1.27	1.05	1.01

$h\lambda$	−.01	−.1	−.5	−.9	−.99	−1
a	2.73	2.87	4.00	12.92	104.76	∞

Once $h\lambda$ is no longer tiny, a is no longer close to $e \approx 2.718$. Moreover, for $h\lambda < -1$, the "approximations" $y^{i+1} = (1 + h\lambda)y^i$ are violently oscillating because $1 + h\lambda$ is negative, and bear no resemblance to the desired solution.

In conclusion, we have qualitatively acceptable solutions for $-\frac{1}{2} \le h\lambda \le 1$, but good approximations only for $|h\lambda| \ll 1$. For $\lambda \gg 0$ it could be expected that the step size h must be very small, in view of the growth behavior of $e^{\lambda t}$. However, it is surprising that a very small step size is also necessary for $\lambda \ll 0$, even though the solution is negligibly small past a certain t-value! For negative values of λ (for example, when $\lambda = -105$ and $h = 0.02$, see Figure 4.3), an

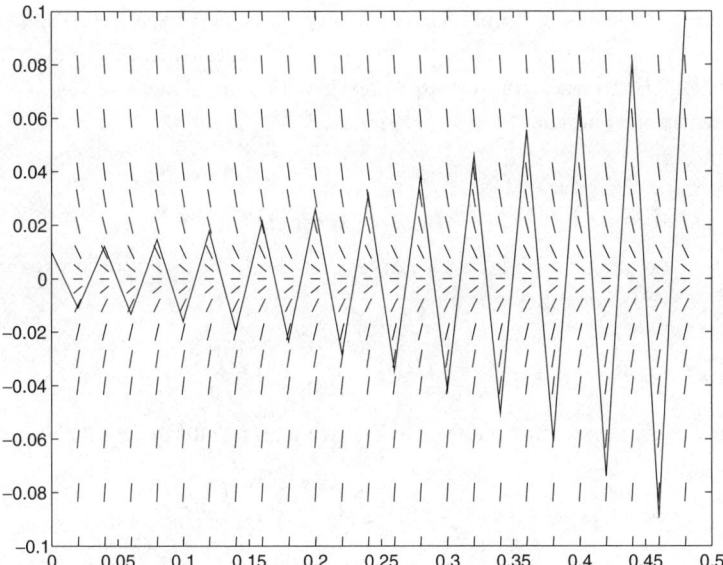

Figure 4.3. Divergence of Euler's method for the stiff equation $y' = -\lambda y$, with $\lambda = -105$ and step size $h = 0.02$.

oscillating sequence occurs that increases in amplitude with t, and the sequence has nothing at all to do with the solution.

The reason for the poor behavior of Euler's method for $\lambda \ll 0$ is the very steep direction field of the corresponding differential equation in the immediate vicinity of the solution. Such differential equations are called *stiff*; to solve such equations effectively, special care must be taken. To improve the accuracy of Euler's method, we may consider the Taylor series expansion

$$y(t+h) = y(t) + hy'(t) + \frac{h^2}{2}y''(t) + O(h^3)$$

$$= y(t) + hF(y(t)) + \frac{h^2}{2}F'(y(t))F(y(t)) + O(h^3).$$

This suggests the method

$$t_{i+1} = t_i + h_i, \quad y^{i+1} = y^i + h_i F(y^i) + \frac{h_i^2}{2}F'(y^i)F(y^i),$$

which can be shown to have a global error of $O(h^2)$. We may similarly take more terms in the series to obtain formulas of arbitrarily high order. A disadvantage of this *Taylor series method* is that, for $F(y) \in \mathbb{R}^n$, we must compute the entries of the Jacobian $F'(y)$ (or even higher derivatives); usually, this additional cost is unwarranted. However, if combined with automatic differentiation, an advantage is an easy rigorous error analysis with interval methods (Moore [63], Lohner [57], Neumaier [70a]).

Multistep Methods

To derive methods with increased order that avoid the use of derivative information, we assume that the values

$$y^j \approx y(t_j), \quad f^j = F(y^j)$$

have already been computed for $j \leq i$. If we integrate the differential equation,

$$y(t_{i+1}) = y(t_i) + \int_{t_i}^{t_{i+1}} F(y(t)) \, dt,$$

we know on the right side the function values $F(y(t_j)) \approx F(y^j) = f^j$ approximately and can therefore approximate the integral with an interpolatory

Table 4.1.   *Coefficients for Adams-Bashforth of order* $p = s + 1$

s	β_0	β_1	β_2	β_3
0	1			
1	$\frac{3}{2}$	$-\frac{1}{2}$		
2	$\frac{23}{12}$	$-\frac{16}{12}$	$\frac{5}{12}$	
3	$\frac{55}{24}$	$-\frac{59}{24}$	$\frac{37}{24}$	$-\frac{9}{24}$

quadrature formula (see Section 4.2). We consider only the equidistant case

$$t_i = a + ih, \quad h = \frac{b-a}{N}.$$

If one uses the last $s + 1$ points $t_{i-s}, \ldots, t_i$ as data points, then

$$\begin{aligned} y(t_{i+1}) &= y(t_i) + \int_{t_i}^{t_i+h} F(y(t))\, dt \\ &= y(t_i) + h \sum_{j=0:s} \beta_j F(y(t_{i-j})) + O(h^{s+2}) \end{aligned}$$

for certain constants β_j depending on s.

The constants β_j may be computed by integrating the Lagrange polynomials; they are given for $s \le 3$ in Table 4.1. They differ from those derived in Section 4.2 because now the interpolation points are not chosen from $[t_i, t_{i+1}]$, but lie to the left of t_i. By dropping the error term, one obtains the following *explicit Adams-Bashforth multistep method* of order $s + 1$:

$$t_{i+1} = t_i + h, \quad y^{i+1} = y^i + h \sum_{j=0:s} \beta_j F(y^{i-j}).$$

The cost per step involves one function evaluation only. The global error can be proved to be $O(h^{s+1})$; as for numerical integration, one loses one power of h for the global estimates because one has to sum over $(b-a)/h$ steps.

If one includes the point to be computed in the set of interpolation points, one finds *implicit* methods with an order that is higher by 1. Indeed,

$$y(t_{i+1}) = y(t_i) + h \sum_{j=-1:s} \beta_j F(y(t_{i-j})) + O(h^{s+3})$$

with s-dependent constants β_j given for $s \le 3$ in Table 4.2. One obtains the

Table 4.2.   *Coefficients for Adams-Moulton formulas of order*
$$p = s + 2$$

s	β_{-1}	β_0	β_1	β_2	β_3
0	$\frac{1}{2}$	$\frac{1}{2}$			
1	$\frac{5}{12}$	$\frac{8}{12}$	$-\frac{1}{12}$		
2	$\frac{9}{24}$	$\frac{19}{24}$	$-\frac{5}{24}$	$\frac{1}{24}$	
3	$\frac{251}{720}$	$\frac{646}{720}$	$-\frac{264}{720}$	$\frac{106}{720}$	$-\frac{19}{720}$

following *implicit Adams-Moulton multi-step method* of order $s + 2$,

$$t_{i+1} = t_i + h, \quad y^{i+1} = y^i + h\left(\beta_{-1}F(y^{i+1}) + \sum_{j=0:s} \beta_j F(y^{i-j})\right),$$

in which the desired function value also appears on the right side.

Instead of solving this nonlinear system of equations, one replaces y^{i+1} on the right side by the result of the corresponding explicit formula; the order, smaller by 1, is restored when one multiplies by h. One thus computes a *predictor* $\tilde{y}^{i+1}$ with an explicit method and then obtains the value y^{i+1} by using an implicit method as a *corrector*, with $\tilde{y}^{i+1}$ on the right side.

For example, one obtains with $s = 3$ the *fifth order Adams-Moulton predictor-corrector method*:

$t_{i+1} = t_i + h;$
% predictor step
$\tilde{y}^{i+1} = y^i + \frac{h}{24}(55f^i - 59f^{i-1} + 37f^{i-2} - 9f^{i-3});$
$\tilde{f}^{i+1} = F(\tilde{y}^{i+1});$
% corrector step
$y^{i+1} = y^i + \frac{h}{720}(251\tilde{f}^{i+1} + 646f^i - 264f^{i-1} + 106f^{i-2} - 19f^{i-3});$
$f^{i+1} = F(y^{i+1});$

Depending on the implementation, one or several corrector iterations may be performed. Compared with the explicit method, the cost per step increased to two function evaluations (if only one corrector step is taken).

It can be shown that the methods of Adams-Bashforth and of Adams-Moulton converge with convergence order $s+1$ (i.e., global error $O(h^{s+1})$) and $s+2$ (i.e., global error $O(h^{s+2})$), respectively. However, a formal analysis is important here because, for general multistep methods where y^{i+1} depends linearly on older values y^{i-j}, $j > 0$, the convergence and stability behavior may change drastically.

4.5.2 Example. The Simpson rule gives

$$y(t_{i+1}) - y(t_{i-1}) = \int_{t_i-h}^{t_i+h} F(y(t)) \, dt$$

$$= \frac{2h}{6} \left(F(y(t_{i+1})) + 4F(y(t_i)) + F(y(t_{i-1})) \right) + O(h^5),$$

hence suggests the implicit multistep formula

$$y^{i+1} = y^{i-1} + \frac{h}{3}(F(y^{i+1}) + 4F(y^i) + F(y^{i-1})). \tag{5.2}$$

For the model problem

$$y' = \lambda y, \quad y(0) = 1,$$

we find

$$y^{i+1} = y^{i-1} + \frac{\lambda h}{3}(y^{i+1} + 4y^i + y^{i-1}).$$

The general solution of this difference equation is

$$y^i = \gamma_1 z_1^i + \gamma_2 z_2^i \tag{5.3}$$

where

$$\left.\begin{matrix} z_1 \\ z_2 \end{matrix}\right\} = \frac{2\lambda h \pm \sqrt{9 + 3\lambda^2 h^2}}{3 - \lambda h} = \begin{cases} 1 + \lambda h + O((\lambda h)^2) \approx e^{\lambda h}, \\ -1 + \frac{\lambda h}{3} + O((\lambda h)^2) \approx -e^{-\lambda h/3} \end{cases}$$

are the solutions of the quadratic equation

$$z^2 = 1 + \frac{\lambda h}{3}(z^2 + 4z + 1).$$

Given y^0 and y^1, we can solve the equations $\gamma_1 + \gamma_2 = y_0 = 1$, $\gamma_1 z_1 + \gamma_2 z_2 = y_1$ to find

$$\gamma_1 = \frac{y_1 - z_2}{z_1 - z_2}, \quad \gamma_2 = \frac{z_1 - y_1}{z_1 - z_2}.$$

Now, if $\lambda < 0$ then the solution $y(t) = e^{\lambda t}$ decays for $t \to \infty$; but unless $y_1 = z_1$ (an unlikely situation), $\gamma_2 \neq 0$ and (5.3) explodes exponentially, even for tiny values of λh.

Solving Stiff Systems

Of considerable practical importance for stiff problems are additional implicit multistep methods obtained by numerical differentiation. Here we approximate $g(\tau) := y(t_i - \tau h)$ by an interpolating polynomial $p_s(\tau)$ of degree $s + 1$ with data points $(\tau, g(\tau)) = (l, y^{i-l})$ $(l = -1, 0, \ldots, s)$. This gives

$$g(\tau) \approx p_{s+1}(\tau) := \sum_{j=-1:s} L_j(\tau) y^{i-j}$$

where

$$L_j(\tau) = \prod_{l \neq j} \frac{\tau - l}{l - j}$$

is a Lagrange polynomial of degree $s + 1$; of course, for $n > 1$, we interpret $p_{s+1}(t)$ as a vector of polynomials. We then have

$$F(y^{i+1}) \approx F(y(t_{i+1})) = y'(t_{i+1}) = \frac{1}{h} g'(-1) \approx \frac{1}{h} p'_{s+1}(-1),$$

$$= \frac{1}{h} \sum_{j=-1:s} L'_j(-1) y^{i-j}.$$

It is natural to fix y^{i+1} implicitly through the condition

$$y^{i+1} = \sum_{j=0:s} \alpha_j y^{i-j} + h\beta_{-1} F(y^{i+1}); \tag{5.4}$$

the coefficients $\beta_{-1} = 1/L'_{-1}(-1)$ and $\alpha_j = -\beta_{-1} L'_j(1)$ are easily computed from the definition. They are gathered for $s \leq 5$ in Table 4.3.

The formulas obtained in this way (by Gear [29]) are called *backwards differentiation formulas*, or BDF-formulas; they have order $s + 1$. Note that here

Table 4.3.   *Coefficients for BDF formulas of order $p = s + 1$*

s	β_{-1}	α_0	α_1	α_2	α_3	α_4	α_5
0	1	1					
1	$\frac{2}{3}$	$\frac{4}{3}$	$-\frac{1}{3}$				
2	$\frac{6}{11}$	$\frac{18}{11}$	$-\frac{9}{11}$	$\frac{2}{11}$			
3	$\frac{12}{25}$	$\frac{48}{25}$	$-\frac{36}{25}$	$\frac{16}{25}$	$-\frac{3}{25}$		
4	$\frac{60}{137}$	$\frac{300}{137}$	$-\frac{300}{137}$	$\frac{200}{137}$	$-\frac{75}{137}$	$\frac{12}{137}$	
5	$\frac{60}{147}$	$\frac{360}{147}$	$-\frac{450}{147}$	$\frac{400}{147}$	$-\frac{225}{147}$	$\frac{72}{147}$	$-\frac{10}{147}$

y^{i+1} is not computed as a sum of y^i and a linear combination of the previous s function values $F(y^{i-j})$, but as a sum of the function value at the new point y_{i+1} and a linear combination of the previous s values y^{i-1}.

For stiff differential equations, both the Adams-Bashforth methods and the Adams-Moulton methods suffer from the same instability as Euler's method so that tiny step sizes are required. However, the BDF-formulas have (for $s \leq 5$ only) better stability properties, and are suitable (and indeed, the most popular choice) for integrating stiff differential equations. However, when using the BDF-formulas in predictor-corrector mode, one loses this stability property; therefore, one must find y^{i+1} directly from (5.4) by solving a nonlinear system of equations (e.g., using the methods of Chapter 6). The increased work per step is more than made good for by the ability to use step sizes that reflect the behavior of the solution instead of being forced by the approximation method to tiny steps.

Properties of and implementation details for the BDF-formulas are discussed in Gear [29]. A survey of other methods for stiff problems can be found in Byrne and Hindmarsh [10].

4.6 Step Size and Order Control

For solving differential equations accurately, step sizes and order must be chosen adaptively. In our discussion of numerical integration, we have already seen the relevance of an adaptive choice of step size; but it is even more important in the present context. To see this, we consider the following.

4.6.1 Example. The initial value problem

$$y' = y^2, \quad y(0) = 3$$

has the solution

$$y(t) = \frac{3}{1 - 3t},$$

only defined for $t < 1/3$. Euler's method with constant step size h gives

$$y^0 = 3, \quad y^{i+1} = y^i + h(y^i)^2 \quad \text{for } i \geq 0.$$

As Table 4.4 with results for constant step sizes $h = 0.1$ and $h = 0.01$ shows, the method with step size $h = 0.1$ continues far beyond the singularity, without any noticeable effect in the approximate solution, whereas with step size $h = 0.01$ the pole is recognized, albeit too late. If one were to choose h still smaller, then

Table 4.4. *Results for constant step sizes using Euler's method*

t_i	$y(t_i)$	y^i $h = 0.1$	y^i $h = 0.01$
0	3	3	3
0.1	4.29	3.9	4.22
0.2	7.5	5.42	7.06
0.3	30	8.36	19.48
0.4	—	15.34	$3.35 \cdot 10^6$
0.47	—			overflow
0.5	—	38.91	—
0.6	—	190.27	—
0.7	—	3810.34	—

one would obtain the position of the pole correspondingly more accurately. In general, one needs small h near singularities, whereas, in order to keep the cost small, one wishes to compute everywhere else with larger step sizes.

4.6.2 Example. The next example shows that an error analysis must also be done along with the step size control. The initial value problem $y' = -y - 2t/y$, $y(0) = 1$ has the solution $y(t) = \sqrt{1 - 2t}$, defined only for $t \leq \frac{1}{2}$. Euler's method, carried through with step size $h = 0.001$, gives the results shown in Table 4.5. Surprisingly, the method can be continued without problems past the singularity at $t = 0.5$. Only a weak hint that the solution might no longer exist is given, by a sudden increase in the size of the increments $y^{i+1} - y^i$ after $t = 0.525$. Exactly the same behavior results for smaller step sizes, except that the large increments occur closer to 0.5.

We conclude that a reliable automatic method for solving initial value problems must include an estimate for the error in addition to the step size control, and must be able to recognize the implications of that estimate.

Table 4.5. *Results for $y' = -y - 2t/y$, $y(0) = 1$, using Euler's method*

t_i	0	0.100	0.200	0.300	0.400	0.499	0.500	0.510	0.520	0.525	0.530	0.600
$y(t_i)$	1	0.894	0.775	0.632	0.447	0.045	0.000	—	—	—	—	—
y^i	1	0.894	0.774	0.632	0.447	0.056	0.038	0.042	−0.023	−0.001	0.777	0.614

In each integration step, one can make several choices because both the order p of the method and the step size h can be chosen arbitrarily. In principle, one could change p and h in each step, but it has been found in practice that doing this produces spurious instability that may accumulate to gross errors in the computed solution. It is therefore recommended that before changing to another step size and a possibly new order $\bar{p}$, one has done at least $\bar{p}$ steps with constant order and step size.

Local Error Estimation

To derive error estimates we must have a precise concept of the order of a method. We say that a multistep method of the form

$$y^{i+1} = \sum_{j=0:s} \alpha_j y^{i-j} + h \sum_{j=0:s} \beta_j F(y^{i-j}); \tag{6.1}$$

has *order p* if, for every polynomial $q(x)$ of degree $\leq p$,

$$q(1) = \sum_{j=0:s} (\alpha_j q(-j) + \beta_j q'(-j)). \tag{6.2}$$

This is justified by the following result for constant step size h.

4.6.3 Theorem. *If the multistep method* (6.1) *has order p, then for all* $(p+2)$-*times continuously differentiable functions y and all* $h \in \mathbb{R}$,

$$y(t + h) = \sum_{j=0:s} \alpha_j y(t - jh) + h \sum_{j=0:s} \beta_j y'(t - jh) + E_h(t) \tag{6.3}$$

with an error term

$$E_h(t) = C_p h^{p+1} y[t, t - h, \ldots, t - (p + 1)h] + O(h^{p+2}) = O(h^{p+1}) \tag{6.4}$$

and

$$C_p = 1 + \sum_{j=0:s} (-j)^p (j\alpha_j - (p + 1)\beta_j).$$

Proof. By Theorem 3.1.8,

$$y[t, t - h, \ldots, t - (p + 1)h] = \frac{y^{(p+1)}(t + \tau h)}{(p + 1)!} = \frac{y^{(p+1)}(t)}{(p + 1)!} + O(h).$$

Taylor expansion of $y(t)$ and $y'(t)$ gives (for bounded j)

$$y(t+jh) = \sum_{k=0:p+1} \frac{y^{(k)}(t)}{k!} h^k j^k + O(h^{p+2})$$

$$= q_h(j) + h^{p+1} j^{p+1} y[t, t-h, \ldots, t-(p+1)h] + O(h^{p+2})$$

$$y'(t+jh) = \sum_{k=0:p} \frac{y^{(k+1)}(t)}{k!} h^k j^k + O(h^{p+1})$$

$$= h^{-1} q_h'(j) + (p+1) h^p j^p y[t, t-h, \ldots, t-(p+1)h]$$
$$+ O(h^{p+1})$$

with a polynomial $q_h(x)$ of degree at most p. Using (6.3) as a definition for $E_h(t)$, plugging in the expressions for $y(t+jh)$ and $y'(t+jh)$ and using (6.2), we find (6.4).   □

If the last step was of order p, it is natural to perform a step of order $\bar{p} \in \{p-1, p, p+1\}$ next. Assuming that one has already performed $\bar{p}$ steps of step size h, the theorem gives a local discretization error of the form

$$y^{i+1} - y(t_{i+1}) = C_{\bar{p}} h^{\bar{p}+1} y[t_i, \ldots, t_{i-\bar{p}-1}] + O(h^{\bar{p}+2}),$$

with $y^{i+1} = \tilde{y}(t_{i+1})$, where $\tilde{y}(t)$ is the Taylor expansion of $y(t)$ of order $\bar{p}$ at t_i. The divided differences are easily computable from the y^{i-j}. If the step size h remains the same, the order $\bar{p}$ with the smallest value of

$$w_{\bar{p}} := \| C_{\bar{p}} h^{\bar{p}+1} y[t_i, \ldots, t_{i-\bar{p}-1}] \|$$

is the natural choice for the order in the next step.

However, if the new step size is changed to $h_i = \alpha h$ (where α is still arbitrary), the discretization error in the next step will be approximately $\alpha^{\bar{p}+1} w_{\bar{p}}$. For this error to be at most as large as the maximum allowable contribution δ to the global error, we must have $\alpha^{\bar{p}+1} w_{\bar{p}} \leq \delta$. Including empirically determined safety factors, one arrives at a recommended value of α given by

$$\alpha = \begin{cases} 0.7(\delta/w_{p-1})^{1/p} & \text{if } \bar{p} = p - 1, \\ 0.8(\delta/w_p)^{1/(p+1)} & \text{if } \bar{p} = p, \\ 0.6(\delta/w_{p+1})^{1/(p+2)} & \text{if } \bar{p} = p + 1. \end{cases}$$

The safety factors for a change of order are taken smaller than those for order p to combat potential instability. One may also argue for multiplying instead by suitable safety factors inside the parentheses, thus making them p dependent.

If the largest of these values for α is not too close to 1, the step size is changed to $h_i = \alpha h$ and the corresponding order $\bar{p}$ is used for the next step. To guarantee accuracy, one must relocate the needed t_{i-j} to their new equidistant positions $t_{i-j} = t_i - jh_i$ and find approximations to $y(t_i - jh_i)$ by interpolating the already computed values. (Take enough points to keep the interpolation error of the same order as the discretization error!)

Initially, one starts any multistep method with the Euler method, which has order 1 and needs no past information. To avoid problems with stability when step sizes change too rapidly (the known convergence results apply only to constant or slowly changing step sizes), it is advisable to restart a multistep method when $\alpha \ll 1$ and truncate large factors α to a suitable threshold value.

The Global Error

As in all numerical methods, it is important to assess the effect of the local approximation errors on the accuracy of the computed solution. Instead of a formal analysis, we simply present some heuristic that covers the principal pitfalls that must be addressed by good software.

The local error is the sum of the rounding errors r_i made in the evaluation of (6.1) and the approximation error $O(h^{p+1})$ (for an order p method) that gives the difference to $y(t_{i+1})$. We assume that rounding errors in computing function values are $O(\varepsilon_F)$ and the rounding error in computing the sum is $O(\varepsilon)$. Then the total local error is $O(h\varepsilon_F) + O(\varepsilon) + O(h^{p+1})$.

For simplicity, let us further assume that the local errors accumulate additively to global errors. (This assumption is often right and often wrong.) If we use constant step size h over the whole interval $[a, b]$, the number of steps needed is $n = (b - a)/h$. The global error therefore becomes

$$y^n - y(t_n) = O(n(h\varepsilon_F + \varepsilon + h^{p+1})) = O\left(\varepsilon_F + \frac{\varepsilon}{h} + h^p\right). \qquad (6.5)$$

Although our derivation was heuristic only and based on assumptions that are often violated, the resulting equation (6.5) can be shown to hold rigorously for all methods that are what is called *asymptotically stable*. In particular, this includes the Adams-Bashforth methods and Adams-Moulton methods of all orders, and the BDF methods of order up to six.

In exact arithmetic, (6.5) shows that the global error remains within a desired bound Δ when $O(\Delta^{1/p})$ steps are taken and F is a p times continuously differentiable function. (Most practical problems need much more flexibility, but it is still an unsolved problem whether there is an efficient control of step size and order such that the same statement holds.)

In finite precision arithmetic, (6.5) exhibits the same qualitative dependency of the total error on the step size h as in the error analysis for numerical differentiation (Section 3.2, Figure 3.2). In particular, the attainable accuracy deteriorates as the step sizes tend to zero. However, as shown later, there are also marked differences between these situations because the cure here is much simpler as in the case of numerical differentiation.

As established in Section 3.2, there is an optimal step size $h_{opt} = O(\varepsilon^{1/(p+1)})$. The total error then has magnitude

$$y_n - y(t_n) = O\left(\varepsilon_F + h_{opt}^p\right) = O\left(\varepsilon_F + \varepsilon^{p/(p+1)}\right).$$

In particular, one sees that not only is a better accuracy attainable with a larger order, but one may also proceed with a larger step size and hence with a smaller total number of steps, thus reducing the total work. We also see that to get optimal accuracy in the final results, one should calculate the linear combination (6.1) with an accuracy $\varepsilon \ll \varepsilon_F^{(p+1)/p}$. In particular, if F is computed to near machine precision, then (6.1) should be calculated in higher precision to make full use of the accuracy provided by F.

In FORTRAN, if F is computed in single precision, one would get the desired accuracy of the y_i by storing double precision versions $yy(s - j)$ of the recent y_{i-j} and proceed with the computation of (6.1) according to the pseudo code

$$F sum = \sum \beta_j F(y(i - j))$$
$$yy(s + 1) = \sum \alpha_j yy(s - j) + \mathrm{dprod}(h, F sum)$$
$$y(i + 1) = \mathrm{real}(yy(s + 1))$$
$$yy(1 : s) = yy(2 : s + 1)$$

(The same analysis and advice is also relevant for adaptive numerical integration, where the number of function evaluations can be large and higher precision may be appropriate in the addition of the contributions to the integral.)

A further error, neglected until now, occurs in computing the t_i. At each step, we actually compute $t_{i+1} = t_i \mp h_i$ with rounding errors that accumulate significantly in a long number of short steps. This can lead to substantial systematic errors when the number of steps becomes large. In order to keep the roundoff error small, it is important to recompute the actual step size $h_i = t_{i+1} - t_i$ used, from the new (approximately computed) argument $t_{i+1} = t_i \mp h_i$; because of cancellation that typically occurs in the computation of $t_{i+1} - t_i$, the step size h_i computed in this way is usually found exactly (cf. Section 1.3). This is one of the few cases in which cancellation is advantageous in numerical computations.

4.7 Exercises

1. Prove the error estimate

$$\int_a^b f(x)\,dx = (b-a)f\left(\frac{a+b}{2}\right) + \frac{(b-a)^3}{24}f''(\xi), \quad \xi \in [a,b]$$

for the *rectangle rule* (also called *midpoint rule*) if f is twice continuously differentiable functions $f(x)$ on the interval $[a,b]$.

2. Because the available function values $\tilde{f}(x_j) = f(x_j) + \varepsilon_j$ $(j = 0,\ldots,N)$ are generally contaminated by roundoff errors, one can ask for weights α_j chosen in such a way that the influence of roundoff errors is small. Under appropriate assumptions (distribution of the ε_j with mean zero and standard deviation σ), the root mean square error of a quadrature formula $Q(f) = \sum_{j=0:N} \alpha_j f(x_j)$ is given by

$$r_N = \sigma \sqrt{\sum_{j=0:N} \alpha_j^2}.$$

(a) Show that r_N for a quadrature formula of order ≥ 1 is minimal when all weights α_j are equal.

(b) A quadrature formula in which all weights are equal and that is exact for polynomials $p(x)$ of degree $\leq n$ is called a *Chebychev quadrature formula*. Determine (for $\omega(x) = 1$) the nodes and weights of the Chebychev formula for $n = 1, 2$, and 3. (*Remark*: The equations for the weights and points have real solutions only for $n \leq 7$ and $n = 9$.)

3. Given a closed interval $[a,b]$ and nodes $x_j := a + jh$, $j = 0,1,2,3$ with $h := (b-a)/3$, derive a quadrature formula $Q(f) = \sum_{i=0:3} \alpha_i f(x_i)$ that matches $I(f) = \int_a^b f(x)\,dx$ exactly for polynomials of degree ≤ 3. Prove the error formula $I(f) - Q(f) = \frac{3}{80}h^5 f^{(4)}(\xi)$, $\xi \in [a,b]$ for $f \in C^4[a,b]$. Is $Q(f)$ also exact for polynomials of degree 4?

4. Let $Q_k(f)$ $(k = 1,2,\ldots)$ be a normalized quadrature formula with non-negative weights and order $k+1$. Show that if $f(x)$ can be expanded in a power series about 0 with radius of convergence $r > 1$, then there is a constant $C_r(f)$ independent of k with

$$\left| \int_{-1}^1 f(x)\,dx - Q_k(f) \right| \leq \frac{C_r(f)}{(r-1)^{k+1}}.$$

Find $C_r(f)$ for $f(x) = 1/(2 - x^2)$, using Remark 4.1.11, and determine r such that $C_r(f)$ is minimal.

5. In order to approximately compute the definite integral $\int_0^1 f(x)\,dx$ of a function $f \in C^6[0, 1]$, the following formulas may be applied ($h := 1/n$):

$$\int_0^1 f(x)\,dx = h\left\{\frac{1}{2}f(0) + \sum_{v=1:n-1} f(vh) + \frac{1}{2}f(1)\right\} - \frac{h^2}{12}f''(\xi_1)$$

(*composite trapezoidal rule*; cf. (3.1)),

$$\int_0^1 f(x)\,dx = \frac{h}{6}\left\{f(0) + 2\sum_{v=1:n-1} f(vh) + 4\sum_{v=1:n} f\left(\frac{(2v-1)h}{2}\right)\right.$$
$$\left. + f(1)\right\} - \frac{h^4}{2880}f^{(4)}(\xi_2)$$

(*composite Simpson rule*; cf. (3.1)),

$$\int_0^1 f(x)\,dx = \frac{h}{90}\left\{7f(0) + 32\sum_{v=1:2n} f\left(\frac{(2v-1)h}{4}\right)\right.$$
$$\left. + 12\sum_{v=1:n} f\left(\frac{(2v-1)h}{2}\right) + 14\sum_{v=1:n-1} f(vh) + 7f(1)\right\}$$
$$- \frac{h^6}{483840}f^{(6)}(\xi_3)$$

(*composite Milne rule*; cf. (4.4)) with $\xi_i \in [0, 1]$, $i = 1, 2, 3$. Let $f(x) := x^{6.5}$.

(a) For each formula, determine the smallest n such that the absolute value of the remainder term is $\leq 10^{-4}$. How many function evaluations would be necessary to complete the numerical evaluation with this n?

(b) Using the formula that seems most appropriate to you, compute an approximation to $\int_0^1 x^{6.5}\,dx$ to the accuracy 10^{-4} (with MATLAB). Compare with the exact value of $\int_0^1 x^{6.5}\,dx$.

6. Let $a = x_0 < x_1 < \cdots < x_N = b$ be an equidistant grid.

(a) Show that if $f(x)$ is monotone in $[a, b]$, then the composite trapezoidal rule satisfies

$$\left|\int_a^b f(x)\,dx - T_N(f)\right| \leq \frac{b-a}{2N}|f(b) - f(a)|.$$

(b) Show that if $Q(f)$ is an arbitrary quadrature formula on the above subdivision of $[a, b]$ and $|\int_a^b f(x)\,dx - Q(f)| \leq R(f)$ for all monotone functions f, where $R(f)$ depends only on the $f(x_j)$, then $R(f) \geq \frac{b-a}{2N}|f(b) - f(a)|$. (Thus, in this sense, the trapezoidal rule is optimal.)

7. Write a MATLAB program to compute an approximation to the integral $\int_a^b f(x)\,dx$ with the composite Simpson rule; use it to compute $\frac{\pi}{4} = \int_0^1 \frac{dx}{1+x^2}$ with an absolute error $\leq 10^{-5}$. How many equidistant points are sufficient? (Assume that there is no roundoff error.)

8. Use MATLAB to compute the integral $I = \int_0^{\pi/2} \sqrt{\cos x}\,dx$ with an absolute error ≤ 0.00005; use Simpson's rule with step size control.

9. Let f be a continuously differentiable function on $[a, b]$.
   (a) Show that

$$\frac{1}{b-a}\int_a^b f(x)\,dx$$
$$\in f([a,b]) \cap \left\{ f\left(\frac{a+b}{2}\right) + [-1, 1]\frac{b-a}{4}\mathrm{rad}(f'([a,b])) \right\}.$$

   (b) Use (a) to write a MATLAB program for the adaptive integration of a given continuous function f where interval extensions for f and f' are available. (Use the range of slopes for f' when the derivative does not exist over some interval, and split intervals recursively until an accuracy specification is met.)

   (c) Compute enclosures for $\int_0^1 \frac{1}{\sqrt{1+x^2}}\,dx$, $\int_{-1}^1 \frac{1}{\sqrt{1-x^2}}\,dx$ and $\int_0^{\pi/2} \sqrt{x}\cos x\,dx$ with an error of at most $5 \times 10^{-i-1}$ $(i = 2, 3, 4)$; give the enclosures and the required number of function evaluations.

10. (a) Show by induction that the *Vandermonde determinant*

$$\det \begin{pmatrix} 1 & 1 & \cdots & 1 \\ x_0 & x_1 & \cdots & x_n \\ x_0^2 & x_1^2 & \cdots & x_n^2 \\ \vdots & \vdots & & \vdots \\ x_0^n & x_1^n & \cdots & x_n^n \end{pmatrix}$$

   does not vanish for pairwise distinct $x_j \in \mathbb{R}$, $j = 0, \ldots, n$.

   (b) Use part (a) to show that for any grid $a \leq x_0 < x_1 < \cdots < x_n \leq b$, there exist uniquely determined numbers $\alpha_0, \alpha_1, \ldots, \alpha_n$ so that

$$\sum_{j=0:n} \alpha_j p(x_j) = \int_a^b p(x)\,dx$$

   for every polynomial $p(x)$ of degree n.

11. (a) Show that the Chebychev polynomials $T_n(x)$ (defined in (3.1.22)) are orthogonal over $[-1, 1]$ with respect to the weight function $\omega(x) :=$ $(1 - x^2)^{-\frac{1}{2}}$.

   (b) Show that $T_n(x) = \operatorname{Re}(x + i\sqrt{1 - x^2})^n$ for $x \in [-1, 1]$.

12. Determine for $N = 0, 1, 2$ the weights α_j and arguments x_j ($j = 0, \ldots, N$) of the Gaussian quadrature formula over the interval $[0, 1]$ with weight function $\omega(x) := x^{-1/2}$. (Find the roots x_j of the orthogonal polynomial $p_3(x)$ with MATLAB's root.)

13. (a) Show that the characteristic polynomials $p_j(x) = \det(xI - A_j)$ of the tridiagonal matrices

$$
A_j = \begin{pmatrix}
a_1 & 1 & & & 0 \\
b_2 & \ddots & \ddots & & \\
& \ddots & \ddots & \ddots & \\
& & \ddots & \ddots & 1 \\
0 & & & b_j & a_j
\end{pmatrix}
$$

satisfy the three-term recurrence relation

$$p_j(x) = (x - a_j)p_{j-1}(x) - b_j p_{j-2}(x) \quad \text{for } j \geq 1, \tag{7.1}$$

started with $p_{-1}(x) = 0$, $p_0(x) = 1$.

   (b) Show that for any eigenvalue λ of A_{n+1},

$$u_j^\lambda = p_{j-1}(\lambda), \quad v_j^\lambda = \frac{p_{j-1}(\lambda)}{b_2 \cdots b_j}$$

define eigenvectors u^λ of A_{n+1} and v^λ of A_{n+1}^T.

   (c) Show that the

$$L_\lambda(x) = c_\lambda^{-1} \sum_j v_j^\lambda p_{j-1}(x), \quad \text{where } c_\lambda = \sum_j b_2 \cdots b_j \left(v_j^\lambda\right)^2$$

are the Lagrange polynomials for interpolation in the eigenvalues of A_{n+1}. (*Hint:* Obtain an identity by computing $(v^\lambda)^T A_{n+1} u^\mu$ in two ways.)

   (d) Deduce from (a)–(c) that the $n + 1$ nodes of a Gaussian quadrature formula corresponding to orthogonal polynomials with three-term recurrence relation (7.1) are given by the eigenvalues $x_j = \lambda$ of A_{n+1}, with corresponding weights $\alpha_j = c_\lambda^{-1} \int \omega(x)\,dx$. (*Hint:* Use the orthogonality relations and $p_0(x) = 1$.)

(e) If all b_j are positive (which is the case for positive weight functions), find a nonsingular diagonal matrix D such that $A'_{n+1} = D^{-1}A_{n+1}D$ is symmetric. Confirm that also $p_{n+1}(x) = \det(xI - A'_{n+1})$. Why does this give another proof of the fact that all nodes are real?

(f) Show that in terms of eigenvectors w^λ of A'_{n+1} of 2-norm $1, c_\lambda = (w_1^\lambda)^2$. Use this to compute with MATLAB's `eig` nodes and weights for the 12th order Gaussian quadrature formula for $\int_{-1}^1 f(x)\,dx$.

14. A quadrature formula $Q(f) := \sum_{i=0:N} \alpha_i f(x_i)$ with $x_i \in [a, b]$ ($i = 0, \ldots, N$) is called *closed* provided $x_0 = a$ and $x_N = b$ and $\alpha_0, \alpha_N \neq 0$. Show that, if there is an error bound of the form

$$\left| \int_a^b f(x)\,dx \right| \leq |E(f)|$$

with $E(f) := \sum_{i=0:N} \beta_i f(x_i)$ for all exponential functions $f = e^{\lambda x}$ ($\lambda \in \mathbb{R}$), then $Q(f)$ must be a closed quadrature formula and $|\beta_i| \geq |\alpha_i|$ for $i = 0, N$.

15. Let $D^k(f) := \sum_{i=0:N} \beta_i^{(k)} f(x_i)$ with $x_0 < x_1 < \cdots < x_N$. Suppose that

$$D^k(f) = \frac{1}{k!}f^{(k)}(s_k(f)) \quad \text{for some } s_k(f) \in [x_0, x_1]$$

holds whenever f is a polynomial of degree $\leq N$.

(a) Show that $D^N(f) = f[x_0, \ldots, x_N]$, and give explicit formulas for the $\beta_i^{(N)}$ in terms of the x_i.

(b) Show that $\beta_i^{(k)} = 0$ for $i \leq N < k$, i.e., that derivatives of order higher than N cannot be estimated with $N + 1$ function values.

16. (a) Write a program for approximate computation of the definite integral $\int_a^b f(x)\,dx$ via Romberg integration. Use $h_1 := (b - a)/2$ as initial step size. Halve the step size until the estimated absolute error is smaller than 10^{-6}, but at most 10 times. Compute and print the values in the "Romberg triangle" as well as the estimated error of the best value.

(b) Test the program on the three functions
   (i) $f(x) := \sin x, a := 0, b := \pi$,
   (ii) $f(x) := \sqrt{x^{13}}, a := 0, b := 1$, and
   (iii) $f(x) := (1 + 15\cos^2 x)^{-1/2}, a := 0, b := \pi/2$.

17. Suppose that the approximation $F(h)$ of the integral I, obtained with step size h, satisfies

$$I = F(h) + ah^{p_1} + bh^{p_2} + ch^{p_3} + \cdots \quad \text{with } p_1 < p_2 < p_3 < \cdots.$$

Give real numbers α, β, and γ such that

$$I = \alpha F(h) + \beta F\left(\frac{h}{2}\right) + \gamma F\left(\frac{h}{4}\right) + O(h^{p_3})$$

holds, to obtain a formula of order h^{p_3}.

18. To approximate the integral $\int_D f(x, y)\, dxdy$, $D \subset \mathbb{R}^2$, via a formula of the form

$$\sum_{v=1:N} A_v f(x_v, y_v) \tag{7.2}$$

with nodes (x_v, y_v) and coefficients A_v, $(v = 1, \ldots, N)$ we need to determine $3N$ parameters. The formula (7.2) is said to have order $m + 1$ provided

$$\sum_{v=1:N} A_v x_v^q y_v^r = \int_D x^q y^r dxdy$$

for all $q, r \geq 0$ with $q + r = 0, 1, \ldots, m$.

(a) What is the one-point formula that integrates all linear functions exactly (i.e., has order 2)?

(b) For integration over the unit square

$$D := \{(x, y) \mid 0 \leq x \leq 1, 0 \leq y \leq 1\},$$

show that there is an integration formula of the form (7.2) with nodes $(0, 0)$, $(0, 1)$, $(1/2, 1/2)$, $(1, 0)$, $(1, 1)$, and order 3, and determine the corresponding A_v, $v = 0, \ldots, 5$.

19. Consider the initial value problem

$$y' = 2y, \quad y(0) = 1.$$

(a) Write down a solution to the initial value problem.

(b) Write down a closed expression for the approximate values y^i obtained with Euler's method with step size h.

(c) How small must h be for the relative error to be $\leq \frac{1}{2} \cdot 10^{-4}$ in $[0, 1]$?

(d) Perform 10 steps with the step size determined in (c).

20. The function $y(t) = t^2/4$ is a solution to the initial value problem

$$y' = \sqrt{y}, \quad y(0) = 0.$$

However, Euler's method gives $y^i = 0$ for all t and $h > 0$. Verify this behavior numerically, and explain it.

21. Consider an initial value problem of the form

$$y' = F(t, y), \quad y(a) = \alpha.$$

Suppose that the function $F(t, z)$ is continuously differentiable in $[a, b] \times \mathbb{R}$. Suppose that

$$F(t, z) \geq 0, \quad F_t(t, z) \leq 0, \quad \text{and} \quad F_y(t, z) \leq 0$$

for every $t \in [a, b]$ and $z \in \mathbb{R}$. Show that the approximations y^i for $y(t_i)$ obtained from Euler's piecewise linear method satisfy

$$y^i \geq y(t_i), \quad i = 0, 1, \dots.$$

22. Compute (by hand or with a pocket calculator) an approximation to $y(0.4)$ to the solution of the initial value problem

$$y''(t) = -y(t), \quad y(0) = 1, \quad y'(0) = 0$$

by
(a) rewriting the differential equation as a system of first order differential equations and using Euler's method with step size $h = 0.1$;
(b) using the following recursion rules derived from a truncated Taylor expansion:

$$y^0 := y(0),$$
$$z^0 := y'(0),$$
$$y^{i+1} := y^i + hz^i + \frac{h^2}{2}F(t_i, y^i),$$
$$z^{i+1} := z^i + hF(t_i, y^i); \quad (i = 0, \dots, 3); \quad (h = 0.1).$$

Interpret the rule explained in (b) geometrically.

23. When f is complex analytic, an initial value problem of the form

$$y' = F(t, y), \quad y(a) = y^0 \qquad (7.3)$$

may be solved via a Taylor expansion. To do this, one expands $y(t)$ about a to obtain

$$y(a + h) = \sum_{v=0:N-1} \frac{y^{(v)}(a)}{v!} h^v + \text{remainder};$$

function values and derivatives at a are computable by repeated differentiation of (7.3). Neglecting the remainder term, one thus obtains an approximation for $y(a + h)$. Expansion of y about $a + h$ gives an approximation at the point $a + 2h$, and so on.

(a) Perform this procedure for $F(t, y) := t - y^2$, $a := 0$ and $y^0 := 1$. In these computations, develop y to the fourth derivative (inclusive) and choose $h = 1/2$, $1/4$, and $1/8$ as step sizes. What is the approximate value at the point $t = 1$?

(b) Use an explicit form of the remainder term and interval arithmetic to compute an enclosure for $y(1)$.

24. To check the effect of different precisions on the global error discussed at the end of Section 4.6, modify Euler's method for solving initial value problems in MATLAB as follows:

Reduce the accuracy of F_h to single precision by multiplying the standard MATLAB result with `1+1e-8*randn` (`randn` produces at different calls normally distribbuted numbers with mean zero and standard deviation 1). Perform the multiplication of F with h and the addition to y^i

(a) in double precision (i.e., normal MATLAB operations);

(b) in simulated single precision (i.e., multiplying after each operation with `1+1e-8*randn`).

Compute an approximate solution at the point $t = 10$ for the initial value problem $y' = -0.1y$, $y(0) = 1$, using constant step sizes $h := 10^{-3}$, 10^{-4} and 10^{-5}. Comment on the results.

25. (a) Verify that the Adams-Bashforth formulas, the Adams-Moulton formulas, and the BDF formulas given in the text have the claimed order.

(b) For the orders for which the coefficients of these formulas are given in the text, find the corresponding error constants C_p. (*Hint:* Use Example 3.1.7.)

5

Univariate Nonlinear Equations

In this chapter we treat the problem of finding a *zero* of a real or complex function f of one variable (a *univariate* function), that is, a number x^* such that $f(x^*) = 0$. This problem appears in various contexts; we give here some typical examples:

(i) Solutions of the equation $p(x) = q(x)$ in one unknown are zeros of functions f such as (among many others) $f := p - q$ or $f := p/q - 1$.

(ii) Interior minima or maxima of continuously differentiable functions f are zeros of the derivative f'.

(iii) Singular points x^* of a continuous function f for which $|f(x)| \to \infty$ as $x \to x^*$ (in particular, poles of f) are zeros of the inverted function

$$g(x) := \begin{cases} 1/f(x), & \text{if } x \neq x^*, \\ 0, & \text{if } x = x^*, \end{cases}$$

continuously completed at $x = x^*$.

(iv) The eigenvalues of a matrix A are zeros of the characteristic polynomial $f(\lambda) := \det(\lambda I - A)$. Because of its importance, we look at this problem more closely (see Section 5.3).

(v) Shooting methods for the solution of boundary value problems also lead to the problem of determining zeros. The boundary-value problem

$$y''(t) = g(t, y(t), y'(t)), \quad y : [a, b] \to \mathbb{R}$$
$$y(a) = y_a, \quad y(b) = y_b$$

can often be solved by solving the corresponding initial-value problem

$$y_s''(t) = g(t, y_s(t), y_s'(t)), \quad y_s : [a, b] \to \mathbb{R}$$
$$y_s(a) = y_a, y_s'(a) = s$$

233

for various values of the parameter s. The solution $y = y_s$ of the initial-value problem is a solution of the boundary-value problem iff s is a zero of the function defined by

$$f(s) := y_s(b) - y_b.$$

We see that, depending on the source of the problem, a single evaluation of f may be cheap or expensive. In particular, in examples (iv) and (v), function values have a considerable cost. Thus one would like to determine a zero with the fewest possible number of function evaluations. We also see that it may be a nontrivial task to provide formulas for the derivative of f. Hence methods that do not use derivative information (such as the secant method) are generally preferred to methods that require derivatives (such as Newton's method).

Section 5.1 discusses the secant method and shows its local superlinear convergence and its global difficulties. Section 5.2 introduces globally convergent bisection methods, based on preserving sign changes, and shows how bisection methods can accomodate the locally fast secant method without sacrificing their good global properties. For computing eigenvalues of real symmetric matrices and other definite eigenvalue problems, bisection methods are refined in Section 5.3 to produce all eigenvalues in a given interval.

In Section 5.4, we show that the secant method has a convergence order of ≈ 1.618, and derive faster methods by Opitz and Muller with a convergence order approaching 2. Section 5.5 considers the accuracy of approximate zeros computable in finite precision arithmetic, the stability problems involved in deflation (a technique for finding several zeros), and shows how to find all zeros in an interval rigorously by employing interval techniques.

Complex zeros are treated in Section 5.6: Local methods can be safeguarded by a spiral search, and complex analysis provides tools (such as Rouché's theorem) for ascertaining the exact number of zeros of analytic functions.

Finally, Section 5.7 explores methods that use derivative information, and in particular Newton's method. Although it is quadratically convergent, it needs more information per step than the secant method and hence is generally slower. However, as a simple prototype for the multivariate Newton method for systems of equations (see Section 6.2), we discuss its properties in some detail – in particular, the need for damping to prevent divergence or cycling.

5.1 The Secant Method

Smooth functions are locally almost linear; in a neighborhood of a simple zero they are well approximated by a straight line. Therefore, if we have two approximations of a zero the zero is likely to be close to the intersection of the interpolating line with the x-axis.

We therefore interpolate the function f in the neighborhood of the zero x^* at two points x_i and x_{i-1} through the linear function

$$p_1(x) := f(x_i) + f[x_i, x_{i-1}](x - x_i).$$

If the slope $f[x_i, x_{i-1}]$ does not vanish, we expect that the zero

$$x_{i+1} := x_i - \frac{f(x_i)}{f[x_i, x_{i-1}]} \tag{1.1}$$

of p_1 represents a better approximation for x^* than x_i and x_{i-1} (Figure 5.1).

The iteration according to (1.1) is known as the *secant method*, and several variants of it are known as *regula falsi methods*. For actual calculations, one

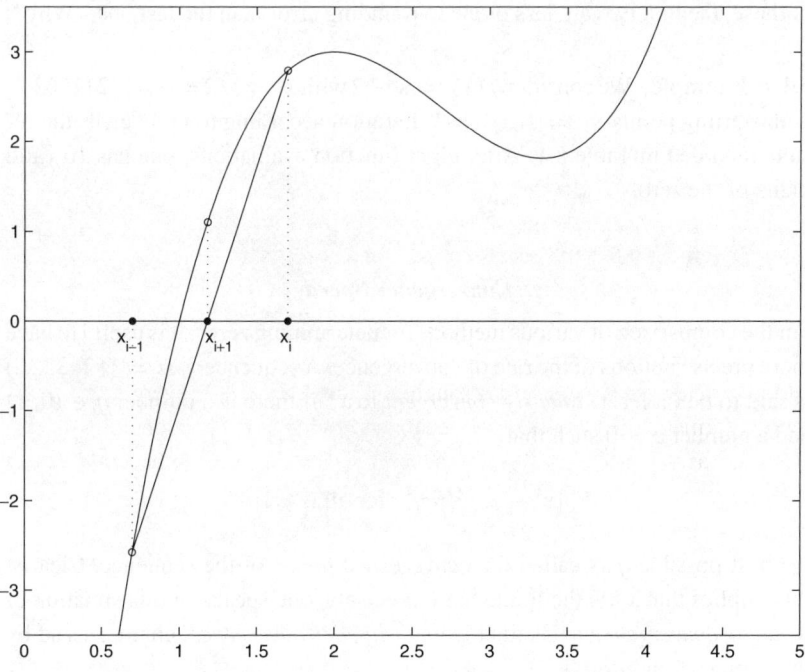

Figure 5.1. A secant step for $f(x) = x^3 - 8x^2 + 20x - 13$.

Table 5.1. *Results of the secant method in* $[1, 2]$ *for* $x^2 - 2$

i	x_i	$\text{sign}(f)$	i	x_i	$\text{sign}(f)$
1	1	$-$	6	1.41421143847487	$-$
2	2	$+$	7	1.41421356205732	$-$
3	1.33333333333333	$-$	8	1.41421356237310	$+$
4	1.40000000000000	$-$	9	1.41421356237310	$+$
5	1.41463414634146	$+$	10	1.41421356237309	$-$

may use one of the equivalent formulas

$$x_{i+1} = x_i - \frac{(x_i - x_{i-1})f(x_i)}{f(x_i) - f(x_{i-1})}$$

$$= x_i - \frac{x_i - x_{i-1}}{1 - f(x_{i-1})/f(x_i)}$$

$$= \frac{x_{i-1}f(x_i) - x_i f(x_{i-1})}{f(x_i) - f(x_{i-1})};$$

of these, the first two are less prone to rounding error than the last one. (Why?)

5.1.1 Example. We consider $f(x) := x^2 - 2$ with $x^* = \sqrt{2} = 1.414213562\ldots$ with starting points $x_1 := 1$, $x_2 := 2$. Iteration according to (1.1) gives the results recorded in Table 5.1. After eight function evaluations, one has 10 valid digits of the zero.

Convergence Speed

For the comparison of various methods for determining zeros, it is useful to have more precise notions of the rate of convergence. A sequence x_i $(i = 1, 2, 3, \ldots)$ is said to be (at least) *linearly convergent* to x^* if there is a number $q \in (0, 1)$ and a number $c > 0$ such that

$$|x^* - x_i| \le cq^i \quad \text{for all } i \ge 1;$$

the best possible q is called the *convergence factor* of the sequence. Clearly, this implies that x^* is the limit. More precisely, one speaks in this situation of *R-linear convergence* as distinct from *Q-linear convergence*, characterized by the stronger condition

$$|x^* - x_{i+1}| \le q|x^* - x_i| \quad \text{for all } i \ge i_0 \ge 1$$

for some *convergence factor* $q \in (0, 1)$. Obviously, each Q-linearly convergent sequence is also R-linearly convergent (with $c = |x^* - x_{i_0}| q^{-i_0}$), but the converse is not necessarily true.

For zero-finding methods, one generally expects a convergence speed that is faster than linear. We say the sequence x_i $(i = 1, 2, 3, \ldots)$ with limit x^* converges *Q-superlinearly* if

$$|x_{i+1} - x^*| \leq q_i |x_i - x^*| \quad \text{for } i \geq 0, \text{ with } \lim_{i \to \infty} q_i = 0,$$

and *Q-quadratically* if a relation of the form

$$|x_{i+1} - x^*| \leq c_0 |x_i - x^*|^2 \quad \text{for } i \geq 0$$

holds. Obviously, each quadratically convergent sequence is superlinearly convergent. To differentiate further between superlinear convergence speeds we introduce later (see Section 5.4) the concept of convergence order.

These definitions carry over to vector sequences in $\mathbb{R}^n$ if one replaces the absolute values with a norm. Because all norms in $\mathbb{R}^n$ are equivalent, the definitions do not depend on the choice of the norm.

We shall apply the convergence concepts to arbitrary iterative methods, and call a method *locally linearly (superlinearly, quadratically) convergent* to x^* if, for all initial iterates sufficiently close to x^*, the method generates a sequence that converges to x^* linearly, superlinearly, or quadratically, respectively. Here the attribute *local* refers to the fact that these measures of speed only apply to an (often small) neighborhood of x^*. In contrast, *global convergence* refers to convergence from arbitrary starting points.

Convergence Speed of the Secant Method

The speed that the secant method displayed in Example 5.1.1 is quite typical. Indeed, usually the convergence of the secant method (1.1) is locally Q-superlinear. The condition is that the limit is a simple zero of f. Here, a zero x^* of f is called a *simple zero* if f is continuously differentiable at x^* and $f'(x^*) \neq 0$.

5.1.2 Theorem. *Let the function f be twice continuously differentiable in a neighborhood of the simple zero x^* and let $c := \frac{1}{2} f''(x^*)/f'(x^*)$. Then the sequence defined by (1.1) converges to x^* for all starting values x_1, x_2 $(x_1 \neq x_2)$ sufficiently close to x^*, and*

$$x_{i+1} - x^* = c_i (x_i - x^*)(x_{i-1} - x^*) \tag{1.2}$$

with

$$\lim_{i \to \infty} c_i = c.$$

In particular, near simple zeros, the secant method is locally Q-superlinearly convergent.

Proof. By the Newton interpolating formula

$$0 = f(x^*)$$
$$= f(x_i) + f[x_i, x_{i-1}](x^* - x_i) + f[x_i, x_{i-1}, x^*](x^* - x_i)(x^* - x_{i-1}).$$

In a sufficiently small neighborhood of x^*, the slope $f[x_i, x_{i-1}]$ is close to $f'(x^*) \neq 0$, and we can "solve" the linear part for x^*. This gives

$$x^* = x_i - \frac{f(x_i)}{f[x_i, x_{i-1}]} - \frac{f[x_i, x_{i-1}, x^*]}{f[x_i, x_{i-1}]}(x^* - x_i)(x^* - x_{i-1})$$
$$= x_{i+1} - c_i(x^* - x_i)(x^* - x_{i-1}),$$

where

$$c_i := \frac{f[x_i, x_{i-1}, x^*]}{f[x_i, x_{i-1}]}. \tag{1.3}$$

So (1.2) holds.

For ξ, ξ' in a sufficiently small closed ball $B[x^*; r]$ around x^*, the quotient

$$c_i(\xi, \xi') := f[\xi, \xi', x^*]/f[\xi, \xi']$$

remains bounded; suppose that $|c_i(\xi, \xi')| \leq \bar{c}$. For starting values x_1, x_2 in the ball $B[x^*; r_0]$, where $r_0 = \min(r, 1/2\bar{c})$, all iterates remain in this ball. Indeed, assuming that $|x_{i-1} - x^*|, |x_i - x^*| \leq r_0$, we find by (1.2) that

$$|x_{i+1} - x^*| \leq \frac{1}{2}|x_i - x^*| \leq r_0.$$

By induction, $|x_i - x^*| \leq 2^{2-i}r_0$, so that x_i converges to x^*. Therefore,

$$\lim_{i \to \infty} c_i = \frac{f[x^*, x^*, x^*]}{f[x^*, x^*]} = \frac{1}{2}\frac{f''(x^*)}{f'(x^*)} = c.$$

With $q_i := |c_i| \, |x_{i-1} - x^*|$ it follows from (1.2) that

$$|x_{i+1} - x^*| = q_i|x_i - x^*|.$$

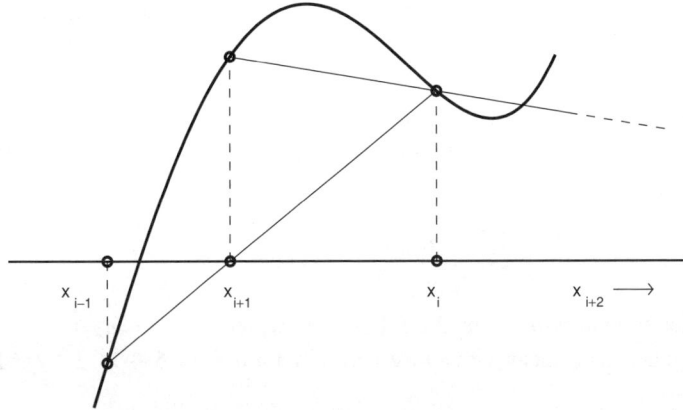

Figure 5.2. Nonconvergence of the secant method.

Now Q-superlinear convergence follows from $\lim q_i = 0$. This proves the theorem. □

5.1.3 Corollary. *If the iterates x_i of the secant method converge to a simple zero x^* with $f''(x^*) \neq 0$, then for large i, the signs of $x_i - x^*$ and $f(x_i)$ in the secant method are periodic with period 3.*

Proof. Let $s_i = \text{sign}(x_i - x^*)$. Because $\lim c_i = c \neq 0$ by assumption, there is an index i_0 such that for $i \geq i_0$, the factors c_i in (1.2) have the sign of c. Let $s = \text{sign}(c)$. Then (1.2) implies $s_{i+1} = ss_is_{i-1}$ and hence $s_{i+2} = ss_{i+1}s_i = sss_is_{i-1}s_i = s_{i-1}$, giving period 3. □

As Table 5.1 shows, this periodic behavior is indeed observed locally, until (after convergence) rounding errors spoil the pattern.

In spite of the good local behavior of the secant method, global convergence is not guaranteed, as Figure 5.2 illustrates. The method may even break down completely if $f[x_i, x_{i-1}] = 0$ for some i because then the next iterate is undefined.

Multiple Zeros and Zero Clusters

It is easy to show that x^* is a simple zero of f if and only if f has the representation $f(x) = (x - x^*)g(x)$ in a neighborhood of x^*, in which g is continuous and $g(x^*) \neq 0$. More generally, x^* is called an *m-fold zero* of f if in a neighborhood of x^* the relation

$$f(x) = (x - x^*)^m g(x), \quad g(x^*) \neq 0 \tag{1.4}$$

holds for some continuous function g.

For multiple zeros $(m > 1)$, the secant method is only locally linearly convergent.

5.1.4 Example. Suppose that we want to determine the double zero $x^* = 0$ of the parabola $f(x) := x^2$. Iteration with (1.1) for $x_1 = 1$ and $x_2 = \frac{1}{2}$ gives the sequence

$$x_i = \frac{1}{3}, \frac{1}{5}, \frac{1}{8}, \frac{1}{13}, \ldots; \quad x_{34} = 1.08_{10} - 7.$$

(The denominators n_i form the Fibonacci sequence $n_{i+1} = n_i + n_{i-1}$.) The convergence is Q-linear with convergence factor $q = (\sqrt{5} - 1)/2 \approx 0.618$.

If one perturbs a function $f(x)$ with an m-fold zero x^* slightly, then the perturbed function $\tilde{f}(x)$ has up to m distinct zeros, a so-called *zero cluster*, in the immediate neighborhood of x^*. (If all these zeros are simple, there is an odd number of them if m is odd, and an even number if m is even.) In particular, for $m = 2$, perturbation produces two closely neighboring zeros (or none) from a double zero. From the numerical point of view, the behavior of closely neighboring zeros and zero clusters is similar to that of multiple zeros; in particular, most zero finding methods converge only slowly towards such zeros, until one finds separating points in between that produce sign changes. If an even number of simple zeros $x^* = x_1^*, \ldots, x_{2k}^*$, well separated from the other zeros, lie in the immediate neighborhood of some point, it is difficult to find a sign change for f because then the product of the linear factors $x - x_i^*$ is positive for all x outside of the hull of the x_i^*. One must therefore hit by chance into the cluster in order to find a sign change. In this case, finding a sign change is therefore essentially the same as finding an "average solution," and is therefore almost as difficult as finding a zero itself.

In practice, zeros of multiplicity $m > 2$ appear only very rarely. For this reason, we consider only the case of a double zero. In a neighborhood of the double zero x^*, $f(x)$ has the sign of $g(x)$ in (1.4); so one does not recognize such a zero through a sign change. However, x^* is a simple zero of the function

$$h(x) := (x - x^*)\sqrt{|g(x)|} = \text{sign}(x - x^*)\sqrt{|f(x)|},$$

and the secant method applied to h gives

$$x_{i+1} = x_i - \frac{x_i - x_{i-1}}{1 - h(x_{i-1})/h(x_i)} = x_i - \frac{x_i - x_{i-1}}{1 \pm \sqrt{f(x_{i-1})/f(x_i)}}.$$

The sign of the root is positive when $x^* \in \bigsqcup\{x_i, x_{i-1}\}$ and negative otherwise.

Table 5.2. *The safeguarded root secant method*

i	x_i	x_{i+1}	$f(x_{i+1})$
1	0	1.20000000000000	0.09760000000000
2	1.74522604659078	1.20000000000000	0.09760000000000
3	1.20000000000000	1.05644633886449	0.00674222757113
4	1.05644633886449	1.00526353445128	0.00005570200768
5	1.00526353445128	1.00014620797161	0.00000004275979
6	1.00014620797161	1.00000038426329	0.00000000000030
7	1.00000038426329	1.00000000002636	0

However, because x^* is not known, we simply choose the negative sign. We call the iteration by means of the formula

$$x_{i+1} = x_i - \frac{x_i - x_{i-1}}{1 - \sqrt{f(x_{i-1})/f(x_i)}} \tag{1.5}$$

the *root secant method*. One can prove (see Exercise 4) that this choice guarantees locally superlinear convergence provided one safeguards the original formula (1.5) by implicitly relabeling the x_i after each iteration such that $|f(x_{i+1})| \le |f(x_i)|$.

5.1.5 Example. The function $f(x) = x^4 - 2x^3 + 2x^2 - 2x + 1$ has a double root at $x^* = 1$. Starting with $x_1 = 0$, $x_2 = 1.2$, the safeguarded root secant method produces the results shown in Table 5.2. Note that in the second step, x_2 and x_3 were interchanged because the new function value was bigger than the so far best function value. Convergence is indeed superlinear; it occurs, as Figure 5.3 shows, from the convex side of $\sqrt{|f(x)|}$. The final accuracy is not extremely high because multiple roots are highly sensitive to rounding errors (see Section 5.5). Of course, for a general f, it may happen that although initially there is no sign change, a sign change is found during the iteration (1.5). We consider this as an asset because it allows us to continue with a globally convergent bracketing method, as discussed in the following section.

5.2 Bisection Methods

A different principle for determining a zero assumes that two points a and b are known at which f has different signs (i.e., for which $f(a)f(b) \le 0$). We then speak of a *sign change* in f and refer to the pair a, b as a *bracket*. If f is continuous in $\square\{a, b\}$ and has there a sign change, then f has at least one zero $x^* \in \square\{a, b\}$ (i.e., some zero is bracketed by a and b).

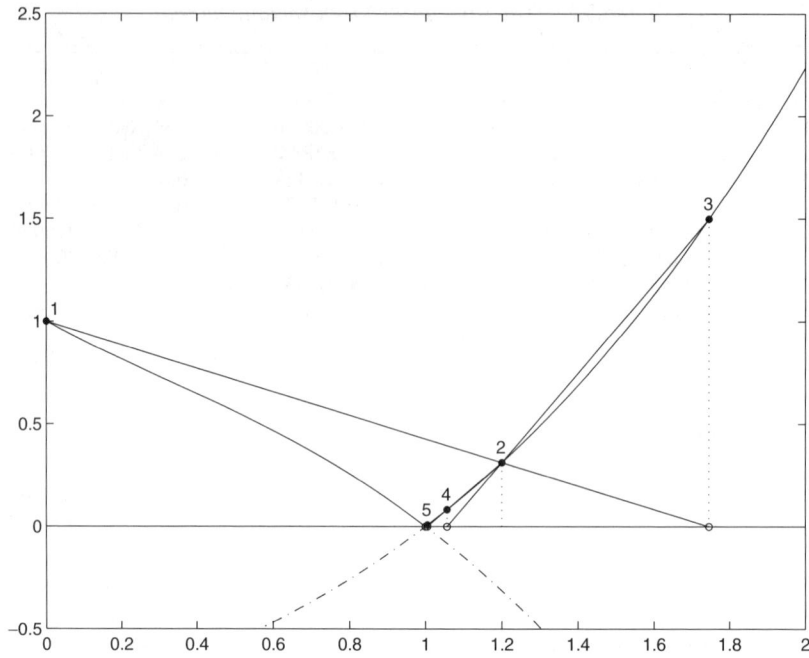

Figure 5.3. The safeguarded root secant method ($\sqrt{|f(x)|}$ against x).

If we evaluate f at some point x between a and b and look at its sign, one finds a sign change in one of the two intervals resulting from splitting the interval at x. We call this process *bisection*. Repeated bisection at suitably chosen splitting points produces a sequence of increasingly narrow intervals, and by a good choice of the splitting points we obtain a sequence of intervals that converges to a zero of f. An advantage of such a *bisection method* is the fact that at each iteration we have (at least in exact arithmetic) an interval in which the zero must lie, and we can use this information to know when to stop.

In particular, if we choose as splitting point always the midpoint of the current interval, the lengths of the intervals are halved at every step, and we get the following algorithm. (0 and 1 are the MATLAB values for *false* and *true*; while 1 starts an infinite loop that is broken by the break in the stopping test.)

5.2.1 Algorithm: Midpoint Bisection to Find a Zero of f in $[a, b]$

```
f=f(a);f2=f(b);
if f*f2>0, error('no sign change'); end;
if f>0, x=a;a=b;b=x; end;    % now f(a)<=0<=f(b)
```

```
i=2;    % counts number of function evaluations
if f==0, x=a;
else
   while 1,
      x=(a+b)/2;
      if abs(b-a)<=delta0, break; end;
      i=i+1;f=f(x);
      if f>=0, b=x; end;
      if f<=0, a=x; end;
   end;
end;
% f has a zero xstar with |xstar-x|<=delta0/2
```

5.2.2 Example. For finding the zero $x^* = \sqrt{2} = 1.414213562\ldots$ of $f(x) :=$ $x^2 - 2$ in the interval $[1, 2]$, Algorithm 5.2.1 gives the results in Table 5.3. Within 10 iterations we get an increase of accuracy of about 3 decimal places, corresponding to a reduction of the initial interval by a factor $2^{10} = 1024$. The accuracy attained at x_9 is temporarily lost since $|x^* - x_{10}| > |x^* - x_9|$. (This may happen with R-linear convergence, but not with Q-linear convergence. Why?)

A comparison with Example 5.1.1 shows that the secant method is much more effective than the midpoint bisection method, which would need 33 function evaluations to reach the same accuracy. This is due to the superlinear convergence speed of the secant method.

In general, if x_i, a_i, and b_i denote the values of x, a, and b after the ith function evaluation, then the maximal error in x_i satisfies the relation

$$|x^* - x_i| \le \frac{1}{2}|b_{i-1} - a_{i-1}| = \frac{1}{2^{i-2}}|b - a| \quad \text{for } i > 2. \tag{2.1}$$

Table 5.3. *The bisection method for*
$f(x) = x^2 - 2 \text{ in } [1, 2]$

i	x_i	i	x_i
3	1.5	8	1.421875
4	1.25	9	1.4140625
5	1.375	10	1.41796875
6	1.4375	11	1.416015625
7	1.40625	12	1.4150390625

5.2.3 Proposition. *The midpoint bisection method terminates after*

$$t = 2 + \left\lceil \log_2 \frac{|b - a|}{\delta_0} \right\rceil$$

function evaluations.

(Here, $\lceil x \rceil$ denotes the smallest integer $\geq x$.)

Proof. Termination occurs for the smallest $i \geq 2$ for which $|b_{i-1} - a_{i-1}| \leq \delta_0$. From (2.1), one obtains $i = t$.   □

5.2.4 Remarks.

(i) Rounding error makes the error estimate (2.1) invalid; the final x_i is a good approximation to x^* in the sense that close to x_i, there is a sign change of the computed version of f.

(ii) Bisection methods make essential use of the continuity of f. The methods are therefore unsuitable for the determination of zeros of rational functions, which may have poles in the interval of interest.

The following proposition is an immediate consequence of (2.1).

5.2.5 Proposition. *The midpoint bisection method converges globally and R-linearly, with convergence factor $q = \frac{1}{2}$.*

Secant Bisection Method

Midpoint bisection is guaranteed to find a zero (when a sign change is known), but it is slow. The secant method is locally fast but may fail globally. To get the best of both worlds, we must combine both methods. Indeed, we can preserve the global convergence of a bisection method if we choose arbitrary bisection points, restricted only by the requirement that the interval widths shrink by at least a fixed factor within a fixed number of iterations.

The convergence behavior of the midpoint bisection method may be drastically accelerated if we manage to place the bisection point close to and on alternating sides of the zero. The secant method is locally able to find points very close to the zero, but as Example 5.1.1 shows, the iterates may fall on the same side of the zero twice in a row.

One can force convergence and sign change in the secant method, given x_1 and x_2 such that $f(x_1)f(x_2) < 0$ if one iterates according to

$$x_{i+1} = x_i - f(x_i)/f[x_i, x_{i-1}];$$
$$\text{if } f(x_{i-1})f(x_{i+1}) < 0, \quad x_i = x_{i-1}; \text{ end};$$

This forces $f(x_i)f(x_{i+1}) < 0$, but in many cases, the lengths of the resulting $\square\{x_i, x_{i-1}\}$ do not approach zero because the x_i converge for large i from one side only, so that the stopping criterion $|x_i - x_{i-1}| \le \delta_0$ is no longer useful. Moreover, the convergence is in such a case only linear.

At the cost of more elaborate programming one can, however, ensure global *and* locally superlinear convergence. It is a general feature of nonlinear problems that there are no longer straightforward, simple algorithms for their solution, and that sophisticated safeguards are needed to make prototypical algorithms robust. The complexity of the following program gives an impression of the difference between a numerical prototype method that works on some examples and a piece of quality software that is robust and generates appropriate warnings.

5.2.6 Algorithm: Secant Bisection Method with Sign Change Search

```
f=feval(func,x);nf=1;
if f==0, ier=0; return; end;
f2=feval(func,x2);nf=2;
if f2==0, x=x2; ier=0; return; end;

% find sign change
while f*f2>0,
   % place best point in position 2
   if abs(f)<abs(f2), x1=x2;f1=f2;x2=x;f2=f;
   else              x1=x;f1=f;
   end;
   % safeguarded root secant formula
   x=x2-(x2-x1)/(1-max(sqrt(f1/f2),2));
   if x==x2 | nf==nfmax,
      % double zero or no sign change found
      x=x2;ier=1; return;
   end;
   f=feval(func,x);nf=nf+1;
   if f==0, ier=0; return; end;
end;
```

```
ier=-1; % we have a sign change f*f2<0
slow=0;
while nf<nfmax,
  % compute new point xx and accuracy
  if slow==0,
    % standard secant step
    if abs(f)<abs(f2),
      xx=x-(x-x2)*f/(f-f2);
      acc=abs(xx-x);
    else
      xx=x2-(x2-x)*f2/(f2-f);
      acc=abs(xx-x2);
    end;
  elseif slow==1,
    % safeguarded secant extrapolation step
    if f1*f2>0,
      quot=max(f1/f2,2*(x-x1)/(x-x2));
      xx=x2-(x2-x1)/(1-quot);
      acc=abs(xx-x2);
    else % f1*f>0
      quot=max(f1/f,2*(x2-x1)/(x2-x));
      xx=x-(x-x1)/(1-quot);
      acc=abs(xx-x);
    end;
  else
    % safeguarded geometric mean step
    if x*x2>0, xx=x*sqrt(x2/x); % preserves the sign!
    elseif x==0, xx=0.1*x2;
    elseif x2==0, xx=0.1*x;
    else          xx=0;
    end;
    acc=max(abs(xx-[x,x2]));
  end;

  % stopping tests
  if acc<=0, x=xx;ier=-1; return; end;
  ff=feval(func,xx);nf=nf+1;
  if ff==0, x=xx;ier=0; return; end;

  % compute reduction factor and update bracket
```

```
if f2*ff<0, redfac=(x2-xx)/(x2-x);x1=x;f1=f;x=xx;f=ff;
else        redfac=(x-xx)/(x-x2);x1=x2;f1=f2;x2=xx;f2=ff;
end;

% force two consecutive mean steps if nonlocal (slow=2)
if redfac>0.7 | slow==2, slow=slow+1;
else                     slow=0;
end;
end;
```

The algorithm is supplied with two starting approximations $-x$ and x_2. If these do not yet produce a sign change, the safeguarded root secant method is applied in the hope that it either locates a double root or that it finds a sign change. (An additional safeguard was added to avoid huge steps when two function values are nearly equal.) If neither happens, the iteration stops after a predetermined number nfmax of function evaluations were used. This is needed because without a sign change, there is no global convergence guarantee.

The function of which zero is to be found must be defined by a function routine whose name is passed in the string func. For example, if func='myfunc', then the expression f=feval(func,x) is interpreted by MATLAB as f=myfunc(x). The algorithm returns in x a zero if ier=0, an approximate zero within the achievable accuracy if ier=-1, and the number with the absolutely smallest function value found in case of failure ier=1. In the last case, it might well be, however, that a good approximation to a double root has been found.

The variable slow counts the number of consecutive times the length of the bracket $[]\{x, x_2\}$ has not been reduced by a factor of at least $1/\sqrt{2}$; if this happened twice in a row, two mean bisection steps are performed in order to speed up the width reduction. However, these mean bisections do not use the arithmetic mean as bisection point (as in midpoint bisection) because for initial intervals ranging over many orders of magnitude, the use of the geometric mean is more appropriate. (For intervals containing zero, the geometric mean is not well defined and an ad hoc recipe is used.)

If only a single iteration was slow, it is assumed that this is due to a secant step that fell on the wrong side of the zero. Therefore, the most recent point x and the last point with a function value of the same sign as $f(x)$ are used to perform a linear extrapolation step. However, because this can lead to very poor points, even outside the current bracket (cf. Figure 5.2), a safeguard is applied that restricts the step in a sensible way (cf. Exercise 7).

The secant bisection method is a robust and efficient method that (for four times continuously differentiable functions) converges locally superlinearly toward simple and double zeros.

The local superlinear and monotone convergence in a neighborhood of double zeros is due to the behavior of the root secant method (cf. Exercise 4); the local superlinear convergence to simple zeros follows because the local sign pattern in the secant method (see Corollary 5.1.3) ensures that locally, the secant bisection method usually reduces to the secant method: In a step with `slow=0`, a sign change occurs, and because the sign pattern has period 3, it must locally alternate as either $-, +, +, -, +, +, -, +, + \ldots$ or $+, -, -, +, -, -, +, -, - \ldots$. Thus, locally, two consecutive slow steps are forbidden. Moreover, locally, after a slow step, the safeguard in the extrapolation formula can be shown to be inactive (cf. Exercise 7).

Of course, global convergence is guaranteed only when a sign change is found (after all, not all functions have zeros). In this case, the method performs globally at most twice as slow as the midpoint bisection method (why?); but if the function is sufficiently smooth, then much fewer function values are needed once the neighborhood of superlinear convergence is reached.

5.2.7 Example. For the determination of a zero of

$$f(x) := 1 - 10x + 0.01e^x$$

(Figure 5.4) with the initial values $x_1 := 5$, $x_2 := 20$, the secant bisection method gives the results in Table 5.4. A sign change is immediate. Sixteen function evaluations were necessary to attain (and check) the full accuracy of 16 decimal digits. Close to the zero, the sign pattern and the superlinear convergence of the pure secant method is noticeable.

Practical experiments on the computer show that generally about 10–15 function evaluations suffice to find a zero. The number is smaller if x and x_2 are not far from a zero, and may be larger if (as in the example) $f(x)$ or $f(x_2)$ is huge. However, because of the superlinear convergence, the number of function evaluations is nearly independent of the accuracy requested, unless the latter is so high that rounding errors produce erratic sign changes. Thus `nfmax=20` is a good choice to prevent useless function evaluations in case of nonconvergence. (To find several zeros, one generally applies a technique called *deflation*; see Section 5.5.)

Unfortunately there are also situations in which the sign change is of no use, namely if the continuity of f in $[a, b]$ is not guaranteed. Thus, for example, it can happen that for rational functions, a bisection method finds instead of a zero, a pole with a sign change (and then with slow convergence). In this case, one must

Table 5.4.   *Locating the zero of* $f(x) := 1 - 10x + 0.01e^x$ *with the secant bisection method*

i	x_i	$f(x_i)$	slow
1	5.0000000000000000	−47.5158684089742350	0
2	20.0000000000000000	4851452.9540979033000000	0
3	5.0001469108434691	−47.5171194663684280	1
4	5.0002938188092605	−47.5183704672215440	2
5	10.0002938144929130	121.3264462602397100	3
6	7.0713794514850772	−57.9360785853060920	0
7	8.0179789309415899	−48.8294182041005270	0
8	8.5868413839981272	−31.2618672757999010	1
9	8.9020135225054009	−14.5526201672795000	2
10	9.4351868431760817	31.8611779028851880	3
11	9.1647237200757186	4.8935773887754692	0
12	9.0986134854694392	−0.5572882149440375	0
13	9.1053724953793740	−0.0183804999496431	1
14	9.1056030246107298	0.0000723790792989	0
15	9.1056021203890367	−0.0000000093485397	0
16	9.1056021205058109	−0.0000000000000568	1

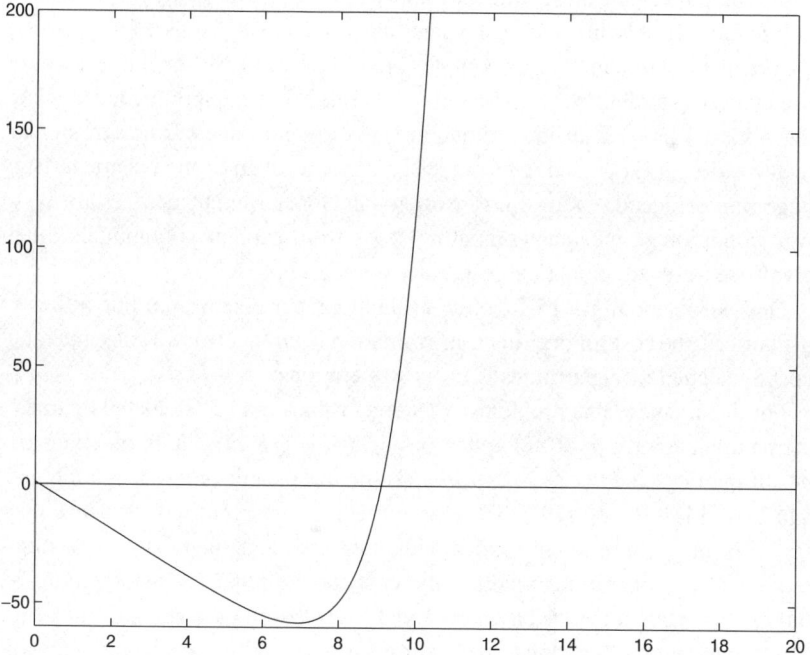

Figure 5.4. Secant bisection method for $f(x) = 1 - 10x + 0.01e^x$.

safeguard the secant method instead by damping the steps taken. The details are similar to that for the damped Newton method discussed in Section 5.7.

5.3 Spectral Bisection Methods for Eigenvalues

Many univariate zero-finding problems arise in the context of eigenvalue calculations. Special techniques are available for eigenvalues, and in this section we discuss elementary techniques related to those for general zero-finding problems. Techniques based on similarity transformations are often superior but cannot be treated here.

The eigenvalues of a matrix A are defined as the zeros of its *characteristic polynomial* $f(\lambda) = \det(\lambda I - A)$. The eigenvalues of a *matrix pencil* (A, B) (the traditional name for a pair of square matrices) are defined as the zeros of its *characteristic polynomial* $f(\lambda) = \det(\lambda B - A)$. And the zeros of $f(\lambda) := \det G(\lambda)$ are the eigenvalues of the *nonlinear eigenvalue problem* associated with a *parameter matrix* $G(\lambda)$, that is, a matrix dependent on the scalar parameter λ. (This includes the ordinary eigenvalue problem with $G(\lambda) = \lambda I - A$ and the general linear eigenvalue problem with $G(\lambda) = \lambda B - A$.) In each case, λ is a possibly complex parameter.

To compute eigenvalues, one may apply any zerofinder to the characteristic polynomial. If only the real eigenvalues are wanted, we can use, for example, the secant bisection method; if complex eigenvalues are also sought, one would use instead a spiral method discussed in Section 5.6. It is important to calculate the value $\det G(\lambda_l)$ at an approximation λ_l to the eigenvalue λ from a triangular factorization of $G(\lambda_l)$ and *not* from the coefficients of an explicit characteristic polynomial. Indeed, eigenvalues are often – and for Hermitian matrices always – well-conditioned, whereas computing zeros from explicit polynomials often gives rise to severe numerical instability (see Example 5.5.1).

One needs about 10–15 function evaluations per eigenvalue. For full $n \times n$-matrices, the cost for one function evaluation is $\frac{2}{3}n^3 + O(n^2)$. Hence the total cost to calculate s eigenvalues is $O(sn^3)$ operations.

For linear eigenvalue problems, *all* n eigenvalues can be calculated by transformation methods, in $O(n^3)$ operations using the QR-*algorithm* for eigenvalues of matrices and the QZ-*algorithm* for matrix pencils (see, e.g., Golub and van Loan [31], Parlett [79]). Thus these methods are preferred for dense matrices. If, however, $G(\lambda)$ is banded with a narrow band, then these algorithms require $O(n^2)$ operations whereas the cost for the required factorizations is only $O(n)$. Thus methods based on finding zeros of the determinant (or poles of suitable rational functions; see Section 5.4) are superior when only a few of the eigenvalues are required.

5.3.1 Example. The matrix

$$A := \begin{pmatrix} 0 & 1 & 0 & -2 \\ 1 & 0 & 2 & 0 \\ 0 & 2 & 0 & 1 \\ -2 & 0 & 1 & 0 \end{pmatrix} \tag{3.1}$$

has the eigenvalues $\pm\sqrt{5} \approx \pm 2.23606797749979$, both of which have multiplicity 2. We compute one of them as a zero of $f(\lambda) := \det(\lambda I - A)$ using the secant bisection method (Algorithm 5.2.6).

With the starting values 3 and 6 we obtain the double eigenvalue $\tilde{\lambda} = 2.23606797749979$ with full accuracy after nine evaluations of the determinant using triangular factorizations.

If we use instead the explicit characteristic polynomial $f(x) = x^4 - 10x^2 + 25$, we obtain after 10 steps $\tilde{x} = 2.2360679866043878$, with only eight valid decimal digits. As shown in Section 5.5, this corresponds to the limit accuracy $O(\sqrt{\varepsilon})$ expected for double zeros.

The surprising order $O(\varepsilon)$ accuracy of the results when the determinant is computed from a matrix factorization is typical for (nondefective) multiple eigenvalues, and can be explained as follows: The determinant is calculated as a product of the diagonal factors in R (and a sign from the permutations), and close to a nondefective m-fold eigenvalue m of these factors have order $O(\varepsilon)$; therefore the unavoidable error in $f(x)$ is only of order $O(\varepsilon^m)$ for $x \approx x^*$ and the error in x^* is $O((\varepsilon^m)^{1/m}) = O(\varepsilon)$.

Definite Eigenvalue Problems

The parameter matrix $G : D \subseteq \mathbb{C} \to \mathbb{C}^{n \times n}$ is called *definite* in a real interval $[\underline{\lambda}, \bar{\lambda}] \subseteq D$ if, for $\lambda \in [\underline{\lambda}, \bar{\lambda}]$, $G(\lambda)$ is Hermitian and (componentwise) continuously differentiable and the (componentwise) derivative $G'(\lambda) := \frac{d}{d\lambda} G(\lambda)$ is positive definite.

5.3.2 Proposition. *The linear parameter matrix $G(\lambda) = \lambda B - A$ is definite (in $\mathbb{R}$) if and only if B and A are Hermitian and B is positive definite. In this case, all the eigenvalues are real. In particular, all eigenvalues of a Hermitian matrix are real.*

Proof. The parameter matrix $G(\lambda) = \lambda B - A$ is Hermitian if and only if B and A are Hermitian. The derivative $G'(\lambda) = B$ is positive definite if and only if

B is positive definite. This proves the first assertion. For each eigenpair (λ, x), $0 = x^H G(\lambda)x = \lambda x^H Bx - x^H Ax$. By hypothesis, $x^H Bx > 0$ and $x^H Ax$ is real; from this it follows that $\lambda = x^H Ax / x^H Bx$ is also real. Therefore all of the eigenvalues are real.

In particular, this applies with $B = I$ to the eigenvalues of a Hermitian matrix A. □

Quadratic eigenvalue problems are often definite after dividing the parameter matrix by λ:

5.3.3 Proposition. *The parameter matrix of the form $G(\lambda) = \lambda B + C - \lambda^{-1} A$ is definite in $(-\infty, 0)$ and $(0, \infty)$ if B, C, and A are Hermitian and A, B are positive definite. All eigenvalues of the parameter matrix are then real.*

Proof. Because B, C, and A are Hermitian, $G(\lambda)$ is also Hermitian. Furthermore, $G'(\lambda) = B + \lambda^{-2} A$ and $x^H G'(\lambda)x = x^H Bx + \lambda^{-2} x^H Ax > 0$ if $x \neq 0$ and $\lambda \in \mathbb{R} \setminus \{0\}$. Consequently, $G(\lambda)$ is definite in $(-\infty, 0)$ and in $(0, \infty)$. For each eigenpair (λ, x),

$$0 = x^H G(\lambda)x = \lambda x^H Bx + x^H Cx - \lambda^{-1} x^H Ax.$$

Because the discriminant $(x^H Cx)^2 + 4(x^H Bx)(x^H Ax)$ of the corresponding quadratic equation in λ is real and positive, all eigenvalues are real. □

For definite eigenvalue problems, the bisection method can be improved to give a method for the determination of *all* real eigenvalues in a given interval.

For the derivation of the method, we need some theory about Hermitian matrices. For this purpose we arrange the eigenvalues of an Hermitian matrix, which by Proposition 5.3.2 are all real, in order from smallest to largest, counting multiple eigenvalues with their multiplicity in the characteristic polynomial

$$\lambda_1(A) \leq \lambda_2(A) \leq \cdots \leq \lambda_n(A).$$

We denote with $p(A)$ the number of nonnegative eigenvalues of A. If the matrix A depends continuously on a parameter t, then the coefficients of the characteristic polynomial $\det(\lambda I - A)$ also vary continuously with t. Because the zeros of a polynomial depend continuously on the coefficients, the eigenvalues $\lambda_i(A)$ depend continuously on t.

Fundamental for spectral bisection methods is the *inertia theorem of Sylvester*. We use it in the following form.

5.3.4 Theorem. *The number $p(A)$ of nonnegative eigenvalues of an Hermitian matrix A is invariant under congruence transformations; that is, if $A_0, A_1 \in C^{n \times n}$ are Hermitian and if S is a nonsingular $n \times n$ matrix with $A_1 = S A_0 S^H$, then $p(A_1) = p(A_0)$.*

Proof. Let σ_0 be a complex number of modulus 1 with $\sigma_0 \neq \text{sign } \sigma$ for every eigenvalue σ of S. Then $S_t := t S - (1-t)\sigma_0 I$ is nonsingular for all $t \in [0, 1]$, and the matrices $A_t := S_t A_0 S_t^H$ are Hermitian for all $t \in [0, 1]$ and have the same rank $r := \text{rank } A_0$. (Because $S_0 A_0 S_0^H = (-\sigma_0) A_0 (-\bar{\sigma}_0) = A_0$ and $S_1 A_0 S_1^H = S A_0 S^H = A_1$, the notation is consistent.) If now A_t has π_t positive and ν_t negative eigenvalues, it follows that $\pi_t + \nu_t = r$ is constant. Because the eigenvalues of a matrix depend continuously on the coefficients, the integers π_t and ν_t must be constant themselves. In particular, $p(A_1) = \pi_1 + n - r = \pi_0 + n - r = p(A_0)$.   □

We note two important consequences of Sylvester's theorem.

5.3.5 Proposition. *Suppose that $A \in \mathbb{C}^{n \times n}$ is Hermitian and that $\sigma \in \mathbb{R}$. Then $p(\sigma I - A)$ is the number of eigenvalues of A with $\lambda \leq \sigma$.*

Proof. The eigenvalues of $\sigma I - A$ are just the numbers $\sigma - \lambda$ in which λ runs through the eigenvalues of A.   □

5.3.6 Proposition. *Suppose that $A, B \in \mathbb{C}^{n \times n}$ are Hermitian and that $A - B$ is positive definite. Then*

$$\lambda_i(A) > \lambda_i(B) \quad \text{for } i = 1, \dots, n. \tag{3.2}$$

In particular, if A has rank r, then

$$p(A) \geq p(B) + n - r. \tag{3.3}$$

Proof. Because $A - B$ is positive definite there exists a Cholesky factorization $A - B = LL^H$. Suppose now that $\sigma := \lambda_i(B)$ is an m-fold eigenvalue of B and that j is the largest index with $\sigma = \lambda_j(B)$. Then $i > j - m$ and $p(\sigma I - B) = j$; and for $C := L^{-1}(\sigma I - B)L^{-H}$ we have $p(C) = p(LCL^H) = p(\sigma I - B) = j$. However, the matrix C has the m-fold eigenvalue zero; therefore $p(C - I) \leq j - m$, whence $p(\sigma I - A) = p(\sigma I - B - LL^H) = p(L(C - I)L^H) = p(C - I) \leq j - m$. Because $i > j - m$ (see previous mention), $\lambda_i(A) > \sigma = \lambda_i(B)$. This proves the assertion (3.2), and (3.3) follows immediately.   □

In order to be able to make practical use of Proposition 5.3.5, we need a simple method for evaluating $p(B)$ for Hermitian matrices B. If an LDL^H factorization of B exists (cf. Proposition 2.2.3 then we obtain $p(B)$ as the number of nonnegative diagonal elements of D; then $p(B) = p(LDL^H) = p(D)$ and the eigenvalues of D are just the diagonal elements.

5.3.7 Example. We are looking for the smallest eigenvalue λ^* in the interval $[0.5, 2.5]$ of the symmetric tridiagonal matrix $T \in \mathbb{R}^{11 \times 11}$ with diagonal entries $1, 2, \ldots, 11$ and subdiagonal entries 1. (The tridiagonal case can be handled very efficiently without computing the entire factorization; see Exercise 11). We initialize the bisection procedure by computing the number of eigenvalues less than 0.5 and 2.5, respectively. We find that two eigenvalues are less than 0.5 and four eigenvalues are less than 2.5. In particular, we see that the third eigenvalue λ_3 is the smallest one in $[0.5, 2.5]$. Table 5.5 shows the result of further bisection steps and the resulting intervals containing λ_3. (The $1.0 + \varepsilon$ in the fourth trial indicates that for $\lambda = 1$, the LDL^H factorization does not exist, and a perturbation must be used.)

A useful property of this *spectral bisection method* is that here an eigenvalue with a particular number can be found. We also obtain additional information about the positions of the other eigenvalues that can subsequently be used to determine them. For example, we can read off Table 5.5 that $[-\infty, 0.5]$ contains the two eigenvalues λ_1 and λ_2, $[0.5, 1.5]$ and $[1.5, 2.5]$ contain one each (λ_3 and λ_4), and $[2.5, \infty]$ contains the remaining seven eigenvalues $\lambda_k, k = 5, \ldots, 11$.

The spectral bisection method is globally convergent, but the convergence is only linear. Therefore, to reduce the number of evaluations of p, it is advisable

Table 5.5. *Example for spectral bisection*

l	σ_l	$p_l = p(\sigma_l I - T)$	$\lambda_3 \in$
1	0.5	2	
2	2.5	4	$[0.5, 2.5]$
3	1.5	3	$[0.5, 1.5]$
4	$1.0 + \varepsilon$	3	$[0.5, 1.0]$
5	0.75	2	$[0.75, 1.0]$
6	0.875	2	$[0.875, 1.0]$
7	0.9375	2	$[0.9375, 1.0]$
8	0.96875	3	$[0.9375, 0.96875]$
9	0.953125	2	$[0.953125, 0.96875]$
10	0.9609375	2	$[0.9609375, 0.96875]$
11	0.96484375	3	$[0.9609375, 0.96484375]$

to switch to a superlinear zerofinder, once an interval containing only one desired eigenvalue has been found. The spectral bisection method is especially effective if triangular factorizations are cheap (e.g., for Hermitian band matrices with small bandwidth) and only a fraction of the eigenvalues (less than about 10%) are wanted.

Problems with the execution of the spectral bisection method arise only if the LDL^H factorization without pivoting does not exist or is numerically unstable. Algorithm 2.2.4 simply perturbs the pivot element to compute the factorization; for the computation of $p(B)$, this seems acceptable as it corresponds to a perturbation of A of similar size, and eigenvalues of Hermitian matrices are not sensitive to small perturbations.

Alternatively, it is recommended by Parlett [79] to simply repeat the function evaluation with a slightly changed value of the argument. One may also employ in place of the LDL^H factorization a so-called Bunch-Parlett factorization with block-diagonal D and symmetric pivoting (see Golub and van Loan [31]), which is numerically stable.

The bisection method demonstrated can also be adopted to solve general definite eigenvalue problems. Almost everything remains the same; however, in place of the inertia theorem of Sylvester, the following theorem is used.

5.3.8 Theorem. *Suppose that $G : D \subseteq \mathbb{C} \to \mathbb{C}^{n \times n}$ is definite in $[\underline{\lambda}, \bar{\lambda}] \subseteq D \cap \mathbb{R}$. Then the numbers $\underline{p} := p(G(\underline{\lambda}))$, $\bar{p} := p(G(\bar{\lambda}))$ satisfy $\underline{p} \leq \bar{p}$, and exactly $\bar{p} - \underline{p}$ real eigenvalues lie in the interval $[\underline{\lambda}, \bar{\lambda}]$. (Here an eigenvalue λ is counted with a multiplicity given by the dimension of the null space of $G(\lambda)$.)*

Proof. Let $g_x(\lambda) := x^H G(\lambda) x$ for $x \in \mathbb{C}^n \setminus \{0\}$. Then

$$\frac{\partial}{\partial \lambda} g_x(\lambda) = x^H G'(\lambda) x > 0 \quad \text{for } \lambda \in [\underline{\lambda}, \bar{\lambda}],$$

that is, g_x is strictly monotone increasing in $[\underline{\lambda}, \bar{\lambda}]$ and

$$x^H G(\lambda_1) x < x^H G(\lambda_2) x$$

for

$$\underline{\lambda} \leq \lambda_1 < \lambda_2 \leq \bar{\lambda}, \quad x \neq 0.$$

We deduce that $G(\lambda_2) - G(\lambda_1)$ is positive definite for $\underline{\lambda} \leq \lambda_1 < \lambda_2 \leq \bar{\lambda}$. From Proposition 5.3.6, we obtain

$$p(G(\lambda_1)) \leq p(G(\lambda_2)) \quad \text{for } \underline{\lambda} \leq \lambda_1 < \lambda_2 \leq \bar{\lambda}.$$

If λ_2 is an m-fold eigenvalue of the parameter matrix $G(\lambda)$, it follows that

$$p(G(\lambda_1)) \leq p(G(\lambda_2)) - m \quad \text{for } \underline{\lambda} \leq \lambda_1 < \lambda_2 \leq \bar{\lambda}.$$

Because of the continuous dependence of the eigenvalues $\lambda_i(G(\lambda))$ on λ, equality holds in this inequality if no eigenvalue of the parameter matrix $G(\lambda)$ lies between λ_1 and λ_2. This proves the theorem. $\qquad\square$

5.4 Convergence Order

To be able to compare the convergence speed of different zero finders, we introduce the concept of *convergence order*. For a sequence of numbers x_i that converge to x^*, we define the quantities

$$e_i := -\log_{10}|x_i - x^*|.$$

Then $|x_i - x^*| = 10^{-e_i}$, and with

$$s_i := \lfloor e_i \rfloor := \max\{s \in \mathbb{Z} \mid s \leq e_i\}$$

we find that

$$10^{-s_i-1} < |x_i - x^*| \leq 10^{-s_i}$$

Therefore x_i has at least e_i correct digits after the decimal point. The growth pattern of e_i characterizes the convergence speed.

For Q-linear convergence,

$$|x_{i+1} - x^*| \leq q|x_i - x^*| \quad \text{with } 0 < q < 1,$$

we find the equivalent relation

$$e_{i+1} \geq e_i + \beta, \quad \text{where } \beta := -\log_{10} q > 0$$

by taking logarithms, and therefore also

$$e_{i+k} \geq k\beta + e_i. \tag{4.1}$$

Thus in k steps, one gains at least $\lfloor k\beta \rfloor$ decimal places. For R-linear convergence,

$$|x_i - x^*| \leq Cq^i, \quad q \in {]}0, 1{[}$$

one similarly obtains the equivalent relation

$$e_i \geq i\beta + \gamma \quad \text{where } \beta := -\log_{10} q, \ \gamma := -\log_{10} C. \tag{4.2}$$

For example, the midpoint bisection method has $q = \frac{1}{2}$, hence $\beta \approx 0.301 >$ 0.3. After any 10 iterations, $10\beta > 3$ further decimal places are ensured. Because (4.1) implies (4.2), we see again that Q-linear convergence implies R-linear convergence.

For Q-quadratic convergence,

$$|x_{i+1} - x^*| \leq C_0 |x_i - x^*|^2,$$

we have the equivalent relation $e_{i+1} \geq 2e_i - \gamma$, where $\gamma := \log_{10} C_0$. Recursively, one finds $e_i - \gamma \geq 2^{i-k}(e_k - \gamma)$. Thus, if we define $\beta := (e_j - \gamma)/2^j$ with the smallest j such that $e_j > \gamma$, we obtain

$$e_i \geq 2^i \beta + \gamma \quad \text{with } \beta > 0,$$

so that the bound for the number of valid digits grows exponentially.

In general, we say that a sequence $x_1, x_2, x_3, \ldots$ with limit x^* converges with (R-)*convergence order* of at least $\kappa > 1$ or $\kappa = 1$ if the numbers

$$e_i := -\log_{10}|x_i - x^*|$$

satisfy

$$e_i \geq \beta \kappa^i + \gamma \quad \text{for all } i \geq 1,$$

or

$$e_i \geq \beta i + \gamma \quad \text{for all } i \geq 1,$$

respectively, for suitable $\beta > 0, \gamma \in \mathbb{R}$.

Thus Q- and R-linearly convergent sequences have convergence order 1, and Q-quadratically convergent sequences have convergence order 2.

To determine the convergence order of methods like the secant method, we need the following.

5.4.1 Lemma. *Let* $p_0, p_1, \ldots, p_s$ *be nonnegative numbers such that*

$$p := p_0 + p_1 + \cdots + p_s - 1 > 0,$$

and let κ be a positive solution of the equation

$$\kappa^{s+1} = p_0 \kappa^s + p_1 \kappa^{s-1} + \cdots + p_s. \qquad (4.3)$$

If e_i is a sequence diverging to $+\infty$ such that

$$e_{i+1} \geq p_0 e_i + p_1 e_{i-1} + \cdots + p_s e_{i-s} + \alpha$$

for some number $\alpha \in \mathbb{R}$ then there are $\beta > 0$ and $\gamma \in \mathbb{R}$ such that

$$e_i \geq \beta \kappa^i + \gamma \quad \text{for all } i \geq 1.$$

Proof. We choose $i_0 \geq s$ large enough such that $e_i + \alpha/p > 0$ for all $i \geq i_0 - s$, and put

$$\beta := \min\{\kappa^{-i}(e_i + \alpha/p) \mid i = i_0 - s, \ldots, i_0\}.$$

Then $\beta > 0$, and we claim that

$$e_i \geq \beta \kappa^i - \alpha/p \qquad (4.4)$$

for all $i \geq i_0 - s$. By construction, (4.4) holds for $i = i_0 - s, \ldots, i_0$. Assuming that it is valid for $i - s, i - s + 1, \ldots, i$ $(i \geq i_0)$ we have

$$
\begin{aligned}
e_{i+1} &\geq p_0 e_i + p_1 e_{i-1} + \cdots + p_s e_{i-s} + \alpha \\
&\geq p_0(\beta \kappa^i - \alpha/p) + p_1(\beta \kappa^{i-1} - \alpha/p) + \cdots + p_s(\beta \kappa^{i-s} - \alpha/p) + \alpha \\
&= \beta \kappa^{i-s}(p_0 \kappa^s + \cdots + p_s) - (p_0 + \cdots + p_s - p)\alpha/p \\
&= \beta \kappa^{i+1} - \alpha/p.
\end{aligned}
$$

So by induction, (4.4) is valid for all $i \geq i_0 - s$. If one chooses

$$\gamma := \min\{-\alpha/p, e_1 - \beta \kappa, e_2 - \beta \kappa^2, \ldots e_{i_0-s-1} - \beta \kappa^{i_0-s-1}\}$$

then one obtains

$$e_i \geq \beta \kappa^i + \gamma \quad \text{for all } i \geq 1.$$

This proves the assertion. $\qquad\qquad\qquad\qquad\qquad\qquad\qquad\qquad\quad \square$

One can show (see Exercise 14) that (4.3) has exactly one positive real solution κ, and this solution satisfies

$$1 < \kappa < 1 + \max\{p_0, p_1, \ldots, p_s\}.$$

5.4.2 Corollary. *For simple zeros, the secant method has convergence order* $\kappa = (1 + \sqrt{5})/2 \approx 1.618\ldots$.

Proof. By Theorem 5.1.2, the secant iterates satisfy the relation

$$|x_{i+1} - x^*| \le \bar{c}|x_i - x^*||x_{i-1} - x^*|,$$

where $\bar{c} = \sup c_i$. Taking logarithms gives

$$e_{i+1} \ge e_i + e_{i-1} + \alpha$$

with $\alpha = -\log_{10} \bar{c}$. Lemma 5.4.1 now implies that

$$e_i \ge \beta\kappa^i + \gamma \quad \text{with } \beta > 0 \text{ and } \gamma \in \mathbb{R}$$

for the positive solution $\kappa = (1 + \sqrt{5})/2$ of the equation

$$\kappa^2 = \kappa + 1. \qquad \square$$

The Method of Opitz

It is possible to increase the convergence order of the secant method by using not only information from the current and the previous iterate, but also information from earlier iterates.

An interesting class of zero-finding methods due to Opitz [77] (and later rediscovered by Larkin [55]) is based on approximating the poles of the function $h := 1/f$ by looking at the special case

$$h(x) = \frac{a}{x^* - x}.$$

For this function, the divided differences

$$h[x_{i-s}, \ldots, x_i] = \frac{a}{(x^* - x_{i-s}) \cdots (x^* - x_i)}$$

were computed in Example 3.1.7. By taking quotients of successive terms, we find

$$x^* = x_i + \frac{h[x_{i-s}, \ldots, x_{i-1}]}{h[x_{i-s}, \ldots, x_i]}.$$

In the general case, we may consider this formula as providing approximations to x^*. For, given $x_1, \ldots, x_{s+1}$ this leads to the iterative *method of Opitz*:

$$x_{i+1} := x_i + \frac{h[x_{i-s}, \ldots, x_{i-1}]}{h[x_{i-s}, \ldots, x_i]} \quad \text{(for } i > s\text{)}. \tag{O_s}$$

In practice, $x_2, \ldots, x_{s+1}$ are generated for $s > 1$ by the corresponding lower order Opitz formulas.

5.4.3 Theorem. *For functions of the form*

$$h(x) = \frac{p_{s-1}(x)}{x - x^*}$$

in which p_{s-1} is a polynomial of degree $\leq s - 1$, the Opitz formula (O_s) gives the exact pole x^* in a single step.

Proof. Indeed, by polynomial division one obtains

$$h(x) = \frac{a}{x^* - x} + p_{s-2}(x) \quad \text{with } a = -p_{s-1}(x^*)$$

and the polynomial p_{s-2} of degree $\leq s - 2$ does not appear in sufficiently higher order divided differences. So the original argument for $h(x) = 1/(x - x^*)$ applies again. $\qquad\square$

5.4.4 Theorem. *Let x^* be a simple pole of h and let $h(x)(x^* - x)$ be at least s times continuously differentiable in a neighborhood of x^*. Then the sequence x_i defined by the method of Opitz converges to x^* for all initial values $x_1, \ldots, x_{s+1}$ sufficiently close to x^*, and*

$$x^* - x_{i+1} = c_i^{(s)}(x^* - x_i)(x^* - x_{i-1}) \cdots (x^* - x_{i-s}) \quad \text{for } i > s,$$

where $c_i^{(s)}$ converges to a constant for $i \to \infty$. In particular, the method (O_s) is superlinearly convergent; the convergence order is the positive solution $\kappa_s = \kappa$ of $\kappa^{s+1} = \kappa^s + \kappa^{s-1} + \cdots + 1$.

Proof. We write

$$h(x) = \frac{a}{x^* - x} + g(x)$$

where g is an $(s - 1)$-times continuously differentiable function. With the abbreviations $\varepsilon_j := x^* - x_j$ $(j = 1, 2, \ldots)$ and $g_{i-s,i} := g[x_{i-s}, \ldots, x_i]$, we obtain

$$h[x_{i-s}, \ldots, x_i] = \frac{a}{\varepsilon_{i-s} \cdots \varepsilon_i} + g_{i-s,i},$$

and

$$\varepsilon_{i+1} = x^* - x_{i+1} = x^* - x_i - \frac{h[x_{i-s}, \ldots, x_{i-1}]}{h[x_{i-s}, \ldots, x_i]},$$

whence

$$\varepsilon_{i+1} = \varepsilon_i - \frac{\frac{a}{\varepsilon_{i-s} \cdots \varepsilon_{i-1}} + g_{i-s,i-1}}{\frac{a}{\varepsilon_{i-s} \cdots \varepsilon_i} + g_{i-s,i}}$$

$$= \varepsilon_{i-s} \cdots \varepsilon_i \frac{g_{i-s,i}\varepsilon_i - g_{i-s,i-1}}{a + g_{i-s,i}\varepsilon_{i-s} \cdots \varepsilon_i}.$$

In the neighborhood of the pole x^* the quantities

$$c_i^{(s)} := \frac{g_{i-s,i}\varepsilon_i - g_{i-s,i-1}}{a + g_{i-s,i}\varepsilon_{i-s} \cdots \varepsilon_i}$$

remain bounded. As in the proof for the secant method, one deduces the locally superlinear convergence and that

$$\lim_{i \to \infty} c_i^{(s)} = c^{(s)} := \frac{-g^{(s-1)}(x^*)}{(s-1)!a}.$$

Taking logarithms of the relation

$$|\varepsilon_{i+1}| \leq C|\varepsilon_{i-s}| \cdots |\varepsilon_i|, \quad C = \sup_{i \geq 1} |c_i^{(s)}|,$$

the convergence order follows from Lemma 5.4.1. □

For the determination of a zero of f, one applies the method of Opitz to the function $h := 1/f$. For $s = 1$, one obtains again the secant method. For $s = 2$, one can show that the method is equivalent to that obtained by *hyperbolic interpolation*,

$$x_{i+1} = x_i - \frac{f(x_i)}{f[x_i, x_{i-1}] - f(x_{i-1})f[x_i, x_{i-1}, x_{i-2}]/f[x_{i-1}, x_{i-2}]}, \quad (4.5)$$

by rewriting the divided differences (see Exercise 15).

The convergence order of (O_s) is monotone increasing with s (see Table 5.6). In particular, (O_2) and (O_3) are superior to the secant method (O_1). The values

Table 5.6. *Convergence order of* (O_s)
for different s

s	κ_s	s	κ_s
1	1.61803	6	1.99196
2	1.83929	7	1.99603
3	1.92756	8	1.99803
4	1.96595	9	1.99902
5	1.98358	10	1.99951

$s = 2$ and $s = 3$ are useful in practice; locally, they give a saving of about 20% of the function evaluations. For a robust algorithm, the Opitz methods must be combined with a bisection method, analogously to the secant bisection method. Apart from replacing in Algorithm 5.2.6 the secant steps by steps computed from (4.5) once enough function values are available, and storing and updating the required divided differences, an asymptotic analysis of the local sign change pattern (that now has period $s + 2$) reveals that to preserve superlinear convergence, one may use mean steps only for `slow` $> s$.

Eigenvalues as Poles

Calculating multiple eigenvalues as in Section 5.3 via the determinant is generally slow because the superlinear convergence of zerofinding methods is lost. This slowdown can be avoided for Hermitian matrices (and more generally for nondefective matrix pencils, i.e., when a basis consisting of eigenvectors exists) if one calculates the eigenvalue as a pole of a suitable function.

To find the eigenvalues of the parameter matrix $G(\lambda)$, one may consider the function

$$h(\lambda) := a^T G(\lambda)^{-1} b$$

for suitable $a, b \in \mathbb{C}^n \backslash \{0\}$. Obviously, each pole of h is an eigenvalue of $G(\lambda)$. Typically, each (simple or multiple) eigenvalue of $G(\lambda)$ is a simple pole of h; however, eigenvalues can be lost by bad choices of a and b, and for *defective* problems (and only for such problems), where the eigenvectors don't span the full space, multiple poles can occur.

5.4.5 Proposition. *Suppose that* $a, b \in \mathbb{C}^n \backslash \{0\}$, *that* $A, B \in \mathbb{C}^{n \times n}$, *and that the matrix* B *is nonsingular. If* $G(\lambda) := \lambda B - A$ *is nondefective, then the function* $h(\lambda) := a^T G(\lambda)^{-1} b$ *has no multiple poles.*

Proof. Because the matrix pencil (A, B) is nondefective, there is a basis $u^1, \ldots,$ $u^n \in \mathbb{C}^n$ of $\mathbb{C}^n$ consisting of eigenvectors, $Au^i = \lambda_i Bu^i$. If we represent $B^{-1}b$ as a linear combination $B^{-1}b = \sum \alpha_i u^i$, where $\alpha_i \in \mathbb{C}$, then

$$x(\lambda) := \sum \frac{\alpha_i}{\lambda - \lambda_i} u^i$$

satisfies the relation

$$G(\lambda)x(\lambda) = \sum \frac{\alpha_i}{\lambda - \lambda_i}(\lambda Bu^i - Au^i)$$

$$= \sum \alpha_i Bu^i = B(B^{-1}b) = b.$$

From this, it follows that

$$h(\lambda) = a^T G(\lambda)^{-1}b = a^T x(\lambda) = \sum \frac{\alpha_i a^T u^i}{\lambda - \lambda_i},$$

that is, h has only simple poles. $\qquad\square$

We illustrate this in the following examples.

5.4.6 Examples.

(i) For $G(\lambda) := (\lambda - \lambda_0)I$, the determinant $f(\lambda) := \det G(\lambda) = (\lambda - \lambda_0)^n$ has the n-fold zero λ_0. However, $h(\lambda) := a^T G(\lambda)^{-1}b = a^T b/(\lambda - \lambda_0)$ has the simple pole λ_0 iff $a^T b \neq 0$.

(ii) For the defective parameter matrix

$$G(\lambda) := \begin{pmatrix} \lambda - 1 & -1 \\ 0 & \lambda - 1 \end{pmatrix},$$

$n = 2$, $f(\lambda) := \det G(\lambda) = (\lambda - 1)^2$ has the double zero $\lambda = 1$, and $h(\lambda) := a^T G(\lambda)^{-1}b = a^T b/(\lambda - 1) + a_1 b_2/(\lambda - 1)^2$ has $\lambda = 1$ as a double pole if $a_1 b_2 \neq 0$. Thus the transformation of zeros to poles brings no advantage for defective problems.

To determine the poles of h by the method of Opitz, one function evaluation is necessary in each iteration. For the calculation of $h(\lambda_l)$, we solve the system of equations $G(\lambda_l)x^l = b$ and find $h(\lambda_l) = a^T x^l$. Thus, as for the calculation of the determinant, one triangular factorization per function evaluation is necessary, but instead of taking the product of the diagonal terms, we must solve two triangular systems.

Table 5.7.   *Evaluations of the function h to find the double eigenvalue*

i	x_i	$1/h(x_i)$
1	6.00000000000000	1.10714285714286
2	3.00000000000000	0.25000000000000
3	2.12500000000000	−0.03875000000000
4	2.23897058823529	0.00100257895440
5	2.23598478789267	−0.00002874150660
6	2.23606785942768	−0.00000004079291
7	2.23606797405966	−0.00000000118853
8	2.23606797740430	−0.00000000003299
9	2.23606797749993	0.00000000000005
10	2.23606797749979	−0.00000000000000
11	2.23606797749979	0.00000000000000

5.4.7 Example. We repeat the calculation of the two double eigenvalues of the matrix (3.1) of Example 5.3.1, this time as the poles of $h(\lambda) := a^T (\lambda I - A)^{-1} b$, where $a = b = (1, 1, 1, 1)^T$. With starting values $x_1 = 6$ and $x_2 = 3$, we do one Opitz step (O_1) and further Opitz steps (O_2), and find results as shown in Table 5.7. After 11 evaluations of the function h, one of the double eigenvalues is found with full accuracy.

Muller's Method

The method of Muller [65] is based on quadratic interpolation. From

$$
\begin{aligned}
0 = f(x^*) \\
\approx f(x_i) + f[x_i, x_{i-1}](x^* - x_i) \\
+ f[x_i, x_{i-1}, x_{i-2}](x^* - x_i)(x^* - x_{i-1}) \\
= f(x_i) + \omega_i(x^* - x_i) + f[x_i, x_{i-1}, x_{i-2}](x^* - x_i)^2,
\end{aligned}
$$

where

$$
\omega_i := f[x_i, x_{i-1}] + f[x_i, x_{i-1}, x_{i-2}](x_i - x_{i-1}),
$$

we deduce that

$$
x^* \approx x_{i+1} := x_i - \frac{2f(x_i)}{\omega_i \pm \sqrt{\omega_i^2 - 4f(x_i)f[x_i, x_{i-1}, x_{i-2}]}}. \tag{4.6}
$$

To ensure that x_{i+1} is the zero closest to x_i of the parabola interpolating in x_i, x_{i-1}, x_{i-2}, the sign of the square root must be chosen such that the magnitude of the denominator is as large as possible (i.e., the correction is as small as possible). For real calculations, therefore, $\pm = \text{sign}(\omega_i)$; for complex calculations, $\pm = \text{sign}(\text{Re}\,\omega_i\,\text{Re}\,q + \text{Im}\,\omega_i\,\text{Im}\,q)$, where q denotes the square root in the denominator.

One can show that the method of Muller has convergence order $\kappa \approx 1.84$, the solution of $\kappa^3 = \kappa^2 + \kappa + 1$ for simple zeros of a three times continuously differentiable function; for double zeros it still converges superlinearly with convergence order $\kappa \approx 1.23$, the solution of $2\kappa^3 = \kappa^2 + \kappa + 1$. A disadvantage is that the method may produce complex iterates from real starting points. To avoid this, it is advisable to replace the square root of a negative number by zero. (However, for finding complex roots, this may be an asset; cf. Section 5.6.)

5.5 Error Analysis

Limiting Accuracy

Suppose that the computed approximation $\tilde{f}(x)$ for $f(x)$ satisfies

$$|\tilde{f}(x) - f(x)| \leq \delta$$

for all x near a zero x^*. Because only $\tilde{f}(x)$ is available in actual computation, the methods considered determine a zero $\tilde{x}^*$ of $\tilde{f}(x)$; and for such a zero, the true function values only satisfy $|f(\tilde{x}^*)| \leq \delta$.

Now suppose that x^* is an m-fold zero of f. Then $f(x) = (x - x^*)^m g(x)$ with $g(x^*) \neq 0$, and it follows that

$$|\tilde{x}^* - x^*| = \sqrt[m]{\left|\frac{f(\tilde{x}^*)}{g(\tilde{x}^*)}\right|} \leq \sqrt[m]{\frac{\delta}{|g(\tilde{x}^*)|}} = O(\sqrt[m]{\delta}). \tag{5.1}$$

For simple zeros, we have more specifically

$$g(\tilde{x}^*) = \frac{f(\tilde{x}^*) - f(x^*)}{\tilde{x}^* - x^*} = f[\tilde{x}^*, x^*] \approx f'(x^*),$$

that is, the absolute error satisfies to a first approximation

$$|\tilde{x}^* - x^*| \approx \leq \frac{\delta}{|f'(x^*)|}.$$

From this, we draw some qualitative conclusions:

(i) For very small $|f'(x^*)|$, that is, for very flat functions, the absolute error in $\tilde{x}^*$ is strongly magnified. In this case, x^* is ill-conditioned.

(ii) In particular, the absolute error in $\tilde{x}^*$ is magnified for multiple zeros; because of (5.1), one finds that the number of correct places is only about the mth part of the mantissa length that is being used.

(iii) A double zero cannot be numerically distinguished from two simple zeros that lie only about $O(\sqrt{\delta})$ apart.

These remarks are valid independently of the method used.

5.5.1 Example. Consider the polynomial

$$f(x) := (x - 1)(x - 2) \cdots (x - 20)$$

of degree 20 in the standard from

$$f(x) = x^{20} - 210x^{19} + \cdots + 20!$$

The coefficient of x^{15} is (check with MATLAB's function `poly`!) -1672280820. Therefore, a relative error of ε in this coefficient produces an absolute perturbation of

$$|\tilde{f}(x) - f(x)| = 1672280820 x^{15} \varepsilon$$

in the function value at x. For the derivative at a zero $x^* = 1, 2, \ldots, 20$,

$$|f'(x^*)| = \prod_{\substack{k=1:20 \\ k \neq x^*}} |x^* - k| = (x^* - 1)!(20 - x^*)!$$

so that the limiting accuracy for the calculation of the zero $x^* = 16$ is about

$$\frac{1672280820 \cdot 16^{15} \varepsilon}{15! \, 4!} \approx 6.14 \cdot 10^{13} \varepsilon \approx 0.014$$

for double precision (machine precision $\varepsilon \approx 2.22 \cdot 10^{-16}$). Thus about two digits of relative accuracy can be expected for this zero. MATLAB's `roots` function for computing all zeros of a polynomial produces for this polynomial the approximation 16.00304396718421 in place of 16, with a relative error of 0.00019; and if we multiply the coefficient of x^{15} by $1 + \varepsilon$, `roots` finds the approximation 15.96922528457543 with a relative error of 0.0019. The slightly higher accuracy obtained is due to our worst-case analysis.

This example is quite typical for the sensitivity of zeros of high degree polynomials in standard form, and explains why, for computing zeros, one should avoid the transformation to the standard form when a polynomial is given in a less explicit form. In particular, this applies to the characteristic polynomial of a matrix whose zeros are the eigenvalues, where, as we have seen in Example 5.3.1, much better accuracy is achievable by using matrix factorizations.

Deflation

To compute several (or all) of the zeros of a function f, the standard technique is called *deflation* and proceeds in analogy to polynomial division, as follows.

If one already knows some zeros $x_1^*, \ldots, x_s^* (s \geq 1)$ with corresponding multiplicities $m_1, \ldots, m_s$, then one may find other zeros by setting

$$g(x) := \frac{f(x)}{\prod\limits_{j=1:s} (x - x_j^*)^{m_j}} \tag{5.2}$$

and seeking solutions of $g(x) = 0$. By definition of a multiple zero, $g(x)$ converges to a number $\neq 0$ as $x \to x_j^*$. Therefore, the solutions $x_1^*, \ldots, x_s^*$ are "divided out" in this way, and one cannot converge to the same zero again.

Although the numerically calculated value of x^* is in general subject to error, the so-called *implicit deflation*, which evaluates $g(x)$ directly by formula (5.2), is harmless from a stability point of view because even when the x_j^* are inaccurate, the other zeros of f are still exactly zeros of g. The only possible problems arise in a tiny neighborhood about the true and the calculated x_j^*, and because there is no further zero, this is usually irrelevant.

Note that when function evaluation is expensive, old function values not very close to a deflated zero can be used again to check for sign changes of the deflated function!

Warning. For polynomials in the power form, it seems natural to perform the division explicitly by each linear factor $x - x^*$ using the Horner scheme, because this again produces a polynomial g of lower degree instead of the expression $f(x)/(x - x^*)$. Unfortunately, this *explicit deflation* may lead to completely incorrect results even for exactly calculated zeros. The reason is that for polynomials with zeros that are stable under small relative changes in the coefficients but very sensitive to small absolute changes, already the deflation of the absolutely largest zero ruins the quality of the other zeros. We show this only by an example; for a more complete analysis of errors in deflation, see Wilkinson [98].

5.5.2 Example. Unlike in the previous example, small relative errors in the coefficients of the polynomial

$$f(x) := (x - 1)(x - 2^{-1}) \cdots (x - 2^{-19})$$
$$= x^{20} - (2 - 2^{-19})x^{19} + \cdots + 2^{-190}$$

cause only small perturbations in the zeros; thus all zeros are well conditioned. Indeed, the MATLAB polynomial root finder `roots` gives the zeros with a maximal absolute error of about $6 \cdot 10^{-15}$. However, explicit deflation of the first zero $x^* = 1$ gives a polynomial $\tilde{g}(x)$ with coefficients differing from those of $g(x) := (x - 2^{-1}) \cdots (x - 2^{-19})$ by $O(\varepsilon)$. The tiny constant term is completely altered, and the computed zeros of $\tilde{g}(x)$ are found to be 0.49999999999390 instead of 0.5, 0.25000106540985 instead of 0.25, and the next largest roots are already complex, $0.13465904883151 \pm 0.01783932277158i$.

5.5.3 Remark. For finding all (real or complex) zeros of *polynomials*, the fastest and most reliable method is perhaps that of Jenkins and Traub [47,48]; it finds all zeros of a polynomial of degree n in $O(n^2)$ operations by reformulating it as an eigenvalue problem. The MATLAB 5 version of `roots` also proceeds that way, but it takes no advantage of the sparsity in the resulting matrix, and hence requires $O(n^3)$ operations.

The Interval Newton Method

For rigorous bounds on the zeros of a function, interval arithmetic may be used. We suppose that the function f is given by an arithmetical expression, that f is twice continuously differentiable in the interval $\mathbf{x} \in \mathbb{IR}$, and that $0 \notin f'(\mathbf{x})$. The latter (holds in reasonably large neighborhoods of simple zeros, and) implies that f is monotone in $\mathbf{x}$ and has there at most one zero x^*.

Rigorous existence and nonexistence tests may be based on sign changes:

$$
\begin{array}{lll}
0 \notin f(\mathbf{x}) & \Rightarrow & \text{there is no zero in } \mathbf{x}. \\
0 \notin f'(\mathbf{x}), \quad f(\underline{\mathbf{x}})f(\bar{\mathbf{x}}) > 0 & \Rightarrow & \text{there is no zero in } \mathbf{x}. \\
0 \notin f'(\mathbf{x}), \quad f(\underline{\mathbf{x}})f(\bar{\mathbf{x}}) \leq 0 & \Rightarrow & \text{there is a unique zero in } \mathbf{x}.
\end{array}
$$

Note that f must be evaluated at the thin intervals $\underline{\mathbf{x}} = [\underline{x}, \underline{x}]$ and $\bar{\mathbf{x}} = [\bar{x}, \bar{x}]$ in order that rounding errors in the evaluation of f are correctly accounted for.

These tests may be used to find *all* real zeros in an interval $\mathbf{x}$ by splitting the interval recursively until one of the conditions mentioned is satisfied. After sufficiently many subdivisions, the only unverified parts are near multiple zeros.

Once the existence test applies, one can refine the box containing the solution as follows. By the mean value theorem,

$$f(\tilde{x}) = f(\tilde{x}) - f(x^*) = f'(\xi)(\tilde{x} - x^*)$$

for $\xi \in \square\{\tilde{x}, x^*\}$. If $x^* \in \mathbf{x}$ then

$$x^* = \tilde{x} - \frac{f(\tilde{x})}{f'(\xi)} \in \tilde{x} - \frac{f(\tilde{x})}{f'(\mathbf{x})}. \tag{5.3}$$

with $\tilde{x} \in \mathbf{x}$. Therefore, the intersection

$$\mathbf{x} \cap \left(\tilde{x} - \frac{f(\tilde{x})}{f'(\mathbf{x})} \right)$$

also contains the zero. Iteration with the choice $\tilde{x} := \check{x}_i = \operatorname{mid} \mathbf{x}_i$ leads to the so-called *interval Newton method*

$$\mathbf{x}_1 = \mathbf{x},$$

$$\mathbf{x}_{i+1} = \mathbf{x}_i \cap \left(\check{\mathbf{x}}_i - \frac{f(\check{\mathbf{x}}_i)}{f'(\mathbf{x}_i)} \right), \quad i = 1, 2, 3, \ldots \tag{5.4}$$

$\check{\mathbf{x}}_i$ is deliberately written in boldface to emphasize that again, in order to correctly account for rounding errors, $\check{\mathbf{x}}_i$ must be a thin interval containing an approximation to the midpoint of $\mathbf{x}_i$. A useful stopping criterion is $\mathbf{x}_{i+1} = \mathbf{x}_i$; because $\mathbf{x}_{i+1} \subseteq \mathbf{x}_i$ and finite precision arithmetic, this holds after finitely many steps.

Because of (5.3), a zero x^* cannot be "lost" by the iteration (5.4); that is, $x^* \in \mathbf{x}$ implies that

$$x^* \in \mathbf{x}_i \quad \text{for all } i.$$

However, it may happen that the iteration (5.4) stops with an empty intersection $\mathbf{x}_i = \emptyset$; then, of course, there was no zero in $\mathbf{x}$.

As shown in Figure 5.5, the radius $\operatorname{rad} \mathbf{x}_i$ is at least halved in each iteration due to the choice $\tilde{x} = \check{x}_i$. Hence, if the iteration does not stop, then $\operatorname{rad} \mathbf{x}_i$ converges to zero, and the sequence $\mathbf{x}_i$ converges (in exact arithmetic) to a thin interval $\mathbf{x}_\infty = x_\infty$. Taking the limit in the iteration formula, we find

$$x_\infty = x_\infty - \frac{f(x_\infty)}{f'(x_\infty)},$$

so that $f(x_\infty) = 0$. Therefore, the limit $x_\infty = x^*$ is the unique zero of f in $\mathbf{x}$.

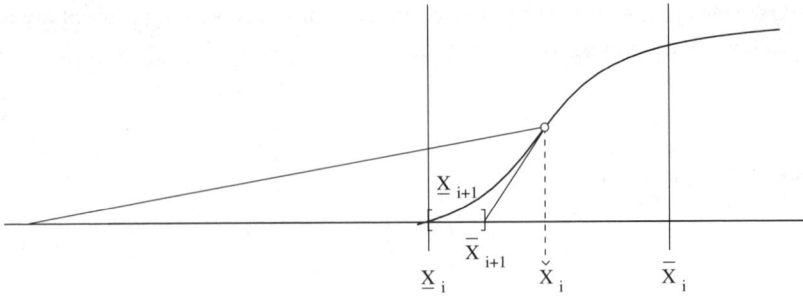

Figure 5.5. The interval Newton method.

5.5.4 Theorem. *If* $0 \notin f'(\mathbf{x})$, *then:*

(i) *The iteration (5.4) stops after finitely many steps with empty* $\mathbf{x}_i = \emptyset$ *if and only if f has no zero in* $\mathbf{x}$.

(ii) *The function f has a (unique) zero* x^* *in* $\mathbf{x}$ *if and only if* $\lim_{i \to \infty} \mathbf{x}_i = x^*$. *In this case,*

$$\operatorname{rad} \mathbf{x}_{i+1} \le \frac{1}{2} \operatorname{rad} \mathbf{x}_i,$$

and

$$\operatorname{rad} \mathbf{x}_{i+1} = O((\operatorname{rad} \mathbf{x}_i)^2). \tag{5.5}$$

In particular, the radii converge quadratically to zero.

Proof. Only the quadratic convergence (5.5) remains to be proved. By the mean value theorem, $f(\check{x}_i) = f'(\xi)(\check{x}_i - x^*)$ with $\xi \in \Box\{\check{x}_i, x^*\}$. Therefore

$$\mathbf{x}_{i+1} = \check{x}_i - (\check{x}_i - x^*) \frac{f'(\xi)}{f'(\mathbf{x}_i)},$$

and

$$\operatorname{rad} \mathbf{x}_{i+1} = |\check{x}_i - x^*| \, |f'(\xi)| \operatorname{rad} \left(\frac{1}{f'(\mathbf{x}_i)} \right). \tag{5.6}$$

Now $|\check{x}_i - x^*| \le \operatorname{rad} \mathbf{x}_i$, f' is bounded on $\mathbf{x}$, and by Theorem 1.5.6(iii),

$$\operatorname{rad} \left(\frac{1}{f'(\mathbf{x}_i)} \right) = O(\operatorname{rad} \mathbf{x}_i).$$

Insertion into (5.6) gives (5.5). $\qquad \Box$

5.5.5 Example. We determine the zero $x^* = \sqrt{2} = 1.41421356\ldots$ of

$$f(x) := 1 - 3/(x^2 + 1)$$

with the interval Newton method. We have $f'(x) = 6x/(x^2+1)^2$. Starting with $\mathbf{x}_1 := [1, 3]$, the first iteration gives

$$f(\check{x}_1) = f([2, 2]) = 1 - \frac{3}{[2, 2]^2 + 1} = 1 - \frac{3}{[5, 5]} = \left[\frac{2}{5}, \frac{2}{5}\right],$$

$$f'(\mathbf{x}_1) = \frac{6 \cdot [1, 3]}{[2, 10]^2} = \frac{[6, 18]}{[4, 100]} = \left[\frac{3}{50}, \frac{9}{2}\right],$$

$$\mathbf{x}_2 = [1, 3] \cap \left([2, 2] - \left[\frac{2}{5}, \frac{2}{5}\right] \Big/ \left[\frac{3}{50}, \frac{9}{2}\right]\right)$$

$$= [1, 3] \cap \left[-\frac{14}{3}, \frac{86}{45}\right] = [1, 1.91111\ldots].$$

Further iteration (with eight-digit precision) gives the results in Table 5.8, with termination because $\mathbf{x}_7 = \mathbf{x}_6$. An optimal inclusion of $\sqrt{2}$ with respect to a machine precision of eight decimal places is attained.

Because interval arithmetic is generally slower than real arithmetic, the splitting process mentioned previously can be speeded up by first locating as many approximate zeros as one can find by the secant bisection method, say. Then one picks narrow intervals containing the approximate zeros, but wide enough that the sign is detectable without ambiguity by the interval evaluation $\mathbf{f}$ at each end point; if the interval evaluation at some end point contains zero, the sign is undetermined, and one must widen the interval adaptively by moving the

Table 5.8. *Interval Newton method for*
$f(x) = 1 - 3/(x^2 + 1)$ *in* $\mathbf{x}_1 = [1, 3]$

i	$\underline{x}_i$	$\bar{x}_i$
1	1.0000000	3.0000000
2	1.0000000	1.9111112
3	1.3183203	1.4422849
4	1.4085591	1.4194270
5	1.4142104	1.4142167
6	1.4142135	1.4142136
7	1.4142135	1.4142136

corresponding end point. The verification procedure can then be restricted to the complements of the already verified intervals.

Error Bounds for Simple Eigenvalues and Associated Eigenvectors

Due to the special structure of eigenvalue problems, it is frequently possible to improve on the error bounds for general functions by using more linear algebra. Here, we consider only the case of simple real eigenvalues of a parameter matrix $G(\lambda)$.

If λ^* is a simple eigenvalue, then we can find a matrix C and vectors $a, b \neq 0$ such that the parameter matrix

$$G_0(\lambda) := CG(\lambda) + ba^H$$

is nonsingular in a neighborhood of λ^*. The following result permits the treatment of eigenvalue problems as the problem of determining a zero of a continuous function.

5.5.6 Proposition. *Let* $G_0(\lambda) := CG(\lambda) + ba^H$ *where C is nonsingular. If* λ^* *is a zero of*

$$f(\lambda) := a^H G_0(\lambda)^{-1} b - 1$$

and $x := G_0(\lambda^*)^{-1} b$, *then* λ^* *is an eigenvalue of* $G(\lambda)$, *and x is an associated eigenvector, that is,* $G(\lambda^*)x = 0$.

Proof. If $f(\lambda^*) = 0$ and $x = G_0(\lambda^*)^{-1}b$, then $a^H x = f(\lambda^*)+1 = 1$. Therefore, we have $CG(\lambda^*)x = G_0(\lambda^*)x - ba^H x = b - b = 0$; because C is nonsingular, it follows that $G(\lambda^*)x = 0$. □

To apply this, we first calculate an approximation $\tilde{s}$ to the unknown (simple) eigenvalue λ^*. Then, to find suitable values for a, b and C, we modify a normalized triangular factorization LR of $G(\tilde{s})$ by replacing the diagonal element R_{ii} of R of least modulus with $\|R\|_\infty$. With the upper triangular matrix R' so obtained, we have $R = R' - \gamma e^{(i)}(e^{(i)})^H$ where $\gamma = R'_{ii} - R_{ii}$, so that

$$(LR')^{-1}G(\tilde{s}) = (LR')^{-1}LR = (R')^{-1}R = I - \gamma(R')^{-1}e^{(i)}(e^{(i)})^H.$$

Thus if we put

$$a = e^{(i)}, \quad b \approx \gamma(R')^{-1}e^{(i)}, \quad C \approx (LR')^{-1},$$

then $G_0(\tilde{s}) \approx I$.

If the components of $G(\lambda)$ are given by arithmetical expressions and $\mathbf{s}$ is a small interval containing $\tilde{s}$, then the interval evaluation $G_0(\mathbf{s})$ contains the matrix $G_0(\tilde{s}) \approx I$. One can therefore expect that $G_0(\mathbf{s})$ is diagonally dominant, or is at least an H-matrix; this can easily be checked in each case. Then for $t \in \mathbf{s}$, $G_0(t) \in G_0(\mathbf{s})$ is nonsingular and f is continuous in $\mathbf{s}$. Therefore, any sign change in $\mathbf{s}$ encloses a zero of f and hence an eigenvalue of $G(\lambda)$. Note that to account for rounding errors, the evaluation of $f(\lambda)$ for finding its sign must be done with a thin interval $[\lambda, \lambda]$ in place of λ.

Once a sign change is verified, we may reduce the interval $\mathbf{s}$ to that defined by the bracket obtained, and obtain the corresponding eigenvector by solving the system of linear interval equations

$$\tilde{B}x = b \quad \text{with } \tilde{B} \in \mathbf{B} = G_0(\mathbf{s})$$

by Krawczyk's method with the matrix $(\text{Diag}\,\underline{B})^{-1}$ as preconditioner.

If during the calculation no sign change is found in $\mathbf{s}$, or if $G_0(\mathbf{s})$ is not diagonally dominant or is not at least an H-matrix, this is an indication that $\tilde{s}$ was a poor eigenvalue approximation, $\mathbf{s}$ was chosen too narrow, or there is a multiple eigenvalue (or several close eigenvalues) near $\tilde{s}$.

5.6 Complex Zeros

In this section, we consider the problem of finding complex zeros $x^* \in D$ of a function f that is analytic in an open and bounded set $D \subseteq \mathbb{C}$ and continuous in its closure $\bar{D}$. In contrast to the situation for $D \subset \mathbb{R}$, the number of zeros (counting their multiplicity) is no longer affected by small perturbations in f, which makes the determination of multiple zeros a much better behaved problem. In particular, there are simple, globally convergent algorithms based on a modified form of damping. The facts underlying such an algorithm are given in the following theorem.

5.6.1 Theorem. *Let $x_0 \in D$ and $|f(x)| > |f(x_0)|$ for all $x \in \partial D$. Then f has a zero in D. Moreover, if $f(x_0) \neq 0$, then every neighborhood of x_0 contains a point x_1 with $|f(x_1)| < |f(x_0)|$.*

Proof. Without loss of generality, we may assume that D is connected. For all α with $x_0 + \alpha$ in a suitable neighborhood of x_0, the power series

$$f(x_0 + \alpha) = f(x_0) + f'(x_0)\alpha + \cdots + \frac{f^{(n)}(x_0)}{n!}\alpha^n + \cdots$$

is convergent.

If all derivatives are zero, then $f(x) = f(x_0)$ in a neighborhood of x_0, so f is a constant in D. This remains valid in $\bar{D}$, contradicting the assumption. Therefore, there is a smallest n with $f^{(n)}(x_0) \neq 0$, and we have

$$f(x_0 + \alpha) = f(x_0) + \alpha^n g_n(\alpha) \tag{6.1}$$

with

$$g_n(\alpha) = \frac{f^{(n)}(x_0)}{n!} + \alpha \frac{f^{(n+1)}(x_0)}{(n+1)!} + \cdots.$$

In particular, $g_n(0) \neq 0$. Taking the square of the absolute value in (6.1) gives

$$\begin{aligned}
|f(x_0 + \alpha)|^2 &= |f(x_0)|^2 + 2\operatorname{Re} \alpha^n g_n(\alpha)\overline{f(x_0)} + |\alpha|^{2n}|g_n(\alpha)|^2 \\
&= |f(x_0)|^2 + 2\operatorname{Re} \alpha^n g_n(0)\overline{f(x_0)} + O(|\alpha|^{2n}) \\
&< |f(x_0)|^2
\end{aligned}$$

if α is small enough and

$$\operatorname{Re} \alpha^n g_n(0)\overline{f(x_0)} < 0. \tag{6.2}$$

To satisfy (6.2), we choose a small $\varepsilon > 0$ and find that

$$\alpha^n g_n(0)\overline{f(x_0)} = -\varepsilon$$

for the choice

$$\alpha = \alpha_\varepsilon = \left(\frac{-\varepsilon}{g_n(0)\overline{f(x_0)}} \right)^{1/n}, \tag{6.3}$$

valid unless $f(x_0) = 0$. Because $\alpha_\varepsilon \to 0$ if $\varepsilon \to 0$, we have $|f(x_0 + \alpha_\varepsilon)| < |f(x_0)|$ for sufficiently small ε. This proves the second part.

Now $|f|$ attains its minimum on the compact set $\bar{D}$, that is, there is some $x^* \in \bar{D}$ such that

$$|f(x^*)| \leq |f(x)| \quad \text{for all } x \in \bar{D}, \tag{6.4}$$

and the boundary condition implies that actually $x^* \in D$. If $f(x^*) \neq 0$, we may use the previous argument with x^* in place of x_0, and find a contradiction to (6.4). Therefore, $f(x^*) = 0$. $\qquad\square$

Spiral Search

Using (6.3), we could compute a suitable α if we knew the nth derivative of f. However, the computation of even the first derivative is usually unnecessarily expensive, and we instead look for a method that works without derivatives. The key is the observation that when $\alpha \neq 0$ makes a full revolution around zero, the left side of (6.2) alternates in sign in $2n$ adjacent sectors of angle π/n. Therefore, if we use a trial correction α and are in the wrong sector, we may correct for it by decreasing α along a spiral toward zero.

A natural choice, originally proposed by Bauhuber [7], is to try the reduced corrections $q^k \alpha$ ($k = 0, 1, \ldots$), where q is a complex number of absolute value $|q| < 1$. The angle of q in polar coordinates must be chosen such that repeated rotation around this angle guarantees that from arbitrary starting points, we land soon in a sector with the correct sign, at least in the most frequent case that n is small. The condition (6.2) that tells us whether we are in a good sector reduces to $\operatorname{Re} q^{kn} \gamma < 0$ with a constant γ that depends on the problem. This is equivalent to

$$kn \arg(q) + \varphi \in \,]2l\pi, (2l + 1)\pi[\quad \text{for some } l \in \mathbb{Z}, \qquad (6.5)$$

where $\varphi = \arg(\gamma) - \pi/2$. If we allow for $n = 1, 2, 3, 4$ at most $2, 2, 3$, resp. 5 consecutive choices in a bad sector, independent of the choice of φ, this restricts the angle to a narrow range, $\arg(q) \in \pm\,]\frac{11}{20}\pi, \frac{5}{9}\pi[$ (cf. Exercise 22). The simplest complex number with an angle in this range is $6i - 1$; therefore, a value

$$q = \lambda(6i - 1) \quad \text{with } \lambda \in [0.05, 0.15],$$

say, may be used. This rotates α in each trial by a good angle and shrinks it by a factor of about $0.3041 - 0.9124$, depending on the choice of λ. Figure 5.6 displays a particular instance of the rotation and shrinking pattern for $\lambda = 0.13$. Suitable starting corrections α can be found by a secant step (1.1), a hyperbolic interpolation step (4.5), or a Muller step (4.6). The Muller step is most useful when there are very close zeros, but a complex square root must be computed.

Rounding errors may cause trouble when f is extremely flat near some point, which typically happens when the above n gets large. (It is easy to construct artificial examples for this.) If the initial correction is within the flat part, the new function value may agree with the old one within their accuracy, and the spiral search never leaves the flat region. The remedy is to expand the

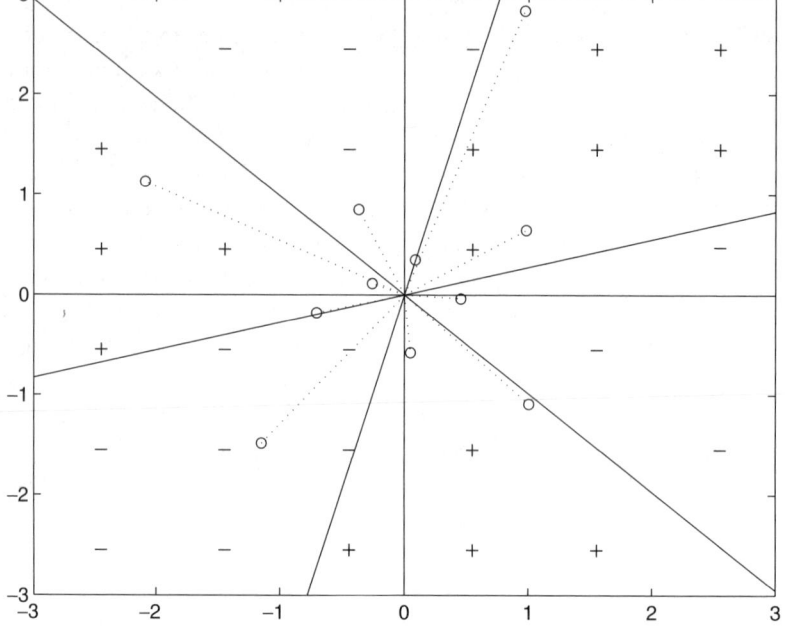

Figure 5.6. Spiral search for a good sign in case $n = 3$. The spiraling factor is $q = 0.13(6i - 1)$. The circles mark the initial α and corresponding rotated and shrinked values. The third trial value gives the required sign; by design, this is the worst possible case for $n = 3$.

size of the correction when the new function value is too close initially to the old one.

Another safeguard is needed to avoid initial step sizes that are too large. One possiblility is to enforce

$$|\alpha| \leq (1 + |x| + |x_{old}|)^p.$$

For $p = 2$, this still allows "local" quadratic convergence to infinite zeros (of rational functions, say). If this is undesirable, one may take $p = 1$.

We formulate the resulting *spiral search*, using a particular choice of constants that specify the conditions when to accept a new point, when to spiral, when to expand and when to stop. ε is the machine precision; *kmax* and p are control parameters that may be set to fixed values. x_0 and an initial α must be provided as input. (In the absence of other information, one may start, e.g., with $x_0 = 0, \alpha = 1$.)

5.6.2 Algorithm: Spiral Search for Complex Zeros

$x = x_0; \ f = |f(x_0)|; \ f_{ok} = \varepsilon * f;$
$q = 0.13 * (6i - 1); \ ql1 = 1 - \sqrt{\varepsilon}; \ qg1 = 1 + \sqrt{\varepsilon};$
while $f > f_{ok}$,
    compute a nonzero correction α to x (e.g. a secant or Muller step);
    if necessary, rescale to enforce $|\alpha| \le (1 + |x| + |x_{old}|)^p$;
    $flat = 1$;
    for $k = 1 : kmax$,
        $f_{new} = |f(x - \alpha)|$;
        if $f_{new} < ql1 * f$,   % good step
            $x = x - \alpha; \ f = f_{new}$; break;
        end;
        if $flat$ & $f_{new} < qg1 * f$,   % flat step
            $\alpha = 10 * \alpha$;
        else   % long step
            $\alpha = q * \alpha; \ flat = 0$;
        end;
        if $k == kmax$, return; end;
    end;
end;

The algorithm stops when $f \le f_{ok}$ or when *kmax* changes of α did not result in a sufficient decrease.

Rigorous Error Bounds

For constructing rigorous error bounds for complex zeros, we use the following tool from complex analysis.

5.6.3 Theorem. *Let f and g be analytic in the interior of a disk D and nonzero and continuous on the boundary ∂D. If*

$$\operatorname{Re} \frac{f(z)}{g(z)} > 0 \quad \text{for all } z \in \partial D \tag{6.6}$$

and each root is counted according to its multiplicity, then f and g have precisely the same number of zeros in D.

Proof. This is a refinement of *Rouché's theorem*, which assumes the stronger condition

$$|f(z)| > |f(z) - g(z)| \quad \text{for all } z \in \partial D.$$

However, the proof of Rouché's theorem in Henrici [43, Theorem 4.10b] uses in fact only the weaker assumption (6.6). □

A simple consequence is the following result (slightly sharper than Neumaier [69]).

5.6.4 Theorem. *Let f be analytic in the disk $D[\tilde{z}; r] = \{z \in \mathbb{C} \mid |z - \tilde{z}| \le r\}$, and let $0 < \delta < r$. If*

$$\text{Re}\, \frac{f^{(k)}(z)}{f^{(k)}(\tilde{z})} > \sum_{l=0:k-1} \frac{k!}{l!} \left| \text{Re}\, \frac{f^{(l)}(\tilde{z})}{f^{(k)}(\tilde{z})} \right| \delta^{l-k} \quad \text{for all } z \in D[\tilde{z}; \delta], \quad (6.7)$$

then f has precisely k roots in $D[\tilde{z}; \delta]$, where each root is counted according to its multiplicity.

Proof. We apply Theorem 5.6.3 with

$$g(z) = \frac{f^{(k)}(\tilde{z})}{k!} (z - \tilde{z})^k$$

to $D = D[\tilde{z}; \delta]$. Hermite interpolation at $\tilde{z}$ gives

$$f(z) = p_{k-1}(z) + f[\tilde{z}, \dots, \tilde{z}, z](z - \tilde{z})^k$$

with

$$p_{k-1}(z) = \sum_{l<k} \frac{f^{(l)}(\tilde{z})}{l!} (z - \tilde{z})^l, \quad f[\tilde{z}, \dots, \tilde{z}, z] = \frac{f^{(k)}(\zeta)}{k!}$$

for some ζ in D. For $z \in \partial D$, we have $|z - \tilde{z}| = \delta$, hence

$$\text{Re}\, \frac{f(z)}{g(z)} = \text{Re}\, \frac{f^{(k)}(\zeta)}{f^{(k)}(\tilde{z})} + \sum_{l<k} \frac{k!}{l!} \text{Re}\, \frac{f^{(l)}(\tilde{z})}{f^{(k)}(\tilde{z})} (z - \tilde{z})^{l-k}$$

$$\ge \text{Re}\, \frac{f^{(k)}(\zeta)}{f^{(k)}(\tilde{z})} - \sum_{l<k} \frac{k!}{l!} \left| \text{Re}\, \frac{f^{(l)}(\tilde{z})}{f^{(k)}(\tilde{z})} \right| \delta^{l-k} > 0$$

by our hypothesis (6.7), and Theorem 5.6.3 applies. □

For polynomials, one may use in place of the kth derivative the divided difference used in the proof, computable by means of the first k steps of the complete Horner scheme. With this change, δ can usually be chosen somewhat smaller.

The theorem may be used as an *a posteriori* test for existence, starting with an approximate root $\tilde{z}$ computed by standard numerical methods. It provides a rigorous existence test, multiplicity count, and enclosure for the root or the root cluster near $\tilde{z}$. The successful application requires that we "guess" the right k and a suitable δ. Because δ should be kept small, a reasonable procedure is the following: Let δ_k be the positive real root of the polynomial

$$q_k(\delta) = \delta^k - \sum_{l<k} \frac{k!}{l!} \left| \mathrm{Re}\, \frac{f^{(l)}(\tilde{z})}{f^{(k)}(\tilde{z})} \right| \delta^l;$$

then (6.7) forces $\delta \geq \delta_k$. If δ_k is small, then $f^{(k)}(z) = f^{(k)}(\tilde{z}) + O(\delta)$ so that it is sufficient to take δ only slightly bigger than δ_k. In practice, it is usually sufficient to calculate δ_k to a relative precision of 10% only and choose $\delta = 2\delta_k$; because k is unknown, one tries $k = 1, 2, \ldots$, until one succeeds or a limit on k is attained.

If there is a k-fold zero z^* and the remaining roots are far away, then Taylor's formula gives

$$\frac{f^{(l)}(\tilde{z})}{l!} \approx \binom{k}{l} \frac{f^{(k)}(\tilde{z})}{k!} (\tilde{z} - z^*)^{k-l}.$$

Thus

$$q_k(\delta) \approx \delta^k - ((\delta + |\tilde{z} - z^*|)^k - \delta^k),$$

and

$$\delta_k \approx |\tilde{z} - z^*|/(\sqrt[k]{2} - 1) < 1.5k|\tilde{z} - z^*|,$$

so that we get with our choice $\delta = 2\delta_k$ an overestimation of roughly $3k$.

5.6.5 Example. Consider $f(x) = x^4 - 2x^3 - x^2 + 2x + 1$ that has a double root $x^* = \frac{1}{2}(1 + \sqrt{5}) \approx 1.618033989$. To verify the double root, we must choose $k = 2$. In this case, a suitable guess for the radius δ is

$$\delta = 2\delta_2 = p + \sqrt{p^2 + 4q},$$

where

$$p = 2\left| \mathrm{Re}\, \frac{f'(\tilde{z})}{f''(\tilde{z})} \right|, \quad q = 2\left| \mathrm{Re}\, \frac{f(\tilde{z})}{f''(\tilde{z})} \right|.$$

Table 5.9. *Approximations $\tilde{z}$ to the root and corresponding enclosures*

$\tilde{z}$	δ (upward rounded)	$\delta/|x^* - \tilde{z}|$
1.5	$6.77 \cdot 10^{-1}$	Enclosure not guaranteed
1.6	$8.91 \cdot 10^{-2}$	4.94
1.61	$3.92 \cdot 10^{-2}$	4.88
1.618	$1.65 \cdot 10^{-4}$	4.83
1.618034	$5.44 \cdot 10^{-8}$	4.83
1.62	$9.48 \cdot 10^{-3}$	4.82
1.65	$1.49 \cdot 10^{-1}$	4.66
1.7	$3.69 \cdot 10^{-1}$	Enclosure not guaranteed

The condition guaranteeing two roots in the disk $D[\tilde{z}; \delta]$ is

$$\inf \left\{ \operatorname{Re} \frac{f''(z)}{f''(\tilde{z})} \,\bigg|\, z \in D[\tilde{z}; \delta] \right\} > (p + q/\delta)/\delta,$$

and this can be verified by means of complex interval arithmetic. For various approximations $\tilde{z}$ to the root, we find the enclosures $|x^* - \tilde{z}| \leq \delta$ shown in Table 5.9. Of course, the test does not guarantee the existence of a double root, but only that of two (possibly coinciding) roots x^* with $|x^* - \tilde{z}| \leq \delta$.

Error Bounds for Polynomial Zeros

The following result, valid for arbitrary $\tilde{x}$, can be used to verify the accuracy of approximations $\tilde{x}$ to simple, real, or complex zeros of polynomials. If rigorous results are required, the error term must be evaluated in interval arithmetic, using as argument the thin interval $[\tilde{x}, \tilde{x}]$.

5.6.6 Theorem. *If f is a polynomial of degree n and $f'(\tilde{x}) \neq 0$, $\tilde{x} \in \mathbb{C}$, then there is at least one zero of f in each disk in the complex plane that contains $\tilde{x}$ and $\tilde{x} - n\frac{f(\tilde{x})}{f'(\tilde{x})}$. In particular, there is a zero x^* with*

$$|\tilde{x} - x^*| \leq n \left| \frac{f(\tilde{x})}{f'(\tilde{x})} \right|. \tag{6.8}$$

The bound is best possible as $f(x) = (x - x^*)^n$ shows.

Proof. For $f(x) := a_0(x - \xi_1) \cdots (x - \xi_n)$,

$$\frac{f'(x)}{f(x)} = \frac{d}{dx} \log |f(x)| = \frac{1}{x - \xi_1} + \cdots + \frac{1}{x - \xi_n}. \qquad (6.9)$$

If x^* is the zero closest to $\tilde{x}$, then

$$|x^* - \tilde{x}| \le |\xi_j - \tilde{x}| \quad \text{for } j = 1, \ldots, n.$$

Therefore,

$$\left| \frac{f'(\tilde{x})}{f(\tilde{x})} \right| \le \frac{n}{|\tilde{x} - x^*|}$$

and

$$|\tilde{x} - x^*| \le n \left| \frac{f(\tilde{x})}{f'(\tilde{x})} \right|.$$

This proves (6.8). A slight generalization of the argument gives the more general assertion (see Exercise 21). □

Much more information about complex zeros can be found in Henrici [43].

5.7 Methods Using Derivative Information

Newton's Method

As $x_{i-1} \to x_i$, the secant slope $f[x_i, x_{i-1}]$ approaches the slope $f'(x_i)$ of the tangent to f at the point x_i. In this limiting case, formula (1.1) yields the formula for *Newton's method*

$$x_{i+1} := x_i - \frac{f(x_i)}{f'(x_i)}, \quad i = 1, 2, 3, \ldots. \qquad (7.1)$$

5.7.1 Example. To compare with the bisection method and with the secant method, the zero $x^* = \sqrt{2} = 1.414215362\ldots$ of the function $f(x) = x^2 - 2$ is approximated by Newton's method. Starting with $x_1 := 1$, we get the results in Table 5.10. After five function evaluations and five derivative evaluations, one has 10 valid digits of x^*.

In the example, the Newton sequence converges faster than the secant method. That this is typical is a consequence of the local Q-quadratic convergence of the Newton method.

Table 5.10. *Results of Newton's*
method for $x^2 - 2$

i	x_i
1	1
2	1.5
3	1.416666667
4	1.414215686
5	1.414213562
6	1.414213562

5.7.2 Theorem. *Let the function f be twice continuously differentiable in a neighborhood of the simple zero x^*, and let $c := \frac{1}{2} f''(x^*)/f'(x^*)$. Then the sequence defined by (7.1) converges to x^* for all x_1 sufficiently close to x^* and*

$$x_{i+1} - x^* = c_i'(x_i - x^*)^2 \qquad (7.2)$$

with

$$\lim_{i \to \infty} c_i' = c.$$

In particular, Newton's method is Q-quadratically convergent to a simple zero, and its convergence order is 2.

Proof. Formula (7.2) is proved just as the corresponding assertion for the secant method. Convergence for initial values sufficiently close to x^* again follows from this. With $c_0 := \sup |c_i'| < \infty$, one obtains from (7.2) the relation

$$|x_{i+1} - x^*| \le c_0 |x_i - x^*|^2$$

from which the Q-quadratic convergence is apparent. By the results in Section 5.4, the convergence order is 2. □

Comparison with the Secant Method

By comparing Theorems 5.7.2 and 5.1.2, we see that locally, Newton's method converges in fewer iterations than the secant method. However, each step is more expensive. If the cost for a derivative evaluation is about the same as that for a function evaluation, it is more appropriate to compare one Newton step with two secant steps. By Theorem 5.1.2, we have for the latter

$$x_{i+2} - x^* = [c_{i+1} c_i (x_{i-1} - x^*)](x_i - x^*)^2,$$

Table 5.11. *A comparison of work versus order of accuracy*

Function evaluation	Newton method $\varepsilon_{i+1} \sim \varepsilon_i^2$	Secant method $\varepsilon_{i+1} \sim \varepsilon_i \varepsilon_{i-1}$
1	ε	ε
2	—	ε
3	ε^2	ε^2
2	—	ε^3
3	ε^4	ε^5
2	—	ε^8
3	ε^8	ε^{13}
2	—	ε^{21}
3	ε^{16}	ε^{34}
2	—	ε^{55}

and because the term in square brackets tends to zero as $i \to \infty$, two secant steps are locally faster than one Newton step. Therefore, the secant method is more efficient when function values and derivatives are equally costly. We illustrate the behaviour in Table 5.11, where the asymptotic order of accuracy after $n = 1, 2, \ldots$ function evaluations is displayed for both Newton's method and the secant method.

In some cases, derivatives are much cheaper than function values when computed together with the latter; in this case, Newton's method may be faster. We use the convergence orders 2 of Newton's method and $(1 + \sqrt{5})/2 \approx 1.618$ of the secant method to compare the asymptotic costs in this case.

Let c and c' denote the cost of calculating $f(x_i)$ and $f'(x_i)$, respectively. As soon as f is not as simple as in our demonstration examples, the cost for the other operations is negligible, so that the cost of calculating x_{i+1} $(i \geq 1)$ is essentially $c + c'$ for Newton's method and c for the secant method. The cost of s Newton steps, namely $s(c + c')$, is equivalent to that of $s(1 + c'/c)$, function evaluations, and the number of correct digits multiplies by a factor 2^s. With the same cost, $s(1 + c'/c)$ secant steps may be performed, giving a gain in the number of accurate digits by a factor $1.618^{s(1+c'/c)}$. Therefore, the secant method is locally more efficient than the Newton method if

$$1.618^{s(1+c'/c)} > 2^s,$$

that is, if

$$\frac{c'}{c} > \frac{\log 2}{\log 1.618} - 1 \approx 0.44.$$

Therefore, locally, Newton's method is preferable to the secant method only when the cost of calculating the derivative is at most 44% of the cost of a function evaluation.

Global Behavior of Newton's Method

As for the secant method, the global behavior of Newton's method must be assessed independent of the local convergence speed.

It can be shown that Newton's method converges for all starting points in some dense subset of $\mathbb{R}$ if f is a polynomial of degree n with real zeros $\xi_1, \ldots, \xi_n$ only. The argument is essentially that if not, it sooner or later generates some $x_i > x^* := \max(\xi_1, \ldots, \xi_n)$ (or $x_i < \min(\xi_1, \ldots, \xi_n)$, which is treated similarly). Then,

$$0 < \frac{1}{x_i - x^*} \leq \frac{f'(x_i)}{f(x_i)} \leq \frac{n}{x_i - x^*},$$

so that $x_{i+1} = x_i - f(x_i)/f'(x_i)$ satisfies the relation

$$x_i - x^* \geq x_i - x_{i+1} \geq \frac{x_i - x^*}{n},$$

that is,

$$0 \leq x_{i+1} - x^* \leq \left(1 - \frac{1}{n}\right)(x_i - x^*).$$

Thus Newton's method converges monotonically for all starting points outside the hull of the set of zeros, with global convergence factor of at least $1 - \frac{1}{n}$.

For large n, a sequence of n Newton steps therefore decreases $|x - x^*|$ by at least a factor $\left(1 - \frac{1}{n}\right)^n < \frac{1}{e} \approx 0.37$. For $f(x) = (x - x^*)^n$, where this bound is asymptotically achieved, convergence is very slow; the same holds initially for general polynomials if the starting point is so far away from all zeros that these "look like a single cluster."

5.7.3 Example. For $x_1 = 100$, Newton's method applied to $f(x) = x^2 - 2$ yields the results in Table 5.12. The sequence initially converges only linearly,

Table 5.12. *Results of Newton's method for* $x^2 - 2$ *with* $x_1 = 100$

i	1	2	3	4	5	6	7	$\ldots$
x_i	100	50.01	25.02	12.55	6.36	3.34	1.97	$\ldots$

Table 5.13. *Results of Newton's method for* $f(x) = 1 - 10x + 0.01e^x$
with $x_1 = 20$

i	x_i	i	x_i
1	20.0000000000000000	10	11.1088835456740740
2	19.0000389558837600	11	10.2610830282869120
3	18.0001392428132970	12	9.5930334887471229
4	17.0003965996736000	13	9.2146744950274755
5	16.0010546324046890	14	9.1119744961101219
6	15.0027299459509860	15	9.1056248937564668
7	14.0069725232866720	16	9.1056021207975100
8	13.0176392772467740	17	9.1056021205058109
9	12.0441649793488760	18	9.1056021205058109

with a convergence factor of $1 - \frac{1}{n} = \frac{1}{2}$. (After iteration 7, quadratic convergence sets in.)

Later in this section we discuss a modified Newton method that overcomes this slow convergence, at least for polynomials.

Slowness of a different kind is observed in the next example.

5.7.4 Example. The function

$$f(x) := 1 - 10x + 0.01e^x$$

has already been considered in Example 5.2.7 for the treatment of the secant bisection method. Table 5.13 shows the convergence behavior of the Newton method for $x_1 := 20$. It takes a long time for the locally quadratic convergence to be noticeable. For the attainment of full accuracy, 17 evaluations of f and f' are necessary. The secant bisection method with $x_1 = 5$, $x_2 = 20$ needs about the same number of function evaluations but no derivative evaluations to achieve the same accuracy.

On nonconvex problems, small perturbations of the starting point may strongly influence the global behavior of Newton's method, if some intermediate iterate gets close to a stationary point.

5.7.5 Example. We demonstrate this with the function

$$f(x) := 1 - 2/(x^2 + 1)$$

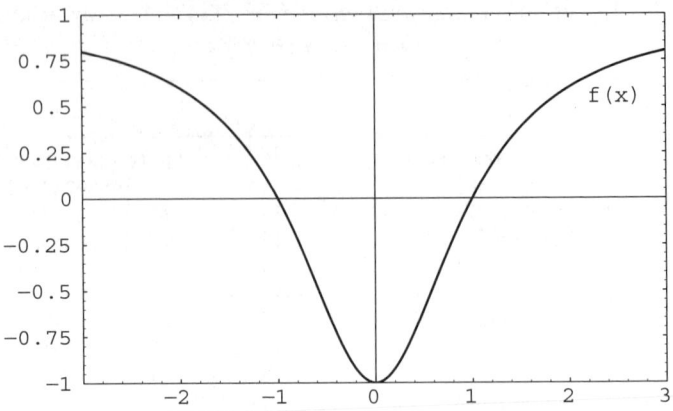

Figure 5.7. Graph of $f(x) = 1 - 2/(x^2 + 1)$.

displayed in Figure 5.7, which has zeros at $+1$ and -1. Table 5.14 shows that three close initial values may result in completely different behavior.

The next example shows that the neighborhood where Newton's method converges to a given zero may be very asymmetric, and very large in some direction.

5.7.6 Example. For the function

$$f(x) := x - 1 - 2/x$$

Table 5.14. *Newton's method for $f(x) = 1 - 2/(x^2 + 1)$*
for different starting points x_1

i	x_i	x_i	x_i
1	1.999500	1.999720	1.999970
2	0.126031	0.125577	0.125062
3	2.109171	2.115887	2.123583
4	−0.118015	−0.134153	−0.152822
5	−2.235974	−1.997092	−1.787816
6	0.446951	−0.130986	−0.499058
7	0.983975	−2.039028	−0.968928
8	0.999874	−0.042251	−0.999533
9	1.000000	−5.959273	−1.000000
10	1.000000	46.906603	−1.000000
11	1.000000	−25754.409557	−1.000000
		↓	
		±∞	

Table 5.15. *Newton's methods for* $f(x) = x - 1 - 2/x$ *with two different starting points* x_1

i	x_i	x_i
1	1000.000000000000000	0.0010000000000000
2	1.0039979920039741	0.0020004989997505
3	1.6702074291915645	0.0040029909876475
4	1.9772917776064771	0.0080139297363667
5	1.9999127426320478	0.0160594553138831
6	1.9999999987309516	0.0322437057559748
7	2.0000000000000000	0.0649734647479344
8		0.1317795480072139
9		0.2698985122099006
10		0.5559697617074131
11		1.0969550112472573
12		1.7454226483923749
13		1.9871575101344159
14		1.9999722751335729
15		1.9999999998718863
16		2.0000000000000000

with zeros -1 and $+2$ and a pole at $x = 0$, Newton's method converges for all values $x_1 \neq 0$. As Table 5.15 shows, the convergence is very slow for starting values $|x_1| \ll 1$ because for tiny x_i we have only $x_{i+1} \approx 2x_i$. Very large starting values, however, give a good approximation to a zero in a few Newton steps.

The Damped Newton Method

If the Newton method diverges, then one can often obtain convergence in spite of this through *damping*. The idea of damping is that, instead of a full *Newton step*,

$$p_i = -f(x_i)/f'(x_i),$$

only a partial step,

$$x_{i+1} := x_i + \alpha_i p_i,$$

is taken, in which the damping factor α_i is chosen by successive halving of an initial trial value $\alpha_i = 1$ until $|f(x_{i+1})| < |f(x_i)|$. This ensures that the values $|f(x_i)|$ are monotonically decreasing and converge (because they are bounded below by zero). Often, but of course not always, the limiting value is zero.

5.7.7 Example. The function

$$f(x) := 10x^5 - 36x^3 + 90x$$

has the derivative

$$f'(x) := 50x^4 - 108x^2 + 90 = 50(x^2 - 1.08)^2 + 31.68 > 0,$$

so that f is monotone in $\mathbb{R}$. The unique real zero of f is $x^* = 0$. Starting from $x_1 = 6$ (Table 5.16), Newton's method damped in this sense accepts always the first trial value $\alpha = 1$ for the damping factor, but the sequence of iterates alternates for $l \geq 8$, and has the limit points $+1$ and -1, and $|f(\pm 1)| = 64$, cf. Figure 5.8. The (mis-)convergence is unusually slow because the improvement factor $|f(x_{i+1})|/|f(x_i)|$ gets closer and closer to 1.

In order to ensure the convergence of the sequence $|f(x_i)|$ to zero, clearly one must require a little more than just $|f(x_{i+1})| < |f(x_i)|$. In practice, one demands that

$$|f(x_i + \alpha_i p_i)| < (1 - q\alpha_i)|f(x_i)| \tag{7.3}$$

for some fixed positive $q < 1$. (One usually takes a small but not tiny value, such as $q = 0.1$.) Because the full Newton step taken for $\alpha_i = 1$ gives local quadratic convergence, one tries in turn $\alpha_i = 1, \frac{1}{2}, \frac{1}{4}, \ldots$ until (7.3) is satisfied. For monotone functions, this indeed guarantees global convergence. (We analyze this in detail in the multidimensional case in Section 6.2.)

Because the step length is now no longer a sign of closeness to the zero, one stops the iteration when $|f(x_i)|$ no longer decreases significantly; for example,

Table 5.16.   *Simple damping does not help in all situations*

i	x_i	$f(x_i)$	$\ldots$	i	x_i	$f(x_i)$
1	6.000000	70524.000000		21	−1.003877	−64.123953
2	4.843907	23011.672462		22	1.003706	64.118499
3	3.926470	7506.913694		23	−1.003550	−64.113517
4	3.198309	2456.653982		24	1.003407	64.108946
5	2.615756	815.686552		25	−1.003276	−64.104738
6	2.133620	284.525312		26	1.003154	64.100850
7	1.685220	115.294770		27	−1.003041	−64.097246
8	1.067196	66.134592		28	1.002936	64.093896
9	−1.009015	−64.287891		29	−1.002839	−64.090774
10	1.008081	64.258095		30	1.002747	64.087856

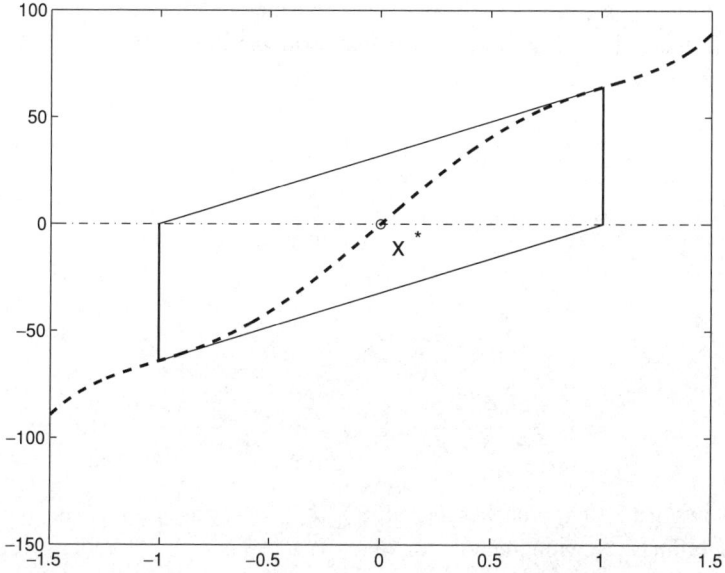

Figure 5.8. Oscillations of simply damped Newton iterates for $f(x) = 10x^5 - 36x^3 + 90x$.

when

$$|f(x_i)| \leq |f(x_{i+1})|(1 + \sqrt{\varepsilon})$$

with the machine precision ε.

A Modified Newton Method

We promised after Example 5.7.3 a modification of the Newton method that, for polynomials, also converges rapidly for large starting values. Let f be a polynomial of degree n; by the fundamental theorem of algebra, there are exactly n zeros $\xi_1, \ldots, \xi_n$ (some of which may coincide), and $f(x) = a_0(x - \xi_1) \cdots (x - \xi_n)$. For a simple zero x^* of f we have

$$\frac{f'(x)}{f(x)} = \frac{d}{dx} \log|f(x)| = \frac{1}{x - x^*} + \sum_{\xi_j \neq x^*} \frac{1}{x - \xi_j} \approx \frac{1}{x - x^*} + \frac{n - 1}{x - c},$$

where $c \in \mathbb{C} \backslash \{x^*\}$, whence

$$x^* \approx x_{i+1} := x_i - \frac{f(x_i)}{f'(x_i) - \frac{n-1}{x_i - c} f(x_i)}. \tag{7.4}$$

The formula (7.4) defines the modified Newton method. Because the additional term in the denominator of the correction term vanishes as $x_i \to x^*$, (7.4) is still locally Q-quadratically convergent. Because

$$x_{i+1} = x_i - \frac{1}{\frac{f'(x_i)}{f(x_i)} - \frac{n-1}{x_i-c}}$$

$$= x_i - \frac{1}{\sum \left(\frac{1}{x_i-\xi_j} - \frac{1}{x_i-c}\right) + \frac{1}{x_i-c}}$$

$$= x_i - \frac{x_i - c}{\sum \frac{\xi_j-c}{x_i-\xi_j} + 1}$$

$$= \left(\sum_{j=1:n} \frac{\xi_j - c}{x_i - \xi_j} x_i + c\right) \Big/ \left(\sum_{j=1:n} \frac{\xi_j - c}{x_i - \xi_j} + 1\right),$$

it follows that x_{i+1} approaches the value $\sum(\xi_j - c) + c$ as $x_i \to \infty$; in contrast to the ordinary Newton method, for which $x_{i+1} = (1 - \frac{1}{n})x_i + O(1) \to \infty$ as $x_i \to \infty$, the modified Newton method provides for very large starting values in only one step a reasonable guess of the magnitude of the zero (unless c is chosen badly). Moreover, the modified Newton method has nice monotonicity properties when the polynomial has only real zeros.

5.7.8 Theorem. *Let f be a real polynomial with real zeros only, and suppose that $x_1 > c$.*

(i) *If all zeros of f lie between c and x_1, then the modified Newton sequence converges and is monotonically decreasing.*
(ii) *If no zero lies between c and x_1, then the modified Newton sequence converges monotonically.*

Proof.

(i) Let x^* be the largest zero of f, and suppose that $x_i > x^*$ (this certainly holds for $i = 1$). Then we have

$$0 < \frac{1}{x_i - x^*} \leq \frac{f'(x_i)}{f(x_i)} - \frac{n-1}{x_i - c} = \frac{1}{x_i - x_{i+1}}$$

and

$$x_i - x^* \geq x_i - x_{i+1} \geq \frac{1}{\frac{n}{x_i-x^*} - \frac{n-1}{x_i-c}} = \frac{(x_i - x^*)(x_i - c)}{x_i - x^* + n(x^* - c)}.$$

Table 5.17. *Results of the modified Newton method for* $f(x) = x^2 - 2$ *with* $x_1 = 100$

i	x_i
1	100.000000000000000
2	2.867949443501232
3	1.536378083259546
4	1.415956475989498
5	1.414213947729941
6	1.414213562373114
7	1.414213562373095
8	1.414213562373095

Therefore,

$$0 \le x_{i+1} - x^* \le (x_i - x^*)\left(1 - \frac{x_i - c}{x_i - x^* + n(x^* - c)}\right)$$

$$= (x_i - x^*)\frac{(n-1)(x^* - c)}{x_i - x^* + n(x^* - c)}$$

$$\le \min\left((n-1)(x^* - c), \left(1 - \frac{1}{n}\right)(x_i - x^*)\right).$$

By induction, it follows that $x^* \le x_{i+1} \le x_i$ for all $i \ge 1$, and the last inequality implies that x_i converges to x^*.

(ii) Similarly proved. $\qquad\square$

5.7.9 Example. Exercise 14 gives the bound $1 + \max\left(\left|\frac{a_1}{a_0}\right|, \left|\frac{a_2}{a_0}\right|\right) = 3$ for the absolute values of the zeros of the polynomial $f(x) := x^2 - 2$. Therefore we may choose $c := -3$. With the starting point $x_1 := 100$, the results of the modified Newton method are given in Table 5.17.

Halley's Method

In the limiting case $x_{i-j} \to x_i (j = 1, \ldots, k)$, the Opitz formula (O_k) derived in Section 5.4 takes the form

$$x_{i+1} := x_i + k\frac{h^{(k-1)}(x_i)}{h^{(k)}(x_i)}. \qquad (H_k)$$

Then

$$|x^* - x_{i+1}| \leq c|x^* - x_i|^{k+1},$$

that is, the method (H_k) has convergence order $k+1$. The derivatives of $h = 1/f$ are calculated from

$$h' = \frac{-f'}{f^2}, \quad h'' = \frac{2f'^2 - ff''}{f^3} \quad \text{etc.}$$

For $k = 1$, one again obtains Newton's method applied to f, with convergence order $\kappa = 2$; the case $k = 2$ leads to *Halley's method*

$$x_{i+1} := x_i - \frac{f(x_i)}{f'(x_i) - \frac{f(x_i)f''(x_i)}{2f'(x_i)}},$$

with cubic convergence ($\kappa = 3$) to simple zeros. It can be shown that, for polynomials with only real roots, Halley's method converges globally and monotonically from arbitrary starting points.

Note that k derivatives of f are needed for the determination of the values $h^{(k-1)}(x_i)$, $h^{(k)}(x_i)$ in (H_k). If for $j = 0, \ldots, k$ these derivatives require a similar computational cost, then a step with (H_k) is about as expensive as $k+1$ secant steps. A method where each step consists of $k+1$ secant steps has convergence order $(1.618\ldots)^{k+1} \gg k+1$; therefore, the secant method is generally much more efficient than any (H_k).

Other variants of the method of Opitz that use derivatives and old function values can be based on partially confluent formulas such as

$$x_{i+1} := x_i + \frac{h[x_i, x_i, x_{i-1}, \ldots, x_{i-s+1}, x_{i-s+1}, x_{i-s}]}{h[x_i, x_i, \ldots, x_{i-s+1}, x_{i-s+1}, x_{i-s}, x_{i-s}]}.$$

In this particular case, the evaluation of h and h' is necessary in each iteration, and we have

$$|x^* - x_{i+1}| \leq c|x^* - x_i|^2 \cdots |x^* - x_{i-s}|^2$$

and the convergence order is bounded by 3. Thus, one step of this variant is already asymptotically less efficient than two steps of (O_2) ($\kappa_2^2 \approx 3.382 > 3$).

A host of other methods for zeros of polynomials (and other univariate functions) is known. The interested reader is referred to the bibliography by McNamee [60]. An extensive numerical comparison of many zerofinders is in Nerinckx and Haegemans [66].

5.8 Exercises

1. What is the maximal length of a cable suspended between two poles of equal height separated by a distance $2d = 100$ m if the height h by which it sags may not exceed 10 m? Can you guarantee that the result is accurate to 1 cm?
   Hint: The form of the cable is described by a catenary $y(x) := a \cosh(x/a)$. First determine the arc length between the poles in terms of a. Then obtain a by finding a zero of a suitable equation.

2. How deeply will a tree trunk in the form of a circular cylinder of radius r (in cm) and density ρ_H (in g/cm^3) be submerged in water (with density $\rho_W = 1$)? The area of the cross-section that is under water is given by

$$F = \frac{r^2}{2}(\alpha - \sin \alpha). \tag{8.1}$$

   The mass of the displaced water is equal to the mass of the tree trunk, whence

$$F = \pi r^2 \rho_H / \rho_W. \tag{8.2}$$

   Using (8.1) and (8.2), establish an equation $f(\alpha) = 0$ for the angle α.
   For $r = 30$ and $\rho_H = 3/4$, determine graphically an approximation α_0 for the zero α^* of $f(\alpha)$. Improve the approximation using the secant method. Use the solution to calculate the depth of immersion.

3. Show that for polynomials of degree n with only real zeros, the secant method, started with two points above the largest zero, converges monotonically, with global convergence factor of at least $\sqrt{1 - \frac{1}{n}}$.

4. Show that for three times continuously differentiable functions f, the root secant method converges locally monotonically and superlinearly toward double zeros x^* of f.
   Hint: You may assume that $f''(x^*) > 0$ (why?). Distinguish two cases depending on whether or not x_1 and x_2 are on different sides of x^*.

5. For the determination of the zero $x^* = \sqrt{2}$ of $f(x) := x^2 - 2$ in the interval $[0, 2]$, use the following variant of the secant method:

```
% Set starting values
x₀ = 0;  x̄₀ = 2; i = 0;
while 1
    x = x̄ᵢ − f(x̄ᵢ)/f[x̄ᵢ, xᵢ]
    if f(xᵢ)f(x) > 0
        xᵢ₊₁ = x;  x̄ᵢ₊₁ = x̄ᵢ₊₁;
```

else

$$x_{i+1} = x_i; \quad \bar{x}_{i+1} = x;$$
end;
$$i = i + 1;$$
end

List i, x_i and $\bar{x}_i$ for $i = 1, 2, \ldots, 6$ to five decimal places, and perform iteration until $|x - \sqrt{2}| \leq 10^{-9}$. Interpret the results!

6. (a) Show that the cubic polynomial $f(x) = ax^3 + bx^2 + cx + d$ ($a \neq 0$) has a sign change at the pair (x_1, x_2) with $x_1 = 0$,

$$x_2 = \begin{cases} -\max(|a|, b, b+c, b+c+d)/a & \text{if } ad \geq 0, \\ -\min(-|a|, b, b+c, b+c+d)/a & \text{otherwise.} \end{cases}$$

Hint: Analyze the Horner form, noting that $|x_2| \geq 1$.

   (b) Based on (a), write a MATLAB program that computes all zeros of a cubic polynomial.

   Hint: After finding one zero, one finds the other two by solving a quadratic equation directly.

   (c) Can one use Cardano's formulas (see Exercise 1.13) to calculate all zeros of $x^3 - 7x + 6 = 0$, using real arithmetic only?

7. (a) Explain the safeguard used in the linear extrapolation step of the secant bisection method.

   (b) Show that in a sufficiently narrow bracket, this safeguard never becomes active.

   Hint: First show that it suffices without loss of generality to consider the case $x_1 < x < x^* < x_2$. Then show that $q = \max(f(x_1)/f(x), 2(x_2 - x_1)/(x_2 - x))$ satisfies $q > 2$ and $(x - x_1)/(1 - q) < -\min(x - x_1, x_2 - x)$. Interpret the behavior of $x_{new} = x - (x - x_1)/(1 - q)$ for the cases $|f(x)| \ll f(x_1)$ and $|f(x)| \gg f(x_1)$, and compare with the original secant formula.

8. (a) Determine the zero of $f(x) := \ln x + e^x - 100x$ in the interval $[1, 10]$ using the original secant method and the secant bisection version with $x_1 := 1$ and $x_2 := 10$. For both methods, list i and x_i ($i = 1, 2, \ldots$) until the error is $\leq \delta_0 = 10^{-9}$. Interpret the results!

   (b) Choose $x_1 := 6$ and $x_2 := 7$ for the above function and iterate with the original secant method until $|x_i - x_{i-1}| \leq 10^{-9}$. List the values of x_i and the corresponding values of i.

9. Which zero of $\sin x$ is found by the secant bisection method started with $x_1 = 1$ and $x_2 = 8$?

10. Let $A \in \mathbb{C}^{n \times n}$ be a symmetric tridiagonal matrix with diagonal elements $\alpha_i, i = 1, \ldots, n$ and subdiagonal elements $\beta_i, i = 1, \ldots, n-1$.

(a) Show that for $x \in \mathbb{C}^n$ with $x_i \neq 0$ for $i = 1, \ldots, n$, (λ, x) is an eigenpair of A if and only if the quotients $\gamma_i = x_{i-1}/x_i, i = 1, \ldots, n$ satisfy the recurrence relations

$$\beta_{i-1} \gamma_i = \lambda - \alpha_i - \beta_i / \gamma_{i+1}, \quad \text{for } i = n-1, \ldots, 1,$$

$$\beta_{n-1} \gamma_n = \lambda - \alpha_n.$$

(b) Show that each zero of the continued fraction

$$q(t) := t - \alpha_1 - \cfrac{\beta_1^2}{t - \alpha_2 - \cfrac{\beta_2^2}{\ddots \\ t - \alpha_{n-1} - \frac{\beta_{n-1}^2}{t - \alpha_n}}}$$

is an eigenvalue of A.

(c) When is the converse true in (b)?

11. Let A be a Hermitian tridiagonal matrix.

(a) Suppose that $\sigma I - A$ has an LDL^H factorization. Derive recurrence formulas for the elements of the diagonal matrix D. (You may proceed directly from the equation $\sigma I - A = LDL^H$, or use Exercise 10.)

(b) Show that the elements of L can be eliminated from the recurrence derived in (a) to yield algorithm that determines $\det(\sigma I - A)$ with a minimal amount of storage.

(c) Use (b) and the spectral bisection method to calculate bounds for the two largest eigenvalues of the 21×21 matrix A with

$$A_{ik} := \begin{cases} 11 - i, & \text{if } k = i \\ 1, & \text{if } |k - i| = 1. \\ 0, & \text{otherwise} \end{cases}$$

Stop as soon as two decimal places are guaranteed.

12. Let $G(\lambda) : \mathbb{C} \to \mathbb{C}^{3 \times 3}$ with

$$G(\lambda) = \begin{pmatrix} -10\lambda^2 + \lambda + 10 & 2\lambda^2 + 2\lambda + 2 & -\lambda^2 + \lambda - 1 \\ 2\lambda^2 + 2\lambda + 2 & -11\lambda^2 + \lambda + 9 & 2\lambda^2 + 2\lambda + 3 \\ -\lambda^2 + \lambda - 1 & 2\lambda^2 + 2\lambda + 3 & -12\lambda^2 + 10 \end{pmatrix}$$

be given.

(a) Why are all eigenvalues of the parameter matrix $G(\lambda)$ real?

(b) $G(\lambda)$ has six eigenvalues in the interval $[-2, 2]$. Find them by spectral bisection.

13. Let $f : \mathbb{R} \to \mathbb{R}$ be continuously differentiable, and let x^* be a simple zero of f. Show for an arbitrary sequence x_i ($i = 0, 1, 2, \ldots$) converging to x^*:

(a) If, for $i = 1, 2, \ldots$,

$$|x_{i+1} - x^*| \leq q|x_i - x^*| \quad \text{with } q < q_0 \leq 1,$$

then there exists an index i_0 such that

$$|f(x_{i+1})| \leq q_0|f(x_i)| \quad \text{for all } i \geq i_0.$$

(b) If

$$\lim_{i \to \infty} \frac{x_{i+1} - x^*}{x_i - x^*} = q \quad \text{with } |q| < q_1 \leq 1,$$

then there exists an index i_1 such that

$$|x_{i+1} - x_i| \leq q_1|x_i - x_{i-1}| \quad \text{for all } i \geq i_1.$$

14. (a) Show that equation (4.3) has exactly one positive real solution κ, and that this solution satisfies $1 < \kappa < \bar{\kappa} = 1 + \max\{p_0, p_1, \ldots, p_s\}$.
Hint: Show first that $(p_0\kappa^s + p_1\kappa^{s-1} + \cdots + p_s)/\kappa^{s+1} - 1$ is positive for $\kappa \leq 1$ and negative for $\kappa = \bar{\kappa}$, and its derivative is negative for all $\kappa > 0$.

(b) Let $p(x) = a_0x^n + a_1x^{n-1} + \cdots + a_{n-1}x + a_n$ be a polynomial of degree n, and let q^* be the uniquely determined positive solution of

$$|a_0|q^n = |a_1|q^{n-1} + \cdots + |a_{n-1}|q + |a_n|.$$

Show that

$$|\xi_\nu| \leq q^* \leq 1 + \max_{\nu=1:n} \left|\frac{a_\nu}{a_0}\right|$$

for all zeros ξ_ν of $p(z)$.
Hint: Find a positive lower bound for $|f(\xi)|$ if $|\xi| > q^*$.

15. (a) Show that formula (4.5) is indeed equivalent to the Opitz formula with $k = 2$ for $h = 1/f$.

(b) Show that for linear functions and hyperbolas $f(x) = (ax + b)/(cx + d)$, formula (4.5) gives the zero in a single step.

16. Let f be $p + 1$ times continuously differentiable and suppose that the iterates x_j of Muller's method converge to a zero x^* of f with multiplicity $p = 1$ or $p = 2$. Show that there is a constant $c > 0$ such that

$$|x_{j+1} - x^*|^p \leq c|x_j - x^*| \, |x_{j-1} - x^*| \, |x_{j-2} - x^*|$$

for $j = 3, 4, \ldots$. Deduce that that the (R-)convergence order of Muller's method is the solution of $\kappa^3 = \kappa^2 + \kappa + 1$ for convergence to simple zeros, and the solution of $2\kappa^3 = \kappa^2 + \kappa + 1$ for convergence to double zeros.

17. The polynomial $p_1(x) := \frac{3}{2}x^4 - 12x^3 + 36x^2 - 48x + 24$ has the 4-fold zero 2, and $p_2(x) := x^4 - 10x^3 + 35x^2 - 50x + 24$ has the zeros 1, 2, 3, 4. What is the effect on the zeros if the constant term is changed by adding $\varepsilon = \pm 0.0024$? Find the change exactly for $p_1(x)$, and numerically to six decimal places for $p_2(x)$.

18. (a) Show that for a factored polynomial

$$f(x) = a_0 \prod_{i=1:n} (x - x_i)^{m_i}$$

we have

$$\frac{f'(x)}{f(x)} = \sum_{i=1:n} \frac{1}{x - x_i}.$$

(b) If f is a polynomial with real roots only, then the zeros of f are the only finite local extrema of the absolute value of the Newton correction $\Delta(x) = -f(x)/f'(x)$.
Hint: Show first that $\Delta(x)^{-1}$ is monotone between any two adjacent zeros of f.

(c) Based on (b) and deflation, devise a damped Newton method that is guaranteed to find all roots of a polynomial with real roots only.

(d) Show that for $f(x) = (x - 1)(x^2 + 3)$, $|\Delta(x)|$ has a local minimum at $x = -1$.

19. (a) Let $x_1^*, \ldots, x_j^* (j < n)$ be known zeros of $f(x)$. Show that for the calculation of the zero x_{j+1}^* of $f(x)$ with implicit deflation, Newton's method is equivalent with the iteration

$$x_{i+1} := x_i - f(x_i) \bigg/ \left(f'(x_i) - \sum_{k=1:j} \frac{f(x_i)}{x_i - x_k^*} \right).$$

(b) To see that explicit deflation is numerically unstable, use Newton's method (with starting value $x_1 = 10$ for each zero) to calculate both with implicit and explicit deflation all zeros of $f(x) := x^7 - 7x^6 + 11x^5 + 15x^4 - 34x^3 - 14x^2 + 20x + 8$, and compare the results with the exact values $1, 1 \pm \sqrt{2}, 1 \pm \sqrt{3}, 1 \pm \sqrt{5}$.

20. Assume that in Exercise 2 you only know that

$$r \in [29.5, \ 30.5], \quad \rho_H \in [0.74, 0.76].$$

Starting with an initial interval that contains α^*, iterate with the interval Newton method until no further improvement is obtained and calculate an interval for the depth of immersion. (You are allowed to use unrounded interval arithmetic if you have no access to directed rounding.)

21. (a) Prove Theorem 5.6.6.

   Hint: Multiply (6.9) by a complex number, take its real part, and conclude that there must be some zero with $\mathrm{Re}(c/(x - \xi_i)) \geq 0$.

   (b) Let $x' = \tilde{x} - \frac{n f(\tilde{x})}{2 f'(\tilde{x})}$. What is the best error bound for $|x' - x^*|$?

22. (a) Let $k_1 = k_2 = 2, k_3 = 3, k_4 = 5$. Show that for $\arg(q) \in \pm \,]\frac{11}{20}\pi, \frac{5}{9}\pi[$, condition (6.5) cannot be violated for all $k = 0, \ldots, k_n$.

   (b) In which sense is this choice of the k_n best possible?

23. (a) Write a MATLAB program that calculates a zero of an analytic function $f(x)$ using Muller's method. If for two iterates x_j, x_{j+1},

$$\left| 1 - \frac{x_{j+1}}{x_j} \right| < \delta$$

   (where δ is an input tolerance), perform one more iteration and terminate. If $x_j = x_{j+1}$, terminate the iteration immediately. Test the program with $f(x) = x^4 + x^3 + 2x^2 + x + 1$.

   (b) Incorporate into the MATLAB program a spiral search and find the zeros again. Experiment with various choices for the contraction and rotation factor q. Compare with the use of the secant formula instead of Muller's.

24. (a) Determine, for $n = 12, 16, 20$, all zeros $x_1^*, \ldots, x_n^*$, of the truncated exponential series

$$p_n(x) := \sum_{\nu=0:n} \frac{x^\nu}{\nu!}.$$

   (The zeros are simple and pairwise conjugate complex.)

   Use the programs of Exercise 23 with $x_0 := -6, x_1 := -5, x_2 := -4$ as starting values and $\delta = 10^{-4}$ as tolerance. On finding a pair of conjugate complex zeros (in MATLAB, `conj` gives the complex conjugate), remove them by implicit deflation and begin again with the above starting points.

(b) Calculate, for each of the estimates $\tilde{x}_k^*$ of the zeros, the error bound

$$|x_k^* - \tilde{x}_k^*| \le n \cdot \left| \frac{p_n(\tilde{x}^*)}{p_n'(\tilde{x}^*)} \right|.$$

Print for each zero the number of iteration required, the last iterate that was calculated, and the error bound.

(c) Plot the positions of the zeros you have found. What do you expect to see for $n \to \infty$?

25. For the functions $f_1(x) := x^4 - 7x^2 + 2x + 2$ and $f_2(x) := x^2 - 1999x + 1000000$, perform six steps of Newton's method with $x_0 = 3$ for $f_1(x)$ and with $x_0 = 365$ for $f_2(x)$. Interpret the results.

26. Let x^* be a zero of the twice continuously differentiable function $f : \mathbb{R} \to \mathbb{R}$, and suppose that f is strictly monotonically increasing and convex for $x \ge x^*$. Show the following:

(a) If $x_1 > x_2 > x^*$, then the sequence defined by the secant method started with x_1 and x_2 is well defined and converges monotonically to x^*.

(b) If $y_1 > x^*$, then the sequence y_i defined by Newton's method $y_{i+1} = y_i - f(y_i)/f'(y_i)$ is well defined and converges monotonically to x^*.

(c) If $x_1 = y_1 > x^*$ then $x^* < x_{2i} < y_i$ for all $i > 1$; that is, two steps of the secant method always give (with about the same computational cost) a better approximation than one Newton step.
Hint: The error can be represented in terms of x^* and divided differences.

27. Investigate the convergence behavior of Newton's method for the polynomial $p(x) := x^5 - 10x^3 + 69x$.

(a) Show that Newton's method converges with any starting point $|x_0| < \frac{1}{3}\sqrt{23}$.

(b) Show that $g(x) = x - \frac{f(x)}{f'(x)}$ is monotone on the interval $\boldsymbol{x} := \left[\frac{1}{3}\sqrt{23}, \sqrt{3}\right]$. Deduce that $g(x)$ maps $\boldsymbol{x}$ into $-\boldsymbol{x}$ and $-\boldsymbol{x}$ into $\boldsymbol{x}$. What does this imply for Newton's method started with $|x_0| \in \boldsymbol{x}$?

28. Let $p(x) = a_0 x^n + a_1 x^{n-1} + \cdots + a_{n-1} x + a_n, a_0 > 0$, be a polynomial of degree n with a real zero ξ^* for which

$$\xi^* \ge \text{Re}\, \xi_\nu \quad \text{for all zeros } \xi_\nu \text{ of } p(x).$$

(a) Show that for $z_0 > \xi^*$ the iterative method

$$z_{j+1} := z_j - \frac{p(z_j)}{p'(z_j)}$$

$$w_{j+1} := z_j - n\frac{p(z_j)}{p'(z_j)}$$

generates two sequences that converge to ξ^* with

$$w_j \leq \xi^* \leq z_j \quad \text{for } j = 1, 2, 3, \ldots$$

and $\{z_j\}$ is monotone decreasing.

Hint: Use Exercise 18(a).

(b) Using Exercise 14, show that the above hypotheses are valid for the polynomial

$$p(x) = x^n - x^{n-1} - x^{n-2} - \cdots - 1.$$

Calculate ξ^* with the method from part (a) for $n = 2, 3, 4, 5, 10$ and $z_0 := 2$.

6

Systems of Nonlinear Equations

In this chapter, we treat methods for finding a zero $x^* \in D$ (i.e., a point $x^* \in D$ with $F(x^*) = 0$) of a continuously differentiable function $F : D \subseteq \mathbb{R}^n \to \mathbb{R}^n$. Such problems arise in many applications, such as the analysis of nonlinear electronic circuits, chemical equilibrium problems, or chemical process design. Nonlinear systems of equations must also frequently be solved as subtasks in solving problems involving differential equations. For example, the solution of initial value problems for stiff ordinary differential equations requires implicit methods (see Section 4.5), which in each time step solve a nonlinear system, and nonlinear boundary value problems for ordinary differential equations are often solved by multiple shooting methods where large banded nonlinear systems must be solved to ensure the correct matching of pieces of the solution. Nonlinear partial differential equations are reduced by finite element methods to solving sequences of huge structured nonlinear systems. Because most physical processes, whether in satellite orbit calculations, weather forecasting, oil recovery, or electronic chip design, can be described by nonlinear differential equations, the efficient solution of systems of nonlinear equations is of considerable practical importance.

Stationary points of scalar multivariate function $f : D \subseteq \mathbb{R}^n \to \mathbb{R}$ lead to a nonlinear system of equations for the gradient, $\nabla f(x) = 0$. The most important case, finding the extrema of such functions, occurs frequently in industrial applications where some practical objective such as profit, quality, weight, cost, or loss must be maximized or minimized, but also in thermodynamical or mechanical applications where some energy functional is to be minimized. (To see how the additional structure of these *optimization problems* can be exploited, consult Fletcher [26], Gill et al. [30] and Nocedal and Wright [73a].)

After the discussion of some auxiliary results in Section 6.1, we discuss the multivariate version of Newton's method in Section 6.2. Emphasis is on obtaining a robust damped version that can be proved to converge under fairly general conditions. Section 6.3 discusses problems of error analysis and interval techniques for the rigorous enclosure of solutions. Finally, some more specialized methods and results are discussed in Section 6.4.

A basic reference for solving nonlinear systems of equations and unconstrained optimization problems is Dennis and Schnabel [18]. See also Allgower and Georg [5] for parametrized or strongly nonlinear systems of equations.

6.1 Preliminaries

We recall here some ideas from multidimensional analysis on norm inequalities and local expansions of functions, discuss the automatic differentiation of multivariate expressions, and state two fixed-point theorems that are needed later. In the following, int D denotes the interior of a set D, and ∂D its boundary.

Integration

Let $g : [a, b] \to \mathbb{R}^n$ be a vector-valued function that is continuous on $[a, b] \subset \mathbb{R}$. The *Riemann integral* $\int_a^b g(t)\, dt$ is defined as the limit

$$\int_a^b g(t)\, dt := \lim_{\max(t_{i+1} - t_i) \to 0} \sum_{i=0:n} (t_{i+1} - t_i) g(t_i)$$

taken over all partitions $a = t_0 < t_1 < \cdots < t_{n+1} = b$ of the interval $[a, b]$, provided this limit exists.

6.1.1 Lemma. *For every norm in $\mathbb{R}^n$,*

$$\left\| \int_a^b g(t)\, dt \right\| \le \int_a^b \|g(t)\|\, dt$$

if both sides are defined.

Proof. This follows from the inequality

$$\left\| \sum_{i=0:n} (t_{i+1} - t_i) g(t_i) \right\| \le \sum_{i=0:n} (t_{i+1} - t_i) \|g(t_i)\|$$

by going to the limit.   □

The Jacobian

The matrix $A \in \mathbb{R}^{m \times n}$ is called the *Jacobian matrix* (or *derivative*) of $F : D \subseteq \mathbb{R}^n \to \mathbb{R}^m$ at the point $x^0 \in D$ if

$$\frac{\|F(x) - F(x^0) - A(x - x^0)\|}{\|x - x^0\|} \to 0 \quad \text{as } x \to x^0;$$

one writes $A = F'(x^0)$. It follows immediately from the definition that

$$F(x) = F(x^0) + F'(x^0)(x - x^0) + o(\|x - x^0\|),$$

and if F' is Lipschitz continuous, the error term is of order $O(\|x - x^0\|^2)$. Because all norms in $\mathbb{R}^n$ are equivalent, the derivative does not depend on the norm used. The function F is called *(continuously) differentiable* in D if the Jacobian $F'(x)$ exists for all $x \in D$ (and is continuous there). If F is continuously differentiable in D, then

$$F'(x)_{ik} = \frac{\partial}{\partial x_k} F_i(x) \quad (i = 1, \ldots, m, \ k = 1, \ldots, n).$$

For example, for $m = n = 2$ and $F(x) = \binom{F_1(x)}{F_2(x)}$, we have

$$F'(x) = \begin{pmatrix} \frac{\partial}{\partial x_1} F_1(x) & \frac{\partial}{\partial x_2} F_1(x) \\ \frac{\partial}{\partial x_1} F_2(x) & \frac{\partial}{\partial x_2} F_2(x) \end{pmatrix}.$$

When programming applications, it is important that derivatives are programmed correctly. Correctness may be checked as in Exercise 1.8.

We generalize the concept of a divided difference from the univariate case (Section 3.1) to multivariate vector-valued functions by defining the *multivariate slope*

$$F[x, x^0] := \int_0^1 F'(x^0 + t(x - x^0)) \, dt$$

if F is differentiable on the line $\overline{xx^0}$.

6.1.2 Lemma. *Let $D \subseteq \mathbb{R}^n$ be convex and let $F : D \to \mathbb{R}^n$ be continuously differentiable in D.*

(i) If $x, x^0 \in D$ then

$$F(x) - F(x^0) = F[x, x^0](x - x^0).$$

(This is a strong form of the mean value theorem.)

(ii) If $\|F'(x)\| \le L$ for all $x \in D$, then

$$\|F[x, x^0]\| \le L \quad \text{for all } x, x^0 \in D$$

and

$$\|F(x) - F(x^0)\| \le L\|x - x^0\| \quad \text{for all } x, x^0 \in D.$$

(This is a weak form of the mean value theorem.)
(iii) If F' is Lipschitz continuous in D, that is, the relation

$$\|F'(x) - F'(y)\| \le \gamma\|x - y\| \quad \text{for all } x, y \in D$$

holds for some $\gamma \in \mathbb{R}$, then

$$\|F(x) - F(x^0) - F'(x^0)(x - x^0)\| \le \frac{\gamma}{2}\|x - x^0\|^2 \quad \text{for all } x, x^0 \in D;$$

in particular

$$F(x) = F(x^0) + F'(x^0)(x - x^0) + O(\|x - x^0\|^2).$$

(This is the truncated Taylor expansion with remainder term.)

Proof. Because D is convex, the function defined by

$$g(t) := F(x^0 + t(x - x^0))$$

is continuously differentiable in $[0, 1]$. Therefore,

$$F(x) - F(x^0) = g(1) - g(0) = \int_0^1 g'(t)\, dt$$

$$= \int_0^1 F'(x^0 + t(x - x^0))(x - x^0)\, dt = F[x, x^0](x - x^0).$$

This proves (i). Now, by Lemma 6.1.1,

$$\|F[x, x^0]\| = \left\| \int_0^1 F'(x^0 + t(x - x^0))\, dt \right\|$$

$$\le \int_0^1 \|F'(x^0 + t(x - x^0))\|\, dt$$

$$\le \int_0^1 L\, dt = L,$$

and

$$\|F(x) - F(x^0)\| = \|F[x, x^0](x - x^0)\| \leq \|F[x, x^0]\|\|x - x^0\| \leq L\|x - x^0\|,$$

whence (ii) follows. The assertion (iii) follows similarly from

$$\|F(x) - F(x^0) - F'(x^0)(x - x^0)\|$$

$$= \left\|\int_0^1 (F'(x^0 + t(x - x^0)) - F'(x^0))(x - x^0)\, dt\right\|$$

$$\leq \int_0^1 \|F'(x^0 + t(x - x^0)) - F'(x^0)\|\|x - x^0\|\, dt$$

$$\leq \int_0^1 \gamma\|x^0 + t(x - x^0) - x^0\|\|x - x^0\|\, dt$$

$$= \gamma\|x - x^0\|^2 \int_0^1 t\, dt = \frac{\gamma}{2}\|x - x^0\|^2. \qquad \square$$

Reverse Automatic Differentiation

For many nonlinear problems, and in particular for those discussed in this chapter, the numerical techniques work better if derivatives of the functions involved are available. Because numerical differentiation is often inaccurate (see Section 3.2), it is advisable to provide programs that compute derivatives analytically. The basic principles were discussed already in Section 1.1 for the univariate case. Here, we look briefly at the multivariate case. For more details, see Griewank and Corliss [33].

Consider the calculation of $y = F(x)$ where $x \in \mathbb{R}^n$, and $y \in \mathbb{R}^m$ is a vector defined in terms of x by arithmetic expressions. We may introduce an auxilary vector $z \in \mathbb{R}^N$ containing all intermediate results from the calculation of y, with y retrievable from suitable coordinates of z, say $y = Pz$ with a $(0, 1)$ matrix P that has exactly one 1 in each row. Then we get each z_i with a single operation from one or two $z_j (j < i)$ or x_j. Therefore, we may write the computation as

$$z = H(x, z), \tag{1.1}$$

where $H_i(x, z)$ depends on one or two of the $z_j (j < i)$ or x_j only. In particular, the partial derivatives $H_x(x, z)$ with respect to x and $H_z(x, z)$ with respect to z are extremely sparse; moreover, $H_z(x, z)$ is strictly lower triangular. Now z depends on x, and its derivative can be found by differentiating (1.1), giving

$$\frac{\partial z}{\partial x} = H_x(x, z) + H_z(x, z)\frac{\partial z}{\partial x}.$$

Bringing the partial derivatives to the left and solving for them, we find

$$\frac{\partial z}{\partial x} = (I - H_z(x, z))^{-1} H_x(x, z). \tag{1.2}$$

Because $F'(x) = \partial y / \partial x = P \partial z / \partial x$, we find

$$F'(x) = P(I - H_z(x, z))^{-1} H_x(x, z). \tag{1.3}$$

Note that the matrix to be inverted is unit lower triangular, so that one can compute (1.2) by solving

$$(I - H_z)\frac{\partial z}{\partial x} = H_x \tag{1.4}$$

by forward substitution: with proper programming, the sparsity can be fully exploited and the computation efficiently arranged. For one-dimensional x, the resulting formulas are equivalent to the approach via differential numbers discussed in Section 1.1. This way of calculating $F'(x)$ is termed *forward automatic differentiation*.

In many cases, especially when $m \ll n$, the formula (1.3) can be evaluated more efficiently. Indeed, we can transpose the formula and find

$$F'(x)^T = H_x(x, z)^T (I - H_z(x, z))^{-T} P^T. \tag{1.5}$$

Now we find $K := (I - H_z(x, z))^{-T} P^T$ by solving with back substitution the associated linear system

$$(I - H_z)^T K = P^T, \tag{1.6}$$

the so-called *adjoint equation* with a unit upper triangular matrix, and obtain

$$F'(x) = K^T H_x(x, z). \tag{1.7}$$

This way of proceeding is called *reverse* or *backward automatic differentiation* because we solve for the components of K in reverse order. An apparent disadvantage of this way of proceeding is that $H_x(x, z)$ must be stored fully while it can be computed on the fly while solving (1.4). However, if $m \ll n$, the advantage in speed is dramatic because (1.4) contains on the right side n columns whereas (1.6) has only m columns. So the number of triangular solves is cut down by a factor of $m/n \ll 1$.

6.1.3 Example. To compute

$$f(x) = \frac{x_1^2 + x_2 x_3}{x_4} e^{x_1 - x_4},$$

we need the intermediate expressions

$$
\begin{aligned}
z_1 &= x_1^2, \\
z_2 &= x_2 x_3, \\
z_3 &= z_1 + z_2, \\
z_4 &= z_3/x_4, \\
z_5 &= x_1 - x_4, \\
z_6 &= e^{z_5}, \\
z_7 &= z_4 z_6
\end{aligned}
\tag{1.8}
$$

to get $f(x) = z_7$. We may write the system (1.8) as $z = H(x, z)$ and have $y = Pz$ with the 1×7-matrix $P = (0\ 0\ 0\ 0\ 0\ 0\ 1)$. The Jacobian of H with respect to z is the strictly lower triangular matrix

$$
H_z(x, z) = \begin{pmatrix}
0 & 0 & 0 & 0 & 0 & 0 & 0 \\
0 & 0 & 0 & 0 & 0 & 0 & 0 \\
1 & 1 & 0 & 0 & 0 & 0 & 0 \\
0 & 0 & 1/x_4 & 0 & 0 & 0 & 0 \\
0 & 0 & 0 & 0 & 0 & 0 & 0 \\
0 & 0 & 0 & 0 & z_6 & 0 & 0 \\
0 & 0 & 0 & z_6 & 0 & z_4 & 0
\end{pmatrix}.
$$

As for any scalar-valued function, $K = k$ is now a vector, and the linear system (1.6) simply becomes

$$
\begin{aligned}
k_1 - k_3 &= 0, \\
k_2 - k_3 &= 0, \\
k_3 - (1/x_4)k_4 &= 0, \\
k_4 - z_6 k_7 &= 0, \\
k_5 - z_6 k_6 &= 0, \\
k_6 - z_4 k_7 &= 0, \\
k_7 &= 1.
\end{aligned}
$$

Thus,

$$k_7 = 1, \quad k_6 = z_4, \quad k_5 = z_6 z_4, \quad k_4 = z_6, \quad k_1 = k_2 = k_3 = z_6/x_4, \tag{1.9}$$

and we get the components of the gradient from (1.7) as

$$g_j = \frac{\partial f(x)}{\partial x_j} = k^T \frac{\partial H(x, z)}{\partial x_j}.$$

Most components of $\partial H / \partial x_j$ vanish, and we end up with

$$
\begin{aligned}
g_1 &= k_1 \cdot 2x_1 + k_5, \\
g_2 &= k_2 x_3, \\
g_3 &= k_2 x_2, \\
g_4 &= k_4\left(-z_3 / x_4^2\right) - k_5 = -k_4 z_4 / x_4 - k_5.
\end{aligned}
\tag{1.10}
$$

The whole gradient is computed from (1.9) and (1.10) with only 11 operations, and because no exponential is involved, computing function value and gradient takes less than twice the work for computing the function value alone. This is not untypical for gradient computations in the reverse mode.

There are several good automatic differentiation programs (such as ADIFOR [2], ADOL-C [34]) based on the formula (1.3) that exploit the forward approach, the backward approach, or a mixture of both. They usually take a program for calculation of $F(x)$ as input and return a program that calculates $F(x)$ and $F'(x)$ simultaneously.

Fixed-Point Theorems

Related to the equation $F(x^*) = 0$ is the fixed point equation

$$x^* = G(x^*) \tag{1.11}$$

and the corresponding fixed point iteration

$$x^{l+1} := G(x^l) \quad (l = 0, 1, 2, \ldots). \tag{1.12}$$

The solutions of (1.11) are called *fixed points* of G. For application to the problem of determining zeros of F, one sets $G(x) = x - C F(x)$ with a suitable $C \in \mathbb{R}^{n \times n}$. There are several important fixed-point theorems giving sufficient conditions for the existence of at least one fixed point. They are useful in proving the existence of zeros in specified regions of space, and relevant to a rigorous error analysis.

Banach's Fixed-Point Theorem

The mapping $G : D \subseteq \mathbb{R}^n \to \mathbb{R}^n$ is called a *contraction* in $K \subseteq D$ if G maps K into itself and if there exists $\beta \in \mathbb{R}$ with $0 \le \beta < 1$ such that

$$\| G(x) - G(y) \| \le \beta \| x - y \| \quad \text{for all } x, y \in K;$$

β is called a *contraction factor* of G (in K). In particular, every contraction in K is Lipschitz continuous.

6.1.4 Theorem. *Let the mapping $G : D \subseteq \mathbb{R}^n \to \mathbb{R}^n$ be a contraction in the closed, convex set $K \subseteq D$. Then G has exactly one fixed point $x^* \in K$ and the iteration $x^{l+1} := G(x^l)$ converges for all $x^0 \in K$ at least linearly to x^*. The relations*

$$\| x^{l+1} - x^* \| \le \beta \| x^l - x^* \|, \tag{1.13}$$

and

$$\frac{\| x^{l+1} - x^l \|}{1 + \beta} \le \| x^l - x^* \| \le \frac{\| x^{l+1} - x^l \|}{1 - \beta} \tag{1.14}$$

hold, in which β denotes a contraction factor of G.

Proof. Let $x^0 \in K$ be arbitrary. Because G maps the set K into itself, x^l is defined for all $l \ge 0$ and for $l \ge 1$,

$$\| x^{l+1} - x^l \| = \| G(x^l) - G(x^{l-1}) \| \le \beta \| x^l - x^{l-1} \|,$$

so by induction,

$$\| x^{l+1} - x^l \| \le \beta^l \| x^1 - x^0 \|.$$

Therefore for $l < m$,

$$\begin{aligned}
\| x^l - x^m \| &\le \| x^l - x^{l+1} \| + \cdots + \| x^{m-1} - x^m \| \\
&\le (\beta^l + \cdots + \beta^{m-1}) \| x^1 - x^0 \| \\
&= \frac{\beta^m - \beta^l}{\beta - 1} \| x^1 - x^0 \|,
\end{aligned}$$

so $\| x^l - x^m \| \to 0$ as $l, m \to \infty$. Therefore x^l is a Cauchy sequence and has a limit x^*. Because G is continuous,

$$x^* = \lim_{l \to \infty} x^l = \lim_{l \to \infty} G(x^{l-1}) = G(x^*),$$

that is, x^* is a fixed point of G. Moreover, because K is closed, $x^* \in K$.

An arbitrary fixed point $x' \in K$ satisfies

$$\|x' - x^*\| = \|G(x') - G(x^*)\| \le \beta \|x' - x^*\|,$$

and we conclude $x' = x^*$ because $\beta < 1$. Therefore x^* is the only fixed point of G in K. The inequality (1.13) now follows from

$$\|x^{l+1} - x^*\| = \|G(x^l) - G(x^*)\| \le \beta \|x^l - x^*\|,$$

and with this result, (1.14) follows from

$$(1 - \beta)\|x^l - x^*\| \le \|x^l - x^*\| - \|x^{l+1} - x^*\| \le \|x^{l+1} - x^l\|$$
$$\le \|x^l - x^*\| + \|x^{l+1} - x^*\| \le (1 + \beta)\|x^l - x^*\|. \qquad \square$$

6.1.5 Remarks.

(i) We know an error bound of the same kind as (1.14) already from Theorem 2.7.2. As seen there, (1.14) is a realistic bound if β does not lie too close to 1.

(ii) Two difficulties in the application of the Banach fixed-point theorem lie in the proof of the property that G maps K into itself and in the determination of a contraction factor $\beta < 1$. By Lemma 6.1.2(ii), one can choose $\beta := \sup\{\|G'(x)\| \mid x \in K\}$. If this supremum is difficult to determine, one can determine a simpler upper bound with the help of interval arithmetic (see Section 6.3).

The Fixed-Point Theorems by Brouwer and Leray-Schauder

Two other important fixed point theorems guarantee the existence of at least one fixed point, but uniqueness can no longer be guaranteed.

6.1.6 Theorem.

(i) (**Fixed-point theorem by Leray and Schauder**) *Suppose that the mapping* $G : D \subseteq \mathbb{R}^n \to \mathbb{R}^n$ *is continuous and that* $K \subseteq D$ *is compact. If* $x^0 \in \operatorname{int} K$ *and*

$$G(x) \ne x^0 + \lambda(x - x^0) \quad \text{for all } x \in \partial K, \ \lambda > 1,$$

then G has at least one fixed point in K.

(ii) (**Fixed point theorem by Brouwer**) *Suppose that the mapping* $G : D \subseteq \mathbb{R}^n \to \mathbb{R}^n$ *is continuous and that* $K \subseteq D$ *is nonempty, convex, and compact. If* G *maps the set* K *into itself, then* G *has at least one fixed point in* K.

Proof. For $n = 1$ the theorem easily follows from the intermediate value theorem. For the general case, we refer to Ortega and Rheinboldt [78] or Neumaier [70] for a proof of (i), and deduce here only (ii) from (i). If $x^0 \in$ int K and $\lambda > 1$ then, assuming $G(x) = x^0 + \lambda(x - x^0)$, the convexity of K implies that

$$x = (1 - \lambda^{-1})x^0 + \lambda^{-1}G(x) \in \text{int } K.$$

Therefore, $G(x) \neq x^0 + \lambda(x - x^0)$ for each $x \in \partial K$, $\lambda > 1$, and by part (i), G has a fixed point in K. $\qquad\square$

Note that Brouwer's fixed-point theorem weakens the hypotheses of Banach's fixed-point theorem, but also gives a weaker conclusion. That the fixed point is no longer uniquely determined can be seen, for example, by choosing the identity for G.

6.2 Newton's Method and Its Variants

The multivariate extension of Newton's method, started at some $x^0 \in \mathbb{R}^n$, is given by

$$x^{l+1} := x^l - F'(x^l)^{-1}F(x^l) = x^l + p^l, \tag{2.1}$$

where p^l is the solution of the system of linear equations

$$F'(x^l)p^l = -F(x^l). \tag{2.2}$$

The solution p^l of (2.2) is called the *Newton correction* or the *Newton direction*. We emphasize that although the notation involving the inverse of the Jacobian is very useful for purposes of analysis, in practice the matrix $F'(x^l)^{-1}$ is never calculated explicitly because as seen in Section 2.2, p^l can be determined more efficiently by solving (2.2), for example, using Gaussian elimination.

Newton's method is the basis of most methods for determining zeros in $\mathbb{R}^n$; many other methods can be considered as variants of this method. To show this, we characterize an arbitrary sequence $x^l (l = 0, 1, 2, \ldots)$ that converges quadratically to a regular zero. Here, a zero $x^* \in$ int D is called a *regular zero* of $F : D \subseteq \mathbb{R}^n \to \mathbb{R}^n$ if $F(x^*) = 0$, if F is differentiable in a neighborhood of x^* and $F'(x^*)$ is Lipschitz continuous and nonsingular in such a neighborhood.

From now on, we frequently omit the iteration index l and simply write x for x^l and $\bar{x}$ for x^{l+1} (and similarly p for p^l, etc.).

6.2.1 Theorem. *Let $x^* \in D$ be a regular zero of the function $F : D \subseteq \mathbb{R}^n \to \mathbb{R}^n$. Suppose that the sequence x^l ($l = 0, 1, 2, \ldots$) converges to x^* and that $p^l := x^{l+1} - x^l$. Then the following statements are equivalent:*

(i) $\|\bar{x} - x^*\| = O(\|x - x^*\|^2)$,
(ii) $\|F(\bar{x})\| = O(\|F(x)\|^2)$,
(iii) $\|F(x) + F'(x)p\| = O(\|F(x)\|^2)$.

Proof. We remark first that by Lemma 6.1.2(iii),

$$F(x) = F(x^*) + F'(x^*)(x - x^*) + O(\|x - x^*\|^2).$$

Because $F(x^*) = 0$ and $F'(x^*)$ is bounded, it follows that

$$\|F(x)\| = O(\|x - x^*\|), \tag{2.3}$$

and for an appropriate $c > 0$

$$\begin{aligned}
\|x - x^*\| &= \|F'(x^*)^{-1}(F(x) - O(\|x - x^*\|^2))\| \\
&\leq \|F'(x^*)^{-1}\|(\|F(x)\| + c\|x - x^*\|^2).
\end{aligned}$$

For large l, $\|F'(x^*)^{-1}\| \, \|x - x^*\| \leq \frac{1}{2c}$, so

$$\|x - x^*\| \leq \|F'(x^*)^{-1}\|\|F(x)\| + \frac{1}{2}\|x - x^*\|,$$

and solving for $\|x - x^*\|$, we find

$$\|x - x^*\| \leq 2\|F'(x^*)^{-1}\|\|F(x)\| = O(\|F(x)\|). \tag{2.4}$$

If now (i) holds, then by (2.3) and (2.4),

$$\|F(\bar{x})\| = O(\|\bar{x} - x^*\|) = O(\|x - x^*\|^2) = O(\|F(x)\|^2),$$

and if (ii) holds, then by (2.4) and (2.3),

$$\|\bar{x} - x^*\| = O(\|F(\bar{x})\|) = O(\|F(x)\|^2) = O(\|x - x^*\|^2).$$

Therefore, (i) and (ii) are equivalent. To show that (ii) and (iii) are equivalent, we introduce the abbreviation

$$r := F(x) + F'(x)p$$

and remark that by Lemma 6.1.2(iii),

$$F(\bar{x}) = F(x) + F'(x)(\bar{x} - x) + O(\|\bar{x} - x\|^2),$$

and

$$F(\bar{x}) = r + O(\|\bar{x} - x\|^2). \qquad (2.5)$$

If (ii) now holds, then by (2.4)

$$\|\bar{x} - x\| \leq \|\bar{x} - x^*\| + \|x - x^*\|$$
$$= O(\|F(\bar{x})\|) + O(\|F(x)\|) = O(\|F(x)\|),$$

and therefore by (2.5) and (ii)

$$\|r\| \leq \|F(\bar{x})\| + O(\|\bar{x} - x\|^2) = O(\|F(x)\|^2)$$

so that (iii) holds. Conversely, if (iii) holds, then

$$\|r\| = O(\|F(x)\|^2) \qquad (2.6)$$

and

$$\|\bar{x} - x\| = \|p\| = \|F'(x)^{-1}(F(x) - r)\|$$
$$\leq \|F'(x)^{-1}\|(\|F(x)\| + \|r\|)$$
$$= O(1)O(\|F(x)\|) = O(\|F(x)\|).$$

By (2.5) and (2.6) we have

$$\|F(\bar{x})\| \leq \|r\| + O(\|\bar{x} - x\|^2) = O(\|F(x)\|^2),$$

so that (ii) holds. Therefore, (ii) and (iii) are also equivalent. $\qquad \square$

Because condition (iii) is obviously satisfied if $p = -F'(x)^{-1}F(x)$ is the Newton correction, we find the following.

6.2.2 Corollary. *Newton's method for a system of equations is locally quadratically convergent if it converges to a regular zero.*

6.2.3 Example. Consider the function

$$F(x) := \begin{pmatrix} x_2^2 - 3x_1^2 \\ x_1^2 + x_1 x_3 + x_3^2 - 3x_2^2 \\ x_2^2 + x_2 + 1 - 3x_3^2 \end{pmatrix}$$

with Jacobian

$$F'(x) := \begin{pmatrix} -6x_1 & 2x_2 & 0 \\ 2x_1 + x_3 & -6x_2 & x_1 + 2x_3 \\ 0 & 2x_2 + 1 & -6x_3 \end{pmatrix}.$$

With the starting value $x^0 := \left(\frac{1}{4}, \frac{1}{2}, \frac{3}{4}\right)^T$, the Newton method gives the results displayed in Table 6.1. Only the seven leading digits are displayed; however, full accuracy is obtained, and the local quadratic convergence is clearly visible until rounding errors start to dominate.

The Discretized Newton Method

If a program calculating $F'(x)$ is not available, one can approximate the partial derivatives by numerical differentiation to obtain an approximation A_l for $F'(x^l)$. If one then determines p^l from the equation $A_l p^l = -F(x^l)$ and again sets $x^{l+1} := x^l + p^l$, then one speaks of a *discretized Newton method*. In order to obtain quadratic convergence, the approximation must satisfy

$$\|A - F'(x)\| = O(\|F(x)\|);$$

Table 6.1. *Simple Newton iteration*

l	x^l			$\|F(x^l)\|_1$	$\|p^l\|_1$
0	0.2500000	0.5000000	0.7500000	$1.88e - 01$	$2.67e - 01$
1	0.3583333	0.6000000	0.8083333	$3.40e - 02$	$4.05e - 02$
2	0.3386243	0.5856949	0.8018015	$1.09e - 03$	$1.28e - 03$
3	0.3379180	0.5852901	0.8016347	$1.57e - 06$	$1.53e - 06$
4	0.3379171	0.5852896	0.8016345	$2.84e - 12$	$2.47e - 12$
5	0.3379171	0.5852896	0.8016345	$5.55e - 16$	$6.52e - 16$
6	0.3379171	0.5852896	0.8016345	$5.00e - 16$	$4.32e - 16$

because $p = -A^{-1}F(x) = O(\|F(x)\|)$, it follows that

$$\|F(x) + F'(x)p\| = \|(A - F'(x))p\|$$
$$\leq \|A - F'(x)\|\|p\| = O(\|F(x)\|^2).$$

The matrix A must therefore approximate $F'(x)$ increasingly well as $\|F(x)\|$ decreases.

In the case that different components of $F(x)$ cannot be calculated one by one, one may determine the approximation for $F'(x)$ columnwise from

$$F'(x)u \approx \frac{F(x + hu) - F(x)}{h} \quad \text{or} \quad \approx \frac{F(x + hu) - F(x - hu)}{2h}, \quad (2.7)$$

in which $h \neq 0$ is chosen suitably and $u \in \mathbb{R}^n$ runs through the unit vectors $e^{(1)}, e^{(2)}, \ldots, e^{(n)}$ in succession.

If the dimension n is large but $F'(x)$ is sparse, that is, each component of F depends only on a few variables, one can make use of this in the calculation of A. If one knows, for example, that $F'(x)$ is tridiagonal, then one obtains from (2.7) with

$$u = (1, 0, 0, 1, 0, 0, 1, 0, 0, \ldots)$$
$$u = (0, 1, 0, 0, 1, 0, 0, 1, 0, \ldots)$$
$$u = (0, 0, 1, 0, 0, 1, 0, 0, 1, \ldots)$$

approximations for the columns

$$1, 4, 7, \ldots$$
$$2, 5, 8, \ldots, \quad \text{and}$$
$$3, 6, 9, \ldots \quad \text{of } F'(x),$$

respectively (why?). In this case, only four evaluations of F are required for the calculation of A by the forward and six evaluations of F for the central difference quotient.

Damping

Under special monotonicity and convexity conditions (see Section 6.4), Newton's method can be shown to converge in its simple form. However, as in the univariate case, in general some sort of damping is necessary to get a robust algorithm.

Table 6.2. *Divergence with undamped Newton method*

l	x^l		$\|F(x^l)\|_1$	$\|p^l\|_1$
0	2.00000	2.00000	$5.1e+00$	$3.3e+01$
1	-18.15888	-10.57944	$6.8e+02$	$1.5e+01$
2	-8.37101	-5.22873	$1.7e+02$	$7.3e+00$
3	-3.55249	-2.71907	$4.1e+01$	$3.6e+00$
4	-1.20146	-1.47283	$1.1e+01$	$2.8e+00$
5	-0.00041	0.09490	$2.4e+01$	$1.9e+03$
6	0.04510	-1865.23122	$3.5e+06$	$9.3e+02$
$\vdots$	$\vdots$	$\vdots$	$\vdots$	$\vdots$
21	0.13650	0.15806	$6.2e+00$	$4.4e+00$
22	0.33360	-4.06082	$2.0e+01$	$2.7e+00$
23	0.19666	-1.49748	$7.3e+00$	$2.6e+00$
24	0.15654	1.03237	$6.6e+00$	$2.4e+00$

6.2.4 Example. The function

$$F(x) := \begin{pmatrix} x_1 + 3\ln|x_1| - x_2^2 \\ 2x_1^2 - x_1 x_2 - 5x_1 + 1 \end{pmatrix}$$

has the Jacobian

$$F'(x) = \begin{pmatrix} 1 + 3/x_1 & -2x_2 \\ 4x_1 - x_2 - 5 & -x_1 \end{pmatrix}.$$

For the starting vector $x^0 := (2, 2)^T$, the Newton method gives the results displayed in Table 6.2, without apparent convergence.

As in the one-dimensional case, *damping* consists in taking only a partial step

$$\bar{x} := x + \alpha p,$$

with a damping factor α chosen such that

$$\|F(x + \alpha p)\| < (1 - q\alpha)\|F(x)\| \tag{2.8}$$

for some fixed $q, 0 < q < 1$. It is important to note that requiring $\|F(x+\alpha p)\| < \|F(x)\|$ is not enough (see Example 5.7.7).

Because $\alpha = 1$ corresponds to a full Newton step, which gives quadratic convergence, one first tests whether (2.8) is satisfied. If this is not the case, then one halves α until (2.8) holds; we shall show that this is possible under very weak conditions. If in step l a smaller damping factor was used, then it is improbable that the next step has a large damping factor. Therefore, one uses in the next iteration an adaptive initial step size, computed, for example, by

$$\bar{\alpha} = \min(1, 4\alpha_{\text{old}}) \tag{2.9}$$

or the more cautious strategy

$$\bar{\alpha} = \begin{cases} \alpha & \text{if } \alpha \text{ decreased in the previous step,} \\ \min(1, 2\alpha) & \text{otherwise.} \end{cases} \tag{2.10}$$

Because in high dimensions, the computation of the exact Newton direction may be expensive, we allow for approximate search directions p by requiring only a bound

$$\|F(x) + F'(x)p\| \le q'\|F(x)\| \tag{2.11}$$

on the residuals. For $F : D \subseteq \mathbb{R}^n \to \mathbb{R}^n$ the following algorithm results, with fixed numbers q, q' satisfying

$$0 < q < 1, \quad 0 < q' < 1 - q. \tag{2.12}$$

6.2.5 Algorithm: Damped Approximate Newton Method

STEP 1: Choose a starting value $x^0 \in D$ and compute $\|F(x^0)\|$. Set $l := 0$; $\alpha = 1$.

STEP 2: Determine p such that (2.11) holds, for example by solving $F'(x)p = -F(x)$.

STEP 3: Update α by (2.9) or (2.10).

STEP 4: Compute $\bar{x} = x + \alpha p$. If $\bar{x} = x$, stop.

STEP 5: If $\bar{x} \notin \text{int } D$ or $\|F(\bar{x})\| \ge (1 - q\alpha)\|F(x)\|$, then replace α with $\alpha/2$ and return to Step 4. Otherwise, replace l with $l + 1$ and return to Step 2.

6.2.6 Example.
We reconsider the determination of a zero of

$$F(x) := \begin{pmatrix} x_1 + 3\ln|x_1| - x_2^2 \\ 2x_1^2 - x_1 x_2 - 5x_1 + 1 \end{pmatrix}$$

from Example 6.2.4 with the same starting point $x^0 = (2, 2)^T$ for which the ordinary Newton method diverged. The damped Newton method (with $q = 0.5$, $q' = 0.4$) converges for $\|\cdot\| = \|\cdot\|_1$ to the solution $\lim_{l \to \infty} x^l = (1.3734784, -1.5249648)^T$, with both strategies for choosing the initial step size (Table 6.3). The use of (2.9) takes fewer iterations and hence Jacobian

Table 6.3. *Damped Newton method with 1-norm*

	l	$\|F(x^l)\|_1$	$\|p^l\|_1$	α^l	nf
With damping (2.9)	0	$5.08e+00$	$3.27e+01$	$6.25e-02$	6
	1	$4.14e+00$	$1.96e+00$	$2.50e-01$	8
	2	$3.31e+00$	$1.95e+00$	$2.50e-01$	12
	3	$2.69e+00$	$2.19e+00$	$2.50e-01$	16
	4	$2.31e+00$	$3.89e+00$	$6.25e-02$	22
	5	$2.21e+00$	$5.44e+00$	$3.12e-02$	27
	6	$2.16e+00$	$6.70e+00$	$3.12e-02$	31
	7	$2.12e+00$	$7.76e+00$	$3.12e-02$	35
	8	$2.09e+00$	$6.91e+00$	$3.12e-02$	39
	9	$2.04e+00$	$4.79e+00$	$1.25e-01$	41
	10	$1.88e+00$	$1.16e+00$	$5.00e-01$	43
	11	$8.59e-01$	$2.55e-01$	$1.00e+00$	45
	12	$3.93e-02$	$1.04e-02$	$1.00e+00$	47
	13	$8.11e-05$	$2.32e-05$	$1.00e+00$	49
	14	$3.92e-10$	$1.04e-10$	$1.00e+00$	51
With damping (2.10)	0	$5.08e+00$	$3.27e+01$	$6.25e-02$	6
	1	$4.14e+00$	$1.96e+00$	$6.25e-02$	8
	2	$3.89e+00$	$1.94e+00$	$1.25e-01$	10
	$\vdots$	$\vdots$	$\vdots$	$\vdots$	$\vdots$
	13	$1.25e+00$	$4.05e-01$	$5.00e-01$	37
	14	$6.00e-01$	$1.36e-01$	$1.00e+00$	39
	15	$3.30e-02$	$1.34e-02$	$1.00e+00$	41
	16	$1.78e-04$	$5.64e-05$	$1.00e+00$	43
	17	$2.91e-09$	$6.65e-10$	$1.00e+00$	45

evaluations, whereas (2.10) takes fewer function evaluations (displayed in the column labeled nf).

Similarly, the 2-norm gives convergence for both strategies, but not as fast as for the 1-norm cf. Table 6.4. However, for the maximum norm (and (2.9)), the iterates stall near $x^{22} = (-0.39215, -0.20505)^T$, although $\|F(x^{22})\|_\infty \approx 3.24$. The size of the correction $\|p^{22}\|_\infty \approx 2 \cdot 10^4$ is a sign that the Jacobian matrix is very ill-conditioned near x^{22}. There is a nearby singularity of the Jacobian, and our convergence analysis does not apply. The other initial step strategy also leads to nonconvergence with the maximum norm.

Convergence

We now investigate the feasibility and convergence of the damped Newton method. First, we show that the loop in Steps 4–5 is finite.

Table 6.4. *Damped Newton method, with damping 2.9*

With 2-norm			
l	$\|F(x^l)\|_2$	$\|p^l\|_2$	α^l	nf
0	$5.00e + 00$	$2.38e + 01$	$6.25e - 02$	6
1	$2.99e + 00$	$1.60e + 00$	$2.50e - 01$	8
2	$2.36e + 00$	$1.72e + 00$	$2.50e - 01$	12
3	$1.91e + 00$	$2.12e + 00$	$2.50e - 01$	16
⋮	⋮	⋮	⋮	⋮
12	$1.16e + 00$	$3.34e - 01$	$1.00e + 00$	50
13	$2.62e - 01$	$9.75e - 02$	$1.00e + 00$	52
14	$1.27e - 02$	$4.67e - 03$	$1.00e + 00$	54
15	$2.39e - 05$	$7.37e - 06$	$1.00e + 00$	56
16	$5.43e - 11$	$1.29e - 11$	$1.00e + 00$	58

With ∞-norm			
l	$\|F(x^l)\|_\infty$	$\|p^l\|_\infty$	α^l	nf
0	$5.00e + 00$	$2.02e + 01$	$1.25e - 01$	5
1	$4.36e + 00$	$4.59e + 00$	$2.50e - 01$	8
2	$3.36e + 00$	$6.30e + 00$	$3.12e - 02$	15
3	$3.28e + 00$	$1.04e + 01$	$7.81e - 03$	21
4	$3.26e + 00$	$1.40e + 01$	$3.91e - 03$	26
5	$3.25e + 00$	$1.83e + 01$	$3.91e - 03$	30
6	$3.25e + 00$	$3.02e + 01$	$9.77e - 04$	36
⋮	⋮	⋮	⋮	⋮
22	$3.24e + 00$	$2.89e + 04$	$1.86e - 09$	119
23	$3.24e + 00$	$5.40e + 04$	$4.66e - 10$	125
24	$3.24e + 00$	$9.07e + 04$	$1.16e - 10$	131
25	$3.24e + 00$	$1.27e + 05$	$5.82e - 11$	136

6.2.7 Proposition. *Let the function $F : D \subseteq \mathbb{R}^n \to \mathbb{R}^n$ be continuously differentiable. Suppose that $x \in$ int D, that $F'(x)$ is nonsingular, and that*

$$\|F(x) + F'(x)p\| \le q'\|F(x)\| \neq 0. \tag{2.13}$$

If (2.12) holds then

$$\|p\| \le (q' + 1)\|F'(x)^{-1}\|\|F(x)\| \tag{2.14}$$

and (2.8) holds for all sufficiently small $\alpha > 0$. If x^* is a regular zero and $\|x - x^*\|$ is sufficiently small, then (2.8) holds even for $0 < \alpha \le 1$.

Proof. From (2.13) we find

$$\|p\| \le \|F'(x)^{-1}\| \|F'(x)p\| \le \|F'(x)^{-1}\| (\|F(x) + F'(x)p\| + \|F(x)\|)$$
$$\le (q' + 1)\|F'(x)^{-1}\| \|F(x)\|;$$

so (2.14) holds. For $0 < \alpha \le 1$ and $x + \alpha p \in D$, it follows from (2.13) that

$$\|F(x + \alpha p)\| = \|F(x) + F'(x)\alpha p + o(\alpha \|p\|)\|$$
$$= \|(1 - \alpha)F(x) + \alpha(F(x) + F'(x)p)\| + o(\alpha \|p\|)$$
$$\le (1 - \alpha)\|F(x)\| + \alpha\|F(x) + F'(x)p\| + o(\alpha \|p\|)$$
$$\le (1 - (1 - q')\alpha)\|F(x)\| + o(\alpha \|p\|).$$

However, for sufficiently small α, we have $o(\alpha \|p\|) < (1 - q - q')\alpha \|F(x)\|$, so that (2.8) holds.

If x^* is now a regular zero, then $F'(x)^{-1}$ is defined and bounded as $x \to x^*$; therefore, (2.14) implies

$$\|p\| = O(\|F(x)\|).$$

Because $\|F(x)\| \to 0$ as $x \to x^*$, one obtains for sufficiently small $\|x - x^*\|$ the relations

$$x + \alpha p \in D$$

and

$$o(\alpha \|p\|) = o(\alpha \|F(x)\|) < (1 - q - q')\alpha \|F(x)\|$$

for $0 < \alpha \le 1$. Therefore, (2.8) now holds in this extended range. $\qquad \square$

It follows that if $F'(x)$ is nonsingular in D, the damped Newton method never cycles. By (2.8), α is changed only a finite number of times in each iteration. Moreover, in the case of convergence to a regular zero, for sufficiently large l, the first α is accepted immediately; and because of Step 3, one takes locally only undamped steps ($\alpha = 1$). In particular, locally quadratic convergence is retained if F' is Lipschitz continuous and p is chosen so that

$$\|F(x) + F'(x)p\| = O(\|F(x)\|^2).$$

In an important special case, the global convergence of the damped Newton method can be shown.

6.2.8 Theorem. *Suppose that the function $F : \mathbb{R}^n \to \mathbb{R}^n$ is uniquely invertible, and F and its inverse function F^{-1} are continuously differentiable. Then the sequence x^l ($l = 0, 1, 2, \ldots$) computed by the damped Newton method either terminates with $x^l = x^*$ or converges to x^*, where x^* is the unique zero of F.*

Proof. Because F^{-1} is continuously differentiable,

$$F'(x)^{-1} = (F^{-1})'(F(x))$$

exists; therefore, Step 2 can always be performed. By Proposition 6.2.7, l is increased after finitely many steps so that – unless by chance $F(x^l) = 0$ and we are done – the sequence x^l ($l = 0, 1, 2, \ldots$) is well defined. It is also bounded because $\|F(x^l)\| \leq \|F(x^0)\|$ and $\{x \in \mathbb{R}^n \mid \|F(x)\| \leq \|F(x^0)\|\}$ is bounded as the image of the bounded closed set $\{y \in \mathbb{R}^n \mid \|y\| \leq \|F(x^0)\|\}$ under the continuous mapping F^{-1}. Therefore, the sequence x^l has at least one limit point, that is, there is a convergent subsequence x^{l_ν} ($\nu = 0, 1, 2, \ldots$) whose limit we denote with x^*. Because of the monotonicity of $\|F(x^l)\|$, we have

$$\|F(x^*)\| = \inf_{l \geq 0} \|F(x^l)\|.$$

Similarly, (2.14) implies that the p^{l_ν} are bounded; so there exists a limit point p^* of the sequence p^{l_ν} ($\nu = 0, 1, 2, \ldots$). By deleting appropriate terms of the sequence if necessary, we may assume that $p^* = \lim_{\nu \to \infty} p^{l_\nu}$.

We now suppose that $F(x^*) \neq 0$ and derive a contradiction. By Proposition 6.2.7 there exists an $\alpha^* \in \left\{ 1, \frac{1}{2}, \frac{1}{4}, \frac{1}{8}, \ldots \right\}$ with

$$\|F(x^* + \alpha^* p^*)\| < (1 - q\alpha^*)\|F(x^*)\|;$$

Therefore, for all sufficiently large $l = l_\nu$ we have

$$\|F(x^l + \alpha^* p^l)\| < (1 - q\alpha^*)\|F(x^l)\|.$$

By construction of the α, it follows that $\alpha \geq \alpha^*$. In particular, $\alpha := \liminf \alpha_{l_\nu} \geq \alpha^* > 0$. If we now take the limit in

$$\|F(x^*)\| \leq \|F(\bar{x})\| = \|F(x + \alpha p)\| < (1 - q\alpha)\|F(x)\|,$$

one obtains the contradiction

$$\|F(x^*)\| \leq (1 - q\alpha)\|F(x^*)\| < \|F(x^*)\|.$$

Therefore, $\|F(x^*)\| = 0$; that is, x^* is the (uniquely determined) zero of F. In particular, there is only one limit point, that is, $\lim_{l \to \infty} x^l = x^*$. ☐

6.2.9 Remarks.

(i) In general, some hypothesis like that in Theorem 6.2.8 is necessary because the damped Newton method may otherwise converge slowly toward the set of points where the Jacobian is singular. Whether this happens depends, of course, a lot on the starting point x^0, and cannot happen if the level set $\{x \in D \mid \|F(x)\| \leq \|F(x^0)\|\}$ is compact and $F'(x)$ is invertible and bounded in this level set.

(ii) Alternative stopping criteria used in practice are (with some error tolerance $\varepsilon > 0$)

$$\|F(\tilde{x})\| \leq \varepsilon$$

or

$$\|\tilde{x} - x\| \leq \varepsilon \|x\|.$$

(iii) If the components of $F(x^0)$ or $F'(x^0)$ have very different orders of magnitude, then it is sensible to minimize the scaled function $\|CF(x)\|$ instead of $\|F(x)\|$, where C is an appropriate diagonal matrix. A useful possibility is to choose C such that $CF'(x^0)$ is equilibrated; if necessary, one can modify C during the computation.

(iv) The principal work for a Newton step consists of the determination of the matrix A as a suitable approximation to $F'(x)$, and the solution of the system of linear equations $Ap = -F(x)$. Compared with this, several evaluations of the function F usually play a subordinate role.

6.3 Error Analysis

Limiting Accuracy

The relation

$$F(\tilde{x}) = F[\tilde{x}, x^*](\tilde{x} - x^*) \tag{3.1}$$

derived in Lemma 6.1.2 gives

$$\|\tilde{x} - x^*\| = \|F[\tilde{x}, x^*]^{-1} F(\tilde{x})\| \leq \|F[\tilde{x}, x^*]^{-1}\| \cdot \|F(\tilde{x})\|.$$

If $\tilde{x}$ is the computed approximation to x^*, then one can expect only that $\|F(\tilde{x})\| \leq \varepsilon_F$ for some bound ε_F of the accuracy of function values that depends on F and on the way in which F is evaluated. This gives for the error in

$\tilde{x}$ the approximate bound

$$\|\tilde{x} - x^*\| \leq \|F[\tilde{x}, x^*]^{-1}\|\varepsilon_F \approx \|F'(x^*)^{-1}\|\varepsilon_F.$$

Therefore, the *limiting accuracy* achievable is of the order of $\|F'(x^*)^{-1}\|\varepsilon_F$, and *the maximal error in $F(\tilde{x})$ can be magnified by a factor of up to $\|F'(x^*)^{-1}\|$.* In the following, we make this approximate argument more rigorous.

Norm-wise Error Bounds

We deduce from Banach's fixed-point theorem a constructive existence theorem for zeros and use it for giving explicit and rigorous error bounds.

6.3.1 Theorem. *Suppose that the function $F : D \subseteq \mathbb{R}^n \to \mathbb{R}^n$ is continuously differentiable and that $B := \{x \in \mathbb{R}^n \,|\, \|x - x^0\| \leq \varepsilon\}$ lies in D.*

(i) If there is a matrix $C \in \mathbb{R}^{n \times n}$ with

$$\|I - CF'(x)\| \leq \beta < 1 \quad \text{for all } x \in B,$$

and

$$\varepsilon_0 := \|CF(x^0)\| \leq \varepsilon(1 - \beta),$$

then F has exactly one zero x^* in B, and

$$\varepsilon_0/(1 + \beta) \leq \|x^0 - x^*\| \leq \varepsilon_0/(1 - \beta) \leq \varepsilon.$$

(ii) Moreover, if $\beta < 1/3$ and $\varepsilon_0 \leq \varepsilon/3$, then Newton's method started from x^0 cannot break down and converges to x^*.

Proof. Because $\|I - CF'(x)\| < 1$, C and $F'(x)$ are nonsingular for all $x \in B$. If we define $G : D \to \mathbb{R}^n$ by $G(x) := x - CF(x)$, then $F(x^*) = 0 \Leftrightarrow CF(x^*) = 0 \Leftrightarrow x^* = G(x^*)$. Thus we must show that G has exactly one fixed point in B. For this purpose, we prove that G has the contraction property. First, $G'(x) = I - CF'(x)$, so $\|G'(x)\| \leq \beta < 1$ for $x \in B$, and $\|G(x) - G(y)\| \leq \beta\|x - y\|$ for all $x, y \in B$. In particular, for $x \in B$,

$$\begin{aligned}
\|G(x) - x^0\| &\leq \|G(x) - G(x^0)\| + \|G(x^0) - x^0\| \\
&\leq \beta\|x - x^0\| + \|CF(x^0)\| \\
&\leq \beta\varepsilon + \varepsilon(1 - \beta) \leq \varepsilon;
\end{aligned}$$

therefore, $G(x) \in B$ for $x \in B$. So G is a contraction with contraction factor β, and (i) follows from the Banach fixed-point theorem.

For the proof of (ii), we consider the function $\tilde{G} : D \to \mathbb{R}^n$ defined by

$$\tilde{G}(x) := x - \tilde{A}^{-1} F(x)$$

in which $\tilde{x} \in B$ and $\tilde{A} := F'(\tilde{x})$ are fixed. Then

$$\|I - C\tilde{A}\| \leq \beta, \quad \|(C\tilde{A})^{-1}\| \leq \frac{1}{1 - \beta},$$

whence for $x \in B$,

$$
\begin{aligned}
\|\tilde{G}'(x)\| &= \|I - \tilde{A}^{-1} F'(x)\| \\
&= \|(C\tilde{A})^{-1}(I - CF'(x) - (I - C\tilde{A}))\| \\
&\leq \|(C\tilde{A})^{-1}\|(\|I - CF'(x)\| + \|I - C\tilde{A}\|) \\
&\leq \frac{1}{1 - \beta}(\beta + \beta) = \frac{2\beta}{1 - \beta}.
\end{aligned}
$$

It follows from this that

$$\|\tilde{G}(x) - \tilde{G}(y)\| \leq \frac{2\beta}{1 - \beta}\|x - y\| \quad \text{for } x, y \in B. \tag{3.2}$$

We apply this inequality to Newton's method,

$$x^{l+1} := x^l - F'(x^l)^{-1} F(x^l).$$

For $l = 0$, (i) and the hypothesis give $\|x^0 - x^*\| \leq \varepsilon_0/(1 - \beta) \leq \frac{3}{2}\varepsilon_0 \leq \frac{1}{2}\varepsilon$. If now $\|x^l - x^*\| \leq \frac{1}{2}\varepsilon$ for some $l \geq 0$, then $\|x^l - x^0\| \leq \|x^l - x^*\| + \|x^0 - x^*\| \leq \varepsilon$, so $x^l \in B$. We can therefore apply (3.2) with $x = \tilde{x} = x^l$, $y = x^*$ to obtain

$$\|x^{l+1} - x^*\| \leq \frac{2\beta}{1 - \beta}\|x^l - x^*\|$$

because $x^{l+1} = \tilde{G}(x^l)$. Because $\beta < \frac{1}{3}$, it follows that $2\beta/(1 - \beta) < 1$ and $\|x^{l+1} - x^*\| \leq \frac{1}{2}\varepsilon$; from this it follows by induction that

$$\|x^l - x^*\| \leq (2\beta/(1 - \beta))^l \|x^0 - x^*\| \to 0 \quad \text{as } l \to \infty.$$

Thus Newton's method is well-defined and convergent. $\quad\square$

6.3.2 Remarks.

(i) The theorem is useful to provide rigorous *a posteriori* error bounds for approximate zeros found by other means. For the maximum norm,

$$B = [x^0 - \varepsilon e, x^0 + \varepsilon e] =: \mathbf{x} \in \mathbb{IR}^n,$$

and β can be determined from $\beta := \|I - CF'(\mathbf{x})\|_\infty$ by means of interval arithmetic.

(ii) The best choice of the matrix C is an approximation to $F'(x^0)^{-1}$ or $(\text{mid } F'(\mathbf{x}))^{-1}$. Indeed, for $C = F'(x^0)^{-1}, x^0 \to x^*$ and $\varepsilon \to 0$, the contraction factor β approaches zero. Therefore, one can regard β as a measure of the nearness of x^0 to x^* relative to the degree of nonlinearity of F.

(iii) In an implementation, one must choose ε and ε_0 appropriately. Ideally, ε should be as a small as possible because β generally becomes larger as B becomes wider. However, because $\varepsilon_0 \le \varepsilon(1 - \beta)$ must be guaranteed, one cannot choose ε too small. A useful way to proceed is as follows: One determines the bounds

$$\|CF(x^0)\|_\infty \le \varepsilon_0, \quad \|I - CF'(x^0)\|_\infty \le \beta_0 < 1$$

and then sets $\varepsilon := 2\varepsilon_0/(1 - \beta_0)$. Then one tests the relation

$$\|I - CF'(\mathbf{x})\|_\infty = \beta \le \frac{1 + \beta_0}{2} \quad \text{for } \mathbf{x} := [x^0 - \varepsilon e, x^0 + \varepsilon e], \quad (3.3)$$

from which the required inequality follows. Generally, if (3.3) fails, the zero is either very ill conditioned or x^0 was not close to a zero.

6.3.3 Example. To show that

$$F(x) = \begin{pmatrix} x_1^2 + 5x_1 + 8x_2 - 5 \\ x_2^2 + 5x_2 - 8x_1 + 1 \end{pmatrix}$$

has a zero in the box $\mathbf{x} = \begin{pmatrix} [0,1] \\ [0,1] \end{pmatrix}$, we apply Theorem 6.3.1 with the maximum norm and $x^0 = \begin{pmatrix} 0.5 \\ 0.5 \end{pmatrix}$, $\varepsilon = 0.5$ (so that $B = \mathbf{x}$). We have

$$F'(x) = \begin{pmatrix} 2x_1 + 5 & 8 \\ -8 & 2x_2 + 5 \end{pmatrix}, \quad F'(\mathbf{x}) = \begin{pmatrix} [5,7] & 8 \\ -8 & [5,7] \end{pmatrix},$$

and for

$$C := (\text{mid } F'(\mathbf{x}))^{-1} = \begin{pmatrix} 0.06 & -0.08 \\ 0.08 & 0.06 \end{pmatrix},$$

we find

$$F(x^0) = \begin{pmatrix} 1.75 \\ -0.25 \end{pmatrix}, \quad CF(x^0) = \begin{pmatrix} 0.125 \\ 0.125 \end{pmatrix},$$

$$I - CF'(\mathbf{x}) = [-1, 1] \begin{pmatrix} 0.06 & 0.08 \\ 0.08 & 0.06 \end{pmatrix}.$$

Thus $\varepsilon_0 = 0.125$, $\beta = 0.14$, and $\varepsilon_0 \leq \varepsilon(1 - \beta)$. Thus there is a unique zero x^* in $\mathbf{x}$, and we have

$$0.109 < \frac{0.125}{1.14} \leq \|x^* - x^0\|_\infty \leq \frac{0.125}{0.86} < 0.146.$$

In particular, $x^* \in \left(\begin{smallmatrix}[0.354, 0.646]\\ [0.354, 0.646]\end{smallmatrix}\right)$. Better approximations would have given sharper bounds.

The Interval Newton Method

In order to obtain simultaneous componentwise bounds that take rounding error into account, one uses an interval version of Newton's method. We suppose for this purpose that $\mathbf{x}$ is an interval that contains the required zero x^*, and we calculate $F'(\mathbf{x})$. Then $\underline{F'(\mathbf{x})} \leq F'(\xi) \leq \overline{F'(\mathbf{x})}$ for all $\xi \in \mathbf{x}$ and $F[\tilde{x}, x^*] = \int_0^1 F'(x^* + t(\tilde{x} - x^*)) \, dt \in F'(\mathbf{x})$ for all $\tilde{x} \in \mathbf{x}$. Therefore, (3.1) gives

$$x^* = \tilde{x} - \tilde{A}^{-1} F(\tilde{x}) \in \tilde{x} - \Box\Sigma(F'(\mathbf{x}), F(\tilde{x})),$$

where $\Sigma(F'(\mathbf{x}), F(\tilde{x}))$ denotes the hull of the solution set of the linear interval equation $\tilde{A}\tilde{z} = F(\tilde{x})$ $(\tilde{A} \in F'(\mathbf{x}), \tilde{x} \in \mathbf{x})$. This gives rise to the *interval Newton* iteration

$$\mathbf{x}^{l+1} := (\tilde{x}^l - \Box\Sigma(F'(\mathbf{x}^l), F(\tilde{x}^l))) \cap \mathbf{x}^l \quad \text{with } \tilde{x}^l \in \mathbf{x}^l. \tag{3.4}$$

By our derivation, each $\mathbf{x}^l$ contains any zero in $\mathbf{x}^0$.

In each step, a system of linear equations must be solved. In practice, one uses in place of the hull a more easily computable enclosure, using an approximate midpoint inverse C as preconditioner.

Following Krawczyk, one can proceed in a slightly simpler way, also using C but saving the solution of the linear interval system. We rewrite (3.1) as

$$CF(\tilde{x}) = CF[\tilde{x}, x^*](\tilde{x} - x^*)$$

and observe that

$$x^* = \tilde{x} - CF(\tilde{x}) - (I - CF[\tilde{x}, x^*])(\tilde{x} - x^*),$$

where $I - CF[\tilde{x}, x^*]$ can be expected to be small by the construction of C. From this it follows that

$$x^* \in K(\mathbf{x}, \tilde{x})$$

for $\tilde{x}, x^* \in \mathbf{x}$, where the *Krawczyk operator* $K : \mathbb{IR}^n \times \mathbb{R}^n \to \mathbb{IR}^n$ is defined by

$$K(\mathbf{x}, \tilde{x}) := \tilde{x} - CF(\tilde{x}) - (I - CF'(\mathbf{x}))(\tilde{x} - \mathbf{x}). \tag{3.5}$$

Therefore, one may use in place of the interval Newton iteration (3.4) the *Krawczyk iteration*

$$\mathbf{x}^{l+1} := K(\mathbf{x}^l, \tilde{x}^l) \cap \mathbf{x}^l \quad \text{with } \tilde{x}^l \in \mathbf{x}^l,$$

again with a guarantee of not losing any zero contained in $\mathbf{x}^0$.

Note that because of the lack of the distributive law for intervals, the definition of K cannot be "simplified" to $\mathbf{x} + C(F'(\mathbf{x})(\tilde{x} - \mathbf{x}) - F(\tilde{x}))$; this would produce poor results only, because the radius would be bigger than that of $\mathbf{x}$.

It is important to remember that in order to take rounding error into account, one *must* use in the interval Newton iteration or the Krawczyk iteration in place of $\tilde{x}^l$ the thin interval $\tilde{\mathbf{x}}^l = [\tilde{x}^l, \tilde{x}^l]$.

The Krawczyk operator can also be used to prove the existence of a zero in $\mathbf{x}$.

6.3.4 Theorem. *Suppose that $\tilde{x} \in \mathbf{x}$.*

(i) If F has a zero $x^* \in \mathbf{x}$ then $x^* \in K(\mathbf{x}, \tilde{x}) \cap \mathbf{x}$.
(ii) If $K(\mathbf{x}, \tilde{x}) \cap \mathbf{x} = \emptyset$ then F has no zero in $\mathbf{x}$.
(iii) If either

$$K(\mathbf{x}, \tilde{x}) \subseteq \text{int}\, \mathbf{x} \tag{3.6}$$

or the matrix C in (3.5) is nonsingular and

$$K(\mathbf{x}, \tilde{x}) \subseteq \mathbf{x} \tag{3.7}$$

then F has a unique zero $x^* \in \mathbf{x}$.

Proof. (i) holds by construction, and (ii) is an immediate consequence of (i). (iii) Suppose first that (3.6) holds. Because $B := I - CF'(\tilde{x}) \subseteq I - CF'(\mathbf{x}) =: B$, we have

$$|B|\, \text{rad}\, \mathbf{x} = |B|\, \text{rad}(\tilde{x} - \mathbf{x}) = \text{rad}(B(\tilde{x} - \mathbf{x}))$$
$$\leq \text{rad}(B(\tilde{x} - \mathbf{x})) = \text{rad}\, K(\mathbf{x}, \tilde{x}) < \text{rad}\, \mathbf{x}$$

by (3.6). For $D = \mathrm{Diag}(\mathrm{rad}\,\mathbf{x})$, we therefore have

$$\|I - D^{-1}CF'(\tilde{x})D\|_\infty = \|D^{-1}BD\|_\infty = \max_i \sum_k \frac{|B|_{ik}\,\mathrm{rad}\,\mathbf{x}_k}{\mathrm{rad}\,\mathbf{x}_i}$$

$$= \max_i \frac{(|B|\,\mathrm{rad}\,\mathbf{x})_i}{\mathrm{rad}\,\mathbf{x}_i} < 1.$$

Thus $CF'(\tilde{x})$ is an H-matrix, hence nonsingular, and as a factor, C is nonsingular.

Thus it suffices to consider the case where C is nonsingular and (3.7) holds. In this case, $K(\mathbf{x}, \tilde{x})$ is (componentwise) a centered form for $G(x) := x - CF(x)$; therefore, $G(x) \in K(\mathbf{x}, \tilde{x}) \subseteq \mathbf{x}$ for all $x \in \mathbf{x}$. By Brouwer's fixed-point theorem, G has a unique fixed point $x^* \in \mathbf{x}$, that is, $x^* = x^* - CF(x^*)$. Because C is nonsingular, it follows that x^* is the unique zero of F in $\mathbf{x}$. $\square$

If $\mathbf{x}^0$ is a narrow (but not too narrow) box symmetric around a good approximation of a well-conditioned zero, Theorem 6.3.4 usually allows one to verify existence in that box, and by refining it with Krawczyk's iteration, we may get very narrow enclosures.

For large and sparse problems, the approximate inverse C is full, which makes Krawczyk's method prohibitively expensive. A modification that respects sparsity is given in Rump [84].

Finding All Zeros

If one has found a zero ξ of F and conjectures that another zero x^* exists, then just as in the univariate case, one can attempt to determine it by deflation. For this, one needs a damped Newton method that minimizes $\|F(x)\|/\|x - \xi\|$ instead of $\|F(x)\|$. However, unlike in dimension 1, this function is no longer smooth near ξ. Therefore, deflation in higher dimension tends to give erratic results, and repeated deflation often finds several but rarely all the zeros.

All zeros in a specified box can be found by interval methods. In analogy to the univariate case, interval methods for finding all zeros scan a given initial box $\mathbf{x} \in \mathbb{IR}^n$ by exhaustive covering with subboxes that contain no zero, are tiny and guaranteed to contain a unique regular zero, or are tiny and likely to contain a nonregular zero. This is done by recursively splitting $\mathbf{x}$ until corresponding tests apply, such as those based on the Krawczyk operator. For details, see Hansen [39], Kearfott [49], and Neumaier [70].

Continuation methods, discussed in the next section, can also find all zeros of *polynomial* systems, at least with high probability (in theory with probability one). However, they cannot guarantee to find real zeros before complex ones. Because high order polynomial systems usually have a number of real or complex zeros that is, exponential in the dimension, continuation methods for finding all zeros are limited to low dimensions.

6.4 Further Techniques for Nonlinear Systems

An Affine Invariant Newton Method

The nonlinear systems arising in solving differential equations, for example, for stiff initial value problems by backward differentiation formulas (Section 4.5), have quite often very ill-conditioned Jacobians. As a consequence, $\|F(x)\|$ has steep and narrow curved valleys and a damped Newton method based on minimizing $\|F(x)\|$ has difficulties because it must follow these valleys with very short steps.

For such problems, the natural merit function for measuring the deviation from a zero is a norm $\|\Delta(x)\|$ of the Newton correction $\Delta(x) = -F'(x)^{-1}F(x)$. Near a zero, this gives the distance to it almost exactly because of the local quadratic convergence of Newton's method. Moreover, it is unaffected by the condition of $F'(x)$ because it is *affine invariant*, that is, any linear transformation $\tilde{F}(x) := CF(x)$ with nonsingular C yields the same $\Delta(x)$. Indeed, the matrix C cancels out because

$$\tilde{F}'(x)^{-1}\tilde{F}(x) = (CF'(x))^{-1}CF(x) = F'(x)^{-1}F(x).$$

However, the affinely equivalent function $\tilde{F}(x) = F'(x^0)^{-1}F(x)$ has well-conditioned $\tilde{F}'(x^0) = I$. Therefore, minimizing $\|\Delta(x)\|$ instead of $\|F(x)\|$ improves the geometry of the valleys and allows one to proceed in larger steps in ill-conditioned problems.

Unfortunately, $\|\Delta(x)\|$ may *increase* along the *Newton path*, defined by the family of solutions $x^*(t)$ of $F(x) = (1 - t)F(x^0)$, $(t \in [0, 1])$ along which a method based on decreasing $\|F(x)\|$ would roughly proceed (look at infinitesimal steps to see this). Hence, this variant may get stuck in a local minimum of $\|\Delta(x)\|$ in many situations where methods based on decreasing $\|F(x)\|$ converge to a zero.

A common remedy is to test for a sufficient decrease by

$$\|F'(x^l)^{-1}F(x^l + \alpha_l p_l)\| \le (1 - q\alpha)\|F'(x^l)^{-1}F(x^l)\| \qquad (4.1)$$

for some positive $q < 1$, so that the matrix is kept fixed during the determination of a good step size (see, e.g., Deuflhard [19, 20], who also discusses other implementation details). Unfortunately, this means that the merit function changes at each step, making a global convergence proof along the old lines impossible. Indeed, the method can cycle for innocuous problems, where methods that use $\|F(x)\|$ have no problems (cf. Exercise 9).

However, a global convergence proof can be given if we mix the two approaches, using an acceptance test similar to (4.1) in regions where $\|\Delta(x)\|$ increases, and switching back to the merit function $\|\Delta(x)\|$ when it has moved sufficiently far over the region of increase.

The algorithm presented next is geared toward large-scale applications where it is often inconvenient to solve exactly for the Newton step. We therefore assume that in each step we have a (possibly crude) approximation $M(x)$ to $F'(x)$ whose structure is simple enough that a factorization of $M(x)$ is available. Often, $M(x) = LR$, where L and R are approximate triangular factors of $F'(x)$, obtained by ignoring terms in the factorization that would destroy sparsity. One speaks in this case of *incomplete triangular factorizations*, and calls $M(x)$ the *preconditioner*. The case $M(x) = F'(x)$ is, of course, admissible and produces an affine invariant algorithm.

We now define the modified Newton correction

$$\Delta(x) = -M(x)^{-1} F(x). \tag{4.2}$$

We use an arbitrary but fixed norm that should be a natural measure for changes in x. The 1-norm is suitable if all components of x have about the same magnitude. The algorithm depends on some constants that must satisfy the restrictions

$$0 < q_0 < q_1 < q_2 < 1, \quad 0 < \mu_1 < \mu_2 < 1 - q_0. \tag{4.3}$$

6.4.1 Algorithm: Affine Invariant Modified Newton Method

STEP 0: Start with $l := 0, k := -1, \delta_{-1} = \infty$.

STEP 1: Calculate a factorization of $M(x^l)$ and $\Delta(x^l) = -M(x^l)^{-1} F(x^l)$. If

$$\|\Delta(x^l)\| \le q_2 \delta_k, \tag{4.4}$$

set $k := k + 1$, $M_k := M(x^l)$ and $\delta_k := \|\Delta(x^l)\|$.

STEP 2: Calculate an approximate Newton direction p^l such that

$$\left\| M_k^{-1}(F(x^l) + F'(x^l)p^l) \right\| \le q_0 \left\| M_k^{-1} F(x^l) \right\|. \tag{4.5}$$

STEP 3: Find α_l such that either

$$\left\| M_k^{-1} F(x^l + \alpha_l p^l) \right\| \le q_1 \delta_k \tag{4.6}$$

or

$$1 - \mu_2 \alpha_l \leq \frac{\|M_k^{-1} F(x^l + \alpha_l p^l)\|}{\|M_k^{-1} F(x^l)\|} \leq 1 - \mu_1 \alpha_l. \qquad (4.7)$$

STEP 4: Set $x^{l+1} := x^l + \alpha_l p^l$, $l := l + 1$. If (4.6) was violated, goto Step 2, else goto Step 1.

We first show that Step 3 can always be completed, at least when F is defined everywhere.

6.4.2 Proposition. *Suppose that* $F(x^l + \alpha p^l)$ *is defined for* $0 \leq \alpha \leq \bar{\alpha}$. *If*

$$\bar{\alpha} > (1 - q_1)/\mu \quad \text{with } \mu_1 \leq \mu \leq \mu_2 \qquad (4.8)$$

then either (4.7) *holds for* $\alpha_l = \bar{\alpha}$, *or the function defined by*

$$\varphi(\alpha) := \frac{1}{\alpha} \left(\frac{\|M_k^{-1} F(x^l + \alpha p^l)\|}{\|M_k^{-1} F(x^l)\|} - 1 \right) + \mu \qquad (4.9)$$

has a zero $\hat{\alpha} \in [0, \bar{\alpha}]$. *In particular, Step 3 can be completed if* $\bar{\alpha} > (1 - q_1)/\mu_2$.

Proof. In general, we have

$$\|M^{-1} F(x + \alpha p)\| = \|M^{-1}(F(x) + \alpha F'(x)p)\| + o(\alpha)$$
$$= \|(1 - \alpha)M^{-1} F(x) + \alpha M^{-1}(F(x) + F'(x)p)\| + o(\alpha)$$
$$\leq |1 - \alpha| \|M^{-1} F(x)\| + |\alpha| \|M^{-1}(F(x) + F'(x)p)\|$$
$$+ o(\alpha).$$

By (4.5) and (4.8), we conclude that for sufficiently small $\alpha > 0$,

$$\frac{\|M_k^{-1} F(x^l + \alpha p^l)\|}{\|M_k^{-1} F(x^l)\|} \leq 1 - \alpha + \alpha q_0 + o(\alpha) < 1 - \alpha \mu, \qquad (4.10)$$

so that $\varphi(\alpha) < 0$ for small $\alpha > 0$. If $\varphi(\bar{\alpha}) < 0$ then at $\alpha = \bar{\alpha}$ this quotient is $\leq 1 - \bar{\alpha} \mu \leq q_1$. However, by (4.7) for the previous steps, we find $\|M_k^{-1} F(x^l)\| \leq \|M_k^{-1} F(x^k)\| = \delta_k$, whence (4.6) holds for $\alpha_l = \bar{\alpha}$.

Also, if $\varphi(\bar{\alpha}) \geq 0$, then φ has a zero $\hat{\alpha} \in [0, \bar{\alpha}]$. In particular, if $\bar{\alpha} > 2(1 - q_1)/(\mu_1 + \mu_2)$, then (4.8) holds for $\mu = (\mu_1 + \mu_2)/2$, and (4.7) is satisfied for $\alpha_l = \hat{\alpha}$. $\qquad \square$

Note that with $\mu = (\mu_1 + \mu_2)/2$, approximate zeros of φ also satisfy (4.7), so that one may adapt any root finding method based on sign changes to check each iterate for (4.6) and (4.7), and quit as soon as one of the two conditions holds. In order to preserve fast local convergence, one should first try $\alpha_l = 1$. (One may also use more sophisticated initial values for α_l as long as the choice $\alpha_l = 1$ results locally.)

If the first guess does not work and $\varphi(\alpha_l) > 0$, one already has a sign change using the artificial value $\varphi(0) = \mu - 1 + q_0 < 0$ that (cf. (4.10)) is an upper bound to the true range of limiting values of $\varphi(\alpha)$ as $\alpha \to 0$. If, however, $\varphi(\alpha_l) < 0$, increase α_l by repeated doubling (or so) until (4.6) or (4.7) holds or a sign change is found.

6.4.3 Remarks.

(i) One must be able to store two factorizations, those of M_k and $M(x^l)$.
(ii) In Step 2, $p^l = \Delta(x^l)$ works if $\|I - M(x^l)^{-1}F'(x^l)\| \le q_0$, that is, if $M(x)$ is a sufficiently good approximation of $F'(x)$. In practice, one may want to choose q_0 adaptively, matching it to the accuracy of x^l.

In general, all sorts of misbehavior may occur besides regular convergence (and, as in the univariate case, this must be the case for any algorithm without additional conditions that guarantee the existence of a solution). The possibilities are classified in the following.

6.4.4 Proposition. *Denote by z^k the values of x^l at each completion of Step 1. Then one of the following holds:*

(i) $F(z^k) \to 0$ as $k \to \infty$.
(ii) For some subsequence, $\|z^{k_v}\| \to \infty$ and $\|M(z^{k_v})\| \to \infty$.
(iii) $F'(x^l)$ is undefined or singular for some l.
(iv) At some iteration, $F(x^l + \alpha p^l)$ is not defined for large α, and Step 3 cannot be completed (cf. Proposition 6.4.2).

Proof. Suppose neither (iii) nor (iv) holds. Then Step 2 can be satisfied by choosing $p^l = -F'(x^l)^{-1}F(x^l)$, and Step 3 can be completed by assumption.

If k remains bounded, then for k fixed at its largest value, (4.6) will always be violated, and therefore (4.7) must hold. Equation (4.7) implies that $\|M_k^{-1}F(x^l)\|$ is monotone decreasing, and because (4.6) is violated, it must converge to a positive limit. Going to the limit in (4.7) therefore gives $\alpha_l \to 0$. However, now (4.10) with $\mu = \mu_2$ gives a contradiction.

Hence, k increases without bound. By Steps 1 and 2, $\delta_{k+1} \leq q_2 \delta_k$, hence $\delta_k \leq q_2^k \delta_0$. Therefore,

$$\|F(z^k)\| = \|M_k \Delta(z^k)\| \leq \|M_k\| \delta_k \leq \|M_k\| q_2^k \delta_0. \tag{4.11}$$

If $\|M_k\|$ remains bounded (and by continuity, this holds in particular when z_k remains bounded), then the right side of (4.11) converges to zero and (i) holds. If not, then (ii) holds. □

In particular, in an important case where all difficulties are absent, we have the following global convergence theorem.

6.4.5 Theorem. *If F is continuously differentiable in $\mathbb{R}^n$ with globally nonsingular $F'(x)$ and bounded $M(x)$, then*

$$\lim_{k \to \infty} F(z^k) = 0.$$

Proof. In Proposition 6.4.4, (ii) contradicts the boundedness of M, (iii) the nonsingularity of F', and (iv) cannot hold because F is defined everywhere. The only alternative left is (i). □

Quasi-Newton Methods

It is possible to generalize the univariate secant method to higher dimensions; the resulting methods are called *quasi-Newton methods*. Frequently, they perform well, although they are generally regarded as somewhat less robust because due to the approximations involved, it is not guaranteed that $\|F(x + \alpha p)\|$ decreases along a computed quasi-Newton direction p. (Indeed, to guarantee that, one needs derivative information.)

To motivate the method, we write the univariate secant method in the form $x_{l+1} = x_l - c_l f(x_l)$, where $c_l = 1/f[x_l, x_{l-1}]$. Noting that $c_l(f(x_l) - f(x_{l-1})) = x_l - x_{l-1}$, we see that a natural multivariate generalization is to use

$$x_{l+1} = x_l - C_l F(x_l),$$

where C_l is a square matrix satisfying the *secant condition*

$$C_l y_l = s_l, \quad \text{with } y_l := F(x_l) - F(x_{l-1}), \quad s_l := x_l - x_{l-1}.$$

Note that no linear system must be solved, so that for problems with a dense Jacobian, the linear algebra work is reduced from $O(n^3)$ operations per step to $O(n^2)$.

The secant condition can be easily enforced by defining

$$C_l = C_{l-1} + \frac{(s_l - C_{l-1}y_l)u_l^T}{u_l^T y_l}$$

with an arbitrary vector $u_l \neq 0$. Among many possibilities, the choice regarded as best for well-scaled x is

$$u_l = C_{l-1}^T s_l.$$

This defines *Broyden's method* and has reasonably good local convergence properties (see Dennis and Schnabel [18] and Gay [28]). Of course, one loses quadratic convergence for such methods, and global convergence proofs are available only under stringent conditions. However, one can ensure local super-linear convergence with convergence order at least $\sqrt[n+1]{2}$. Therefore, the number of significant digits doubles about every n iterations. This implies that, for problems with dense Jacobians, quasi-Newton methods are locally not much faster than a discrete Newton method, which uses n function values to get a Jacobian approximation. For sufficiently sparse problems, much fewer function values suffice and discrete Newton methods are usually faster.

However, far from a solution, the convergence speed is much less predictable. One again needs a damping strategy, and hence uses the search direction $p_l = -C_l F(x_l)$ in place of the Newton direction.

Continuation Methods

For functions that satisfy only weak smoothness conditions there are special algorithms that are based on topological concepts: homotopies (that generate a smooth path to a solution), and simplicial decompositions (that generate piecewise linear paths). An in-depth discussion is given in Allgower and Georg [5]; see also Morgan [64] and Zangwill and Garcia [100]. Here we only outline the basic idea of homotopy methods.

For $t \in [0, 1]$ one considers curves $x = x(t)$ (called *homotopies*) for which $x(0) = x^0$ is a known point and $x(1) = x^*$ is a zero of F. Such curves are implicitly defined by the solution set of $F(x, t) = 0$, with an augmented expression $F(x, t)$ that satisfies $F(x^0, 0) = 0$ and $F(x, 1) = F(x)$ for all x. The basic examples are

$$F(x, t) = F(x) - (1 - t)F(x^0)$$

and

$$F(x, t) = t F(x) - (1 - t)F'(x^0)(x - x^0).$$

Both choices are natural in that they are affine invariant, that is, the curves on which the zeros lie are not affected by linear transformations of x or F.

The task of solving $F(x) = 0$ is now replaced by tracing the curves from the known starting point at $t = 0$ to the point given by $t = 1$. Because the curve guides the way to go, this may be the easiest way to arrive at a solution for many hard zerofinding problems. However, it is possible that the path defined by $F(x, t) = 0$ never arrives at $t = 1$, or turns back, or bifurcates into several paths, and all this must be handled by a robust code. Existence of the path can often be guaranteed by a fixed-point theorem.

The path-following problem is the special case $m = 1$, $E = [0, 1]$ of the more general problem of solving a parametrized system of equations

$$F(x, y) = 0$$

for x in dependence on a vector y of parameters from a set $E \subseteq \mathbb{R}^m$, where $F : D \times E \to \mathbb{R}^n, x \in D \subseteq \mathbb{R}^n$. This system of equations determines an m-dimensional manifold of solutions, and often a functional dependence $H : E \to D$ such that $F(H(y), y) = 0$ is sought. (Of course, in general, the manifold need not have a one-to-one projection onto E, and this function may have to be pieced together from several solution branches.)

The well-known theorem on implicit functions gives sufficient conditions for the existence of H in a sufficiently small neighborhood of a solution. A global statement about the existence of H is made in the following theorem, whose proof (see Neumaier [68]) depends on a detailed topological analysis of the situation. As in the fixed-point theorem of Leray and Schauder, a boundary condition is the main thing to be established.

6.4.6 Theorem. *Let $D \subseteq \mathbb{R}^n$ be closed, let $E \subseteq \mathbb{R}^m$ be simply connected, and let $F : D \times E \to \mathbb{R}^n$ be such that*

(i) the derivative $\partial_1 F(x, y)$ of F with respect to x exists in $D \times E$ and is continuous and nonsingular there;
(ii) there exist $x^0 \in \text{int } D$ and $y^0 \in E$ such that

$$F(x, y) \neq t F(x^0, y^0) \quad \text{for all } x \in \partial D, y \in E, t \in [0, 1].$$

Then there is a continuous function $H : E \to D$ with $F(H(y), y) = 0$ for $y \in E$.

To trace the curve determined by

$$F(x(t), t) = 0 \tag{4.12}$$

there are a variety of *continuation methods*. The smooth continuation methods assume that F is continuously differentiable and are based on the differential equation

$$F_x(x(t), t)\dot{x}(t) + F_t(x(t), t) = 0 \qquad (4.13)$$

obtained by differentiating (4.12), using the chain rule. This is an implicit differential equation, but as long as F_x is nonsingular, we can solve for $\dot{x}(t)$,

$$\dot{x}(t) = -F_x(x(t), t)^{-1} F_t(x(t), t). \qquad (4.14)$$

(A closer analysis of the singular case leads to turning points and bifurcation points; see, e.g., Chow and Hale [11] and Seydel [88]. Now, (4.14) can be solved by methods for ordinary differential equations. Alternatively, one may solve (4.13) by methods for differential-algebraic equations; cf. Brenan, Campbell, and Petzold [9].)

If F_x is well conditioned, (4.14) is easy to solve. However, for the application to solving systems of equations by a homotopy, the well-conditioned case is less relevant because one solves such problems easily with damped Newton methods. In the ill-conditioned case, (4.14) is stiff, but advantage can be taken of good predictors computed by extrapolating from past values.

Due to rounding errors and discretization errors in the solution of the differential equation, a numerical solution based on (4.14) alone tends to wander away from the manifold. To avoid this, one must monitor the residual size $\|F(\tilde{x}(t), t)\|$ of the approximate solution curve $\tilde{x}(t)$, and reduce the step size if the residual size is deemed to be too large. For implementation details, see, for example, Rheinboldt [81].

Global Monotone Convergence of Newton's Method

Newton's method is globally and monotonically convergent in an important special case, namely when the derivative $F'(x)$ is an M-matrix and, in addition, either F is convex or F' is isotone in x. (M-matrices were defined in Section 2.4; see also Exercises 2.19 and 2.20.) The assumptions generalize the monotonicity and convexity assumption in the univariate case; they are, for example, met in systems of equations resulting from the discretization of certain nonlinear ordinary or partial differential equations.

6.4.7 Theorem. *Let* $F : D \subseteq \mathbb{R}^n \to \mathbb{R}^n$ *be continuously differentiable. Let* $F'(x)$ *be an M-matrix in a convex subset* $D_0 \subseteq D$, *and suppose that there are* *vectors $u, v > 0$ such that*

$$F'(x)u \geq v \quad \text{for all } x \in D_0. \tag{4.15}$$

Suppose that $x^0 \in D$, that $F(x^0) \geq 0$, and that $[x^0 - uw^T F(x^0), x^0] \subseteq D_0$ for a vector w with $w_i v_i \geq 1$ ($i = 1, \ldots, n$). Then

(i) F has exactly one zero x^* in D_0.
(ii) Each sequence of the form

$$x^{l+1} = x^l - A_l^{-1} F(x^l) \quad (l = 0, 1, 2, \ldots) \tag{4.16}$$

in which the nonsingular matrices A_l remain bounded as $l \to \infty$ and satisfy the ('subgradient') relations

$$A_l^{-1} \geq 0, \quad F(x) \geq F(x^l) + A_l(x - x^l) \quad \text{for all } x \in D_0 \text{ with } x \leq x^l \tag{4.17}$$

remains in D_0 and converges monotonically to x^*:

$$\lim_{l \to \infty} x^l = x^* \leq x^{l+1} \leq x^l \leq \cdots \leq x^0.$$

Also, the error estimate

$$z^l := x^l - uw^T F(x^l) \in D_0 \quad \Rightarrow \quad x^* \in [z^l, x^l] \tag{4.18}$$

holds.

Proof. We proceed in several steps.

STEP 1: For all $x, y \in D_0$, the matrix $\tilde{A} = F[x, y]$ is an M-matrix with

$$F(x) - F(y) = \tilde{A}(x - y), \quad \tilde{A}u \geq v. \tag{4.19}$$

Indeed, $F(x) - F(y) = \tilde{A}(x - y)$ holds by Lemma 6.1.2. Because $F'(x)$ is an M-matrix for $x \in D_0$,

$$\tilde{A}_{ik} = \int_0^1 F'_{ik}(y + t(x - y)) \, dt \leq 0$$

for $i \neq k$, and by (4.15),

$$\tilde{A}u = \int_0^1 F'(y + t(x - y))u \, dt \geq \int_0^1 v \, dt = v;$$

in particular (Exercise 2.20), $\tilde{A}$ is an M-matrix.

STEP 2: We use induction to show that

$$x^l \in D_0, \quad F(x^l) \geq 0 \tag{4.20}$$

and

$$z^k \in D_0 \quad \Rightarrow \quad z^k \leq x^l \quad \text{for } k \leq l \tag{4.21}$$

for all $l \geq 0$. Because this holds for $l = 0$, we assume that (4.20) and (4.21) hold for an index l and for all smaller indices. Then

$$x^{l+1} \leq x^l \tag{4.22}$$

because $A_l^{-1} \geq 0$. If now $z^k \in D_0$ and $k \leq l$, then $z^k \leq x^k$ because $F(x^k) \geq 0$, and by Step 1

$$F(z^k) = F(x^k) + \tilde{A}_k(z^k - x^k) = F(x^k) - \tilde{A}_k u w^T F(x^k)$$
$$\leq F(x^k) - v w^T F(x^k) \leq 0.$$

Therefore, (4.21) and (4.17) imply

$$0 \geq F(z^k) \geq F(x^l) + A_l(z^k - x^l)$$
$$= A_l(x^l - x^{l+1}) + A_l(z^k - x^l) = A_l(z^k - x^{l+1}).$$

Because $A_l^{-1} \geq 0$, it follows that $0 \geq z^k - x^{l+1}$, so $z^k \leq x^{l+1}$. In particular, $x^{l+1} \in [z^0, x^0] \subseteq D_0$, whence by (4.17) and (4.16),

$$F(x^{l+1}) \geq F(x^l) + A_l(x^{l+1} - x^l) = 0.$$

Therefore, (4.20) and (4.21) hold with $l + 1$ replacing l, whence they hold in general. Thus (4.22) also holds in general.

STEP 3: The limit $x^* = \lim_{l \to \infty} x^l$ exists and is a zero of F.

Because the sequence x^l is monotone decreasing and is bounded below by z^0, it has a limit x^*. From (4.16) it follows that $F(x^l) = A_l(x^l - x^{l+1})$, and because of the boundedness of the A_l it follows on taking the limit that $F(x^*) = 0$.

STEP 4: The point x^* is the unique zero of F in D_0.

Indeed, if y^* is an arbitrary zero of F in D_0, then $0 = F(x^*) - F(y^*) = \tilde{A}(x^* - y^*)$ where $\tilde{A}$ is an M-matrix. Multiplication with $\tilde{A}^{-1}$ gives $y^* = x^*$.

STEP 5: The error estimate (4.18) holds: Because of the monotonically decreasing convergence of the x^l to x^*, we have $x^* \leq x^l$, and because of (4.21), we have $z^l \leq x^*$. Therefore, $x^* \in [z^l, x^l]$. $\square$

6.4.8 Remarks.

(i) If $A_l = F'(x^l)$ is an M-matrix, then $A_l^{-1} \geq 0$ (see Exercise 2.19).

(ii) If $A_l^{-1} \geq 0$ and $F'(x) \leq A_l$ for all $x \in D_0$ such that $x \leq x^l$, then (4.17) holds, because then $\tilde{A} \leq A_l$ for the matrix in (4.19) with $y = x^l$, and $F(x) = F(x^l) + \tilde{A}(x - x^l) \geq F(x^l) + A_l(x - x^l)$ because $x \leq x^l$.

(iii) If, in particular, $F'(x)$ $(x \in D_0)$ is an M-matrix with components that are increasing in each variable, then (4.17) holds with $A_l = F'(x^l)$. If $F'(x)$ is an M-matrix, but its components are not monotone, one can often find bounds for the range of values of $F'(x)$ for x in a rectangular box $[x, x^l] \subseteq D_0$ by means of interval arithmetic, and hence find an appropriate A_l in (4.17).

(iv) If F is convex in D_0, that is, if the relation

$$F(\lambda x + (1 - \lambda)y) \leq \lambda F(x) + (1 - \lambda)F(y) \quad \text{for } 0 \leq \lambda \leq 1$$

holds for all $x, y \in D_0$, then

$$F(x) \geq F(y) + (F(y + \lambda(x - y)) - F(y))/\lambda$$

and as $\lambda \to 0$, the inequality

$$F(x) \geq F(y) + F'(y)(x - y) \quad \text{for all } x, y \in D_0 \tag{4.23}$$

is obtained. If in addition, $F'(x^l)$ is an M-matrix, then again, (4.17) holds with $A_l = F'(x^l)$.

(v) The hypothesis $F(x^0) \geq 0$ may be fairly easily satisfied as follows:

   (a) If (4.15) holds, and both x and $x^0 := x + uw^T |F(x)|$ lie in D_0, then $F(x^0) = F(x) + \tilde{A}(x^0 - x) = F(x) + \tilde{A}uw^T |F(x)| \geq F(x) + vw^T |F(x)| \geq F(x) + |F(x)| \geq 0$.

   (b) If F is convex and both x and $x^0 := x - F'(x)^{-1} F(x)$ lie in D_0, then $F(x^0) \geq F(x) + F'(x)(x^0 - x) = 0$ because of (4.23).

(vi) If x^* lies in the interior of D_0, then the hypothesis of (4.18) is satisfied for sufficiently large l. From the enclosure (4.18), the estimate $|x^l - x^*| \leq uw^T F(x^l)$ is obtained, which permits a simple analysis of the error.

(vii) In particular, in the special case $D_0 = \mathbb{R}^n$ it follows from the previous remarks that the Newton method is convergent for all starting values if F is convex in the whole of $\mathbb{R}^n$ and $F'(x)$ is bounded above and below by M-matrices. The last hypothesis may be replaced with the requirement that $F'(x)^{-1} \geq 0$ for $x \in \mathbb{R}^n$, as can be seen by looking more closely at the arguments used in the previous proof.

6.5 Exercises

1. Write a MATLAB program that determines a zero of a function $F : \mathbb{R}^2 \to \mathbb{R}^2$
   (a) by Newton's method, and
   (b) by the damped Newton method (with the 1-norm).
   In both cases, terminate the iteration when $\|x^{l+1} - x^l\|_1 \leq 10^{-6}$. Use

$$F(x, y) := \begin{pmatrix} \exp(x^2 + y^2) - 3 \\ x + y - \sin(3(x + y)) \end{pmatrix}$$

as a test function. For (a), use the starting values $x^0 = \begin{pmatrix} 0 \\ -1 \end{pmatrix}, \begin{pmatrix} 0.5 \\ 0 \end{pmatrix}, \begin{pmatrix} 0 \\ 0 \end{pmatrix}, \begin{pmatrix} 0.5 \\ -0.5 \end{pmatrix}, \begin{pmatrix} 1 \\ -1 \end{pmatrix}, \begin{pmatrix} 1 \\ 1 \end{pmatrix}, \begin{pmatrix} 10 \\ -10 \end{pmatrix}, \begin{pmatrix} 1 \\ -0.5 \end{pmatrix}, \begin{pmatrix} 1 \\ 0 \end{pmatrix}, \begin{pmatrix} 0.5 \\ -1 \end{pmatrix}, \begin{pmatrix} 2 \\ 1.5 \end{pmatrix}, \begin{pmatrix} 0.08 \\ -0.08 \end{pmatrix}, \begin{pmatrix} 0.07 \\ -0.07 \end{pmatrix}$, and $\begin{pmatrix} 0 \\ -0.5 \end{pmatrix}$. Which ones of these are reasonable? For (b), use the starting values $x^0 = \begin{pmatrix} 0 \\ -1 \end{pmatrix}$ and $\begin{pmatrix} 2 \\ 1.5 \end{pmatrix}$.

2. Let $(x_1(s), y_1(s))$ and $(x_2(t), y_2(t))$ be parametric representations of two twice continuously differentiable curves in the real plane that intersect at (x^*, y^*) at an angle $\varphi^* \neq 0$.
   (a) Find and justify a locally quadratically convergent iterative method for the calculation of the point of intersection (x^*, y^*). The iteration formula should depend only on the quantities t, x_i, y_i and the derivatives x_i', y_i'.
   (b) Test it on the pair of curves (s, s^2) $(s \in \mathbb{R})$ and (t^2, t) $(t \in \mathbb{R})$.

3. The surface area M and the volume V of a bucket are given by the formulas

$$M = \pi(sx + sy + x^2), \quad V = \frac{\pi}{3}(x^2 + xy + y^2)h,$$

where x, y, h, and s are shown in Figure 6.1. Find the bucket that has the greatest volume V^* for the given surface area $M = \pi$. The radii of this bucket are denoted by x^* and y^*.

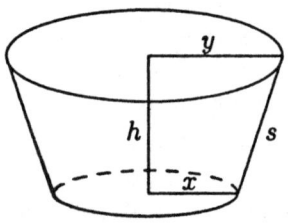

Figure 6.1. Model of a bucket.

(a) Show that these radii are zeros of the function F given by

$$F(x, y)$$
$$= \left(\begin{array}{c} (1 - 4x^2)(x + 2y - 2y^3) + 4xy^2(x^2 - 1) + 3xy^4 \\ 2x(1 - x^2)^2 + y(1 - 2x^2 + 4x^4) + 6xy^2(x^2 - y^2) - 3y^5 \end{array} \right).$$

(b) Write a MATLAB program that computes the optimal radii using Newton's method. Choose as starting values (x^0, y^0) those radii that give the greatest volume under the additional condition $x = y$ (cylindrical bucket; what is its volume?). Determine x^*, y^* and the corresponding volume V^* and height h^*. What is the percentage gain compared with the cylindrical bucket?

4. A pendulum of length L oscillates with a period of $T = 2$ (seconds). How big is its maximum angular displacement? It satisfies the boundary value problem

$$\varphi''(t) + \frac{g}{L} \sin \varphi(t) = 0, \quad \varphi(0) = \varphi(1) = 0;$$

here, t is the time and g is the gravitational acceleration. In order to solve this equation by discretization, one replaces $\varphi''(t)$ with

$$\frac{\varphi(t - h) - 2\varphi(t) + \varphi(t + h)}{h^2}$$

and obtains approximate values $x_i \approx \varphi(ih)$ for $i = 0, \ldots, n$ and $h := 1/n$ from the following system of nonlinear equations:

$$f_i(x) := x_{i-1} - 2x_i + x_{i+1} + h^2 \frac{g}{L} \sin x_i = 0, \qquad (5.1)$$

where $i = 1, \ldots, n - 1$ and $x_0 = x_n = 0$.

(a) Solve (5.1) by Newton's method for $\frac{g}{L} = 39.775354$ and initial vector with components $x_i^{(0)} := 12h^2 i(n - i)$, taking into account the tridiagonal form of $f''(x)$. (Former exercises can be used.)

(b) Solve (5.1) with the discretized Newton method. You should only use three directional derivatives per iteration.

Compare the results of (a) and (b) for different step sizes ($n = 20, 40, 100$). Plot the results and list for each iteration the angular displacement $x_{\frac{n}{2}} \approx \varphi(\frac{1}{2})$ (which is the maximum) in degrees. (The exact maximum angular displacement is $160°$.)

5. To improve an approximation $y^l(x)$ to a solution $y(x)$ of the nonlinear boundary-value problem

$$y'(x) = F(x, y(x)), \quad r(y(a), y(b)) = 0, \qquad (5.2)$$

where $F : \mathbb{R}^2 \to \mathbb{R}$ and $r : \mathbb{R}^2 \to \mathbb{R}$ are given, one can calculate

$$y^{l+1}(x) = y^l(x) - \delta^l(x)$$

with (∂_i denotes the derivative with respect to the ith argument)

$$\delta^l(x) := c + \int_a^x (\partial_2 F(t, y^l(t))\delta^l(t) - F(t, y^l(t)) + (y^l)'(t)) \, dt, \quad (5.3)$$

where the constant c is determined such that

$$\partial_1 r(y^l(a), y^l(b))\delta^l(a) + \partial_2 r(y^l(a), y^l(b))\delta^l(b) = -r(y^l(a), y^l(b)). \tag{5.4}$$

(a) Find an explicit expression for c.

(b) Show that the solution of (5.3) is equivalent to a Newton step for an appropriate zero-finding problem in a space of continuously differentiable functions.

6. (a) Prove the fixed-point theorem of Leray-Schauder for the special Banach space $\mathbb{B} := \mathbb{R}$.

(b) Prove the fixed-point theorem of Brouwer for $\mathbb{B} := \mathbb{R}$ without using (a).

7. Let the mapping $G : D \subseteq \mathbb{R}^n \to \mathbb{R}^n$ be a contraction in $K \subseteq D$ with fixed point $x^* \in K$ and contraction factor $\beta < 1$. Instead of the iterates x^l defined by the fixed-point iteration $x^{l+1} := G(x^l)$, the inaccurate values

$$\tilde{x}^{l+1} := G(x^l) + \Delta G^l$$

are actually calculated because of rounding error. Suppose that the error is bounded according to

$$\|\Delta G^l\| \le \delta \quad \text{for all } l \ge 0.$$

The iteration begins with $x^0 \in K$ and is terminated with the first value of the index $l = l_0$ such that

$$\|\tilde{x}^{l+1} - \tilde{x}^l\| \ge \|\tilde{x}^l - \tilde{x}^{l-1}\|.$$

Show that for the error in $\tilde{x} := \tilde{x}^{l_0}$, a relation of the form

$$\|\tilde{x} - x^*\| \leq C\delta$$

holds, and find an explicit value for the constant C.

8. Using Theorem 6.3.1, give an *a posteriori* error estimate for one of the approximate zeros $\tilde{x}$ found in Exercise 1. Take rounding errors into account by using interval arithmetic. Why must one evaluate $F(\tilde{\mathbf{x}})$ with the point interval $\tilde{\mathbf{x}} = [\tilde{x}, \tilde{x}]$?

9. Let

$$F(x) = \begin{pmatrix} x_1^3 - x_1 x_2 + x_1 - 1 \\ x_2 - x_1^2 \end{pmatrix}.$$

   (a) Show that $F'(x)$ is nonsingular in the region $D = \{x \in \mathbb{R}^2 \mid x_2 < x_1^2 + 1\}$ and that F has there a unique zero at $x^* = \begin{pmatrix} 1 \\ 1 \end{pmatrix}$.

   (b) Show that the Newton paths with starting points in D remain in D and end at x^*.

   (c) Modify a damped Newton method that uses $\|F(x)\|$ as merit function by basing it on (4.1) instead, and compare for various random starting points its behavior on D with that of the standard damped Newton method, and with an implementation of Algorithm 6.4.1.

   (d) Has the norm of the Newton correction a nonzero local minimum in D?

10. Let $F : D \subseteq \mathbb{R}^n \to \mathbb{R}^n$ be continuously differentiable. Let $F'(x)$ be an M-matrix and let the vectors $u, v > 0$ be such that $F'(x)u \geq v$ for all $x \in D$. Also let $w \in \mathbb{R}^n$ be the vector with $w_i = v_i^{-1}$ ($i = 1, \ldots, n$). Show that for arbitrary $\tilde{x} \in D$:

   (a) If D contains all $x \in \mathbb{R}^n$ such that $|x - \tilde{x}| \leq u \cdot w^T |F(\tilde{x})|$ then there is exactly one zero x^* of F such that

$$|x^* - \tilde{x}| \leq u \cdot w^T |F(\tilde{x})|.$$

   (b) If, in addition, D is convex, then x^* is the unique zero of F in D.

11. (a) Let

$$F(x) := \begin{pmatrix} x_1^3 + 2x_1 - x_2 \\ x_2^3 + 3x_2 - x_1 \end{pmatrix}$$

   Show that Newton's method for the calculation of a solution of $F(x) - c = 0$ converges for all $c \in \mathbb{R}^2$ and all starting vectors $x^0 \in \mathbb{R}^2$.

(b) Let

$$F(x) := \begin{pmatrix} \sin x_1 + 3x_1 - x_2 \\ \sin x_2 + 3x_2 - x_1 \end{pmatrix}.$$

Show that Newton's method for $F(x) = 0$ with the starting value $x^0 = \pi \begin{pmatrix} 1 \\ 1 \end{pmatrix}$ converges, but not to a zero of F.

(c) For F given in (a) and (b), use Exercise 10 to determine *a priori* estimates for the solutions of $F(x) = c$ with c arbitrary.

References

The number(s) at the end of each reference give the page number(s) where the reference is cited.

[1] H. Abramowitz and I. A. Stegun, *Handbook of Mathematical Functions*, Dover, New York, 1985. [193, 194]

[2] ADIFOR, Automatic Differentiation of Fortran, WWW-Document, http://www-unix.mcs.anl.gov/autodiff/ADIFOR/ [308]

[3] T. J. Aird and R. E. Lynch, Computable accurate upper and lower error bounds for approximate solutions of linear algebraic systems, *ACM Trans. Math. Software* **1** (1975), 217–231. [113]

[4] G. Alefeld and J. Herzberger, *Introduction to Interval Computations*, Academic Press, New York, 1983. [42]

[5] E. L. Allgower and K. Georg, *Numerical Continuation Methods: An Introduction*, Springer, Berlin, 1990. [302, 334]

[6] E. Anderson et al., *LAPACK Users' Guide*, 3rd ed., SIAM, Philadelphia 1999. Available online: http://www.netlib.org/lapack/lug/ [62]

[7] F. Bauhuber, Direkte Verfahren zur Berechnung der Nullstellen von Polynomen, *Computing* **5** (1970), 97–118. [275]

[8] A. Björck, *Numerical Methods for Least Squares Problems*, SIAM, Philadelphia, 1996. [61, 78]

[9] K. E. Brenan, S. L. Campbell, and L. R. Petzold, *Numerical Solution of Initial-Value Problems in Differential Algebraic Equations, Classics in Applied Mathematics 14*, SIAM, Philadelphia, 1996. [336]

[10] G. D. Byrne and A. C. Hindmarsh, Stiff ODE Solvers: A Review of Current and Coming Attractions, *J. Comput. Phys.* **70** (1987), 1–62. [219]

[11] S.-N. Chow and J. K. Hale, *Methods of Bifurcation Theory*, Springer, New York, 1982. [336]

[12] S. D. Conte and C. de Boor, *Elementary Numerical Analysis*, 3rd ed., McGraw-Hill, Auckland, 1981. [145]

[13] I. Daubechies, *Ten Lectures on Wavelets*, SIAM, Philadelphia, 1992. [178]

[14] P. J. Davis and P. Rabinowitz, *Methods of Numerical Integration*, 2nd ed., Academic Press, Orlando, 1984. [179]

[15] C. de Boor, *A Practical Guide to Splines*, 2nd ed., Springer, New York, 2000. [155, 163, 169]

[16] C. de Boor, Multivariate piecewise polynomials, pp. 65–109, in: *Acta Numerica 1993* (A. Iserles, ed.), Cambridge University Press, Cambridge, 1993. [155]

345

[17] J. Demmel, *Applied Numerical Linear Algebra*, SIAM, Philadelphia, 1997. [61]

[18] J. E. Dennis and R. B. Schnabel, *Numerical Methods for Unconstrained Optimization and Nonlinear Equations*, Prentice-Hall, Englewood Cliffs, NJ, 1983. [302, 334]

[19] P. Deuflhard, A modified Newton method for the solution of ill-conditioned systems of non-linear equations with application to multiple shooting, *Numer. Math.* **22** (1974), 289–315. [329]

[20] P. Deuflhard, Global inexact Newton methods for very large scale nonlinear problems, *IMPACT Comp. Sci. Eng.* **3** (1991), 366–393. [329]

[21] R. A. De Vore and B. L. Lucier, Wavelets, pp. 1–56, in: *Acta Numerica 1992* (A. Iserles, ed.), Cambridge University Press, Cambridge, 1992. [178]

[22] I. S. Duff, A. M. Erisman, and J. K. Reid, *Direct Methods for Sparse Matrices*, Oxford University Press, Oxford, 1986. [64, 75]

[23] J.-P. Eckmann, H. Koch, and P. Wittwer, A computer-assisted proof of universality for area-preserving maps, *Amer. Math. Soc. Memoir* **289**, AMS, Providence (1984). [38]

[24] M. C. Eiermann, Automatic, guaranteed integration of analytic functions, *BIT* **29** (1989), 270–282. [187]

[25] H. Engels, *Numerical Quadrature and Cubature*, Academic Press, London, 1980. [179]

[26] R. Fletcher, *Practical Methods of Optimization*, 2nd ed., Wiley, New York, 1987. [301]

[27] M. R. Garey and D. S. Johnson, *Computers and Intractability: A Guide to the Theory of NP-Completeness*, Freeman, San Francisco, 1979. [115]

[28] D. M. Gay, Some convergence properties of Broyden's method, *SIAM J. Numer. Anal.* **16** (1979), 623–630. [334]

[29] C. W. Gear, *Numerical Initial Value Problems in Ordinary Differential Equations*, Prentice-Hall, Englewood Cliffs, NJ, 1971. [218, 219]

[30] P. E. Gill, W. Murray, and M. H. Wright, *Practical Optimization*, Academic Press, London, 1981. [301]

[31] G. H. Golub and C. F. van Loan, *Matrix Computations*, 2nd ed., Johns Hopkins University Press, Baltimore, 1989. [61, 70, 73, 76, 250, 255]

[32] A. Griewank, *Evaluating Derivatives: Principles and Techniques of Automatic Differentiation*, SIAM, Philadelphia, 2000. [10]

[33] A. Griewank and G. F. Corliss, *Automatic Differentiation of Algorithms*, SIAM, Philadelphia, 1991. [10, 305]

[34] A. Griewank, D. Juedes, H. Mitev, J. Utke, O. Vogel, and A. Walther, ADOL-C: A package for the automatic differentiation of algorithms written in C/C++, *ACM Trans Math. Software* **22** (1996), 131–167. http://www.math.tu-dresden.de/wir/project/adolc/ [308]

[35] E. Hairer, S. P. Nørsett, and G. Wanner, *Solving Ordinary Differential Equations*, Vol. 1, Springer, Berlin, 1987. [179, 212]

[36] E. Hairer and G. Wanner, *Solving Ordinary Differential Equations*, Vol. 2, Springer, Berlin, 1991. [179, 212]

[37] T. C. Hales, The Kepler Conjecture, Manuscript (1998). math.MG/9811071 http://www.math.lsa.umich.edu/~hales/countdown/ [38]

[38] D. C. Hanselman and B. Littlefield, *Mastering MATLAB 5: A Comprehensive Tutorial and Reference*, Prentice-Hall, Englewood Cliffs, NJ, 1998. [1]

[39] E. Hansen, *Global Optimization Using Interval Analysis*, Dekker, New York, 1992. [38, 328]

[40] P. C. Hansen, *Rank-Deficient and Discrete Ill-Posed Problems*, SIAM, Philadelphia, 1997. [81, 125]

[41] J. F. Hart, *Computer Approximations*, Krieger, Huntington, NY, 1978. [19, 22]

[42] J. Hass, M. Hutchings, and R. Schlafli, The double bubble conjecture, *Electron. Res. Announc. Amer. Math. Soc.* **1** (1995), 98–102. http://math.ucdavis.edu/~hass/bubbles.html [38]

[43] P. Henrici, *Applied and Computational Complex Analysis*, Vol. 1. Power Series – Integration – Conformal Mapping – Location of Zeros, Wiley, New York, 1974. [141, 278, 281]

[44] N. J. Higham, *Accuracy and Stability of Numerical Algorithms*, SIAM, Philadelphia, 1996. [17, 77, 93]

[45] IEEE Computer Society, IEEE standard for binary floating-point arithmetic, *IEEE Std* 754-1985, (1985). [15, 42]

[46] E. Isaacson and H. B. Keller, *Analysis of Numerical Methods*, Dover, New York, 1994. [143]

[47] M. A. Jenkins, Algorithm 493: Zeros of a real polynomial, *ACM Trans. Math. Software* **1** (1975), 178–189. [268]

[48] M. A. Jenkins and J. F. Traub, A three-stage variable-shift iteration for polynomial zeros and its relation to generalized Rayleigh iteration, *Numer. Math.* **14** (1970), 252–263. [268]

[49] R. B. Kearfott, *Rigorous Global Search: Continuous Problems*, Kluwer, Dordrecht, 1996. [38, 328]

[50] R. B. Kearfott, Algorithm 763: INTERVAL_ARITHMETIC: A Fortran 90 module for an interval data type, *ACM Trans. Math. Software* **22** (1996), 385–392. [43]

[51] R. Klatte, U. Kulisch, C. Lawo, M. Rauch, and A. Wiethoff, *C-XSC, a C++ Class Library for Extended Scientific Computing*, Springer, Berlin, 1993. [42]

[52] R. Klatte, U. Kulisch, M. Neaga, D. Ratz, and C. Ullrich, *PASCAL-XSC – Language Reference with Examples*, Springer, Berlin, 1992. [42]

[53] R. Krawczyk, Newton-Algorithmen zur Bestimmung von Nullstellen mit Fehlerschranken, *Computing* **4** (1969), 187–201. [116]

[54] A. R. Krommer and C. W. Ueberhuber, *Computational Integration*, SIAM, Philadelphia, 1998. [179]

[55] F. M. Larkin, Root finding by divided differences, *Numer. Math.* **37** (1981) 93–104. [259]

[56] W. Light (ed.), *Advances in Numerical Analysis: Wavelets, Subdivision Algorithms, and Radial Basis Functions*, Clarendon Press, Oxford, 1992. [170]

[57] R. Lohner, Enclosing the solutions of ordinary initial- and boundary-value problems, pp. 255–286, in: E. Kaucher et al., eds., *Computerarithmetic*, Teubner, Stuttgart, 1987. [214]

[58] MathWorks, *Student Edition of MATLAB Version 5 for Windows*, Prentice-Hall, Englewood Cliffs, NJ, 1997. [1]

[59] MathWorks, Matlab online manuals (in PDF), WWW-document, http://www.mathworks.com/access/helpdesk/help/fulldocset.shtml [1]

[60] J. M. McNamee, A bibliography on roots of polynomials, *J. Comput. Appl. Math.* **47** (1993), 391–394. Available online at http://www.elsevier.com/homepage/sac/cam/mcnamee [292]

[61] K. Mehlhorn and S. Näher, *The LEDA Platform of Combinatorial and Geometric Computing*, Cambridge University Press, Cambridge, 1999. [38]

[62] K. Mischaikow and M. Mrozek, Chaos in the Lorenz equations: A computer as-sisted proof. Part II: Details, *Math. Comput.* **67** (1998), 1023–1046. [38]

[63] R. E. Moore, *Methods and Applications of Interval Analysis*, SIAM, Philadelphia, 1981. [42, 214]

[64] A. Morgan, *Solving Polynomial Systems Using Continuation for Engineering and Scientific Problems*, Prentice-Hall, Englewood Cliffs, NJ, 1987. [334]

[65] D. E. Muller, A method for solving algebraic equations using an automatic computer, *Math. Tables Aids Comp.* **10** (1956), 208–215. [264]

[66] D. Nerinckx and A. Haegemans, A comparison of non-linear equation solvers, *J. Comput. Appl. Math.* **2** (1976), 145–148. [292]

[67] NETLIB, A repository of mathematical software, papers, and databases. http://www.netlib.org/ [15, 62]

[68] A. Neumaier, Existence regions and error bounds for implicit and inverse functions, *Z. Angew. Math. Mech.* **65** (1985), 49–55. [335]

[69] A. Neumaier, An existence test for root clusters and multiple roots, *Z. Angew. Math. Mech.* **68** (1988), 256–257. [278]

[70] A. Neumaier, *Interval Methods for Systems of Equations*, Cambridge University Press, Cambridge, 1990. [42, 48, 115, 117, 118, 311, 328]

[70a] A. Neumaier, Global, vigorous and realistic bounds for the solution of dissipative differential equations, *Computing* **52** (1994), 315–336. [214]

[71] A. Neumaier, Solving ill-conditioned and singular linear systems: A tutorial on regularization, *SIAM Rev.* **40** (1998), 636–666. [81]

[72] A. Neumaier, A simple derivation of the Hansen-Bliek-Rohn-Ning-Kearfott enclosure for linear interval equations, *Rel. Comput.* **5** (1999), 131–136. Erratum, *Rel. Comput.* **6** (2000), 227. [118]

[73] A. Neumaier and T. Rage, Rigorous chaos verification in discrete dynamical systems, *Physica D* **67** (1993), 327–346. [38]

[73a] J. Nocedal and S. J. Wright, *Numerical Optimization*, Springer, Berlin, 1999. [301]

[74] GNU OCTAVE. A high-level interactive language for numerical computations. http://www.che.wisc.edu/octave [2]

[75] W. Oettli and W. Prager, Compatibility of approximate solution of linear equations with given error bounds for coefficients and right-hand sides, *Numer. Math.* **6** (1964), 405–409. [103]

[76] M. Olschowka and A. Neumaier, A new pivoting strategy for Gaussian elimination, *Linear Algebra Appl.* **240** (1996), 131–151. [85]

[77] G. Opitz, Gleichungsauflösung mittels einer speziellen Interpolation, *Z. Angew. Math. Mech.* **38** (1958), 276–277. [259]

[78] J. M. Ortega and W. C. Rheinboldt, *Iterative Solution of Nonlinear Equations in Several Variables*, Academic Press, New York, 1970. [311]

[79] B. N. Parlett, *The Symmetric Eigenvalue Problem*, Prentice Hall, Englewood Cliffs, NJ, 1990. (Reprinted by SIAM, Philadelphia, 1998.) [250, 255]

[80] K. Petras, Gaussian versus optimal integration of analytic functions, *Const. Approx.* **14** (1998), 231–245. [187]

[81] W. C. Rheinboldt, *Numerical Analysis of Parameterized Nonlinear Equations*, Wiley, New York, 1986. [336]

[82] J. R. Rice, *Matrix Computation and Mathematical Software*, McGraw-Hill, New York, 1981. [84]

[83] S. M. Rump, Verification methods for dense and sparse systems of equations, pp. 63–136, in: J. Herzberger (ed.), *Topics in Validated Computations – Studies in Computational Mathematics*, Elsevier, Amsterdam, 1994. [100]

[84] S. M. Rump, Improved iteration schemes for validation algorithms for dense and sparse non-linear systems, *Computing* **57** (1996), 77–84. [328]

[85] S. M. Rump, INTLAB – INTerval LABoratory, pp. 77–104, in: *Developments in Reliable Computing* (T. Csendes, ed.), Kluwer, Dordrecht, 1999. http://www.ti3.tu-harburg.de/rump/intlab/index.html [10, 42]

[86] R. B. Schnabel and E. Eskow, A new modified Cholesky factorization, *SIAM J. Sci. Stat. Comput.* **11** (1990), 1136–1158. [77]

[87] SCILAB home page. http://www-rocq.inria.fr/scilab/scilab.html [2]

[88] R. Seydel, *Practical Bifurcation and Stability Analysis*, Springer, New York, 1994. [336]

[89] G. W. Stewart, *Matrix Algorithms*, SIAM, Philadelphia, 1998. [61]

[90] J. Stoer and R. Bulirsch, *Introduction to Numerical Analysis*, Springer, Berlin, 1987. [207]

[91] A. H. Stroud, *Approximate Calculation of Multiple Integrals*, Prentice-Hall, Englewood Cliffs, NJ, 1971. [179]

[92] F. Stummel and K. Hainer, *Introduction to Numerical Analysis*, Longwood, Stuttgart, 1982. [17, 94]

[93] A. van der Sluis, Condition numbers and equilibration of matrices, *Numer. Math.* **14** (1969), 14–23. [101]

[94] R. S. Varga, *Matrix Iterative Analysis*, Prentice-Hall, Englewood Cliffs, NJ, 1962. [98]

[95] W. V. Walter, FORTRAN–XSC: A portable Fortran 90 module library for accurate and reliable scientific computing, pp. 265–285, in: R. Albrecht et al. (eds.), *Validation Numerics – Theory and Applications*. Computing Supplementum 9, Springer, Wien, 1993. [43]

[96] J. Waldvogel, *Pronunciation of Cholesky, Lanczos and Euler*, NA-Digest v90n10 (1990). http://www.netlib.org/cgi-bin/mfs/02/90/v90n10.html#2 [64]

[97] R. Weiss, *Parameter-Free Iterative Linear Solvers*, Akademie-Verlag, Berlin, 1996. [64]

[98] J. H. Wilkinson, *Rounding Errors in Algebraic Processes*, Dover Reprints, New York, 1994. [267]

[99] S. J. Wright, A collection of problems for which Gaussian elimination with partial pivoting is unstable, *SIAM J. Sci. Statist. Comput.* **14** (1993), 231–238. [122]

[100] W. I. Zangwill and C. B. Garcia, *Pathways to Solutions, Fixed Points, and Equilibria*, Prentice-Hall, Englewood Cliffs, NJ, 1981. [334]

Index

(O_s), Opitz method, 260
$A_{:k} = k$th column of A, 62
$A_{i:} = i$th row of A, 62
$D[c; r]$, closed disk in complex plane, 141
df, differential number, 7
e, all-one vector, 62
$e^{(k)}$, unit vectors, 62
$f[x_0, x_1]$, divided difference, 132
$f[x_0, \ldots, x_i]$, divided difference, 134
I, unit matrix, 62
$I(f)$, (weighted) integral of f, 180
J, all-one matrix, 62
$L_i(x)$, Lagrange polynomial, 131
$O(\cdot)$, Landau symbol, 34
$o(\cdot)$, Landau symbol, 34
$Q(f)$, quadrature formula, 180
QR-algorithm, 250
QZ-algorithm, 250
$R_N(f)$, rectangle rule, 204
$T_N(f)$, equidistant trapezoidal rule, 197
$T_{i,k}$, entry of Romberg triangle, 206
$x \gg y$, much greater than, 24
$x \ll y$, much smaller than, 24
$\approx$, approximately equal to, 24
$\mathbb{C}^{m \times n}$, 62
$\check{x}$, midpoint of $\mathbf{x}$, 39
Conv S, closed convex hull, 141
int D, interior of D, 141
mid $\mathbf{x}$, midpoint of $\mathbf{x}$, 39
rad $\mathbf{x}$, radius of $\mathbf{x}$, 39
$\omega(x)$, weight function, 180
∂D, boundary of D, 141
$\mathbb{R}^{m \times n}$, 62
$\square S$, interval hull, 40
$\Sigma(\mathbf{A}, \mathbf{b})$, solution set of linear interval system, 114
$\mathcal{A}(x_1, \ldots, x_n)$, set of arithmetic expressions, 3

absolute value, 40, 98
accuracy, 23
algorithm
    QR, 250
    QZ, 250
Algorithm
    Affine Invariant Modified Newton Method, 330
    Damped Approximate Newton Method, 317
analytic, 4
approximation, 19
    of closed curves, 170
    by Cubic Splines, 165
    expansion in a power series, 19
    by interpolation, 165
    iterative methods, 20
    by least squares, 166
    order
        linear, 45
    rational, 22
argument reduction, 19
arithmetic
    interval, 38
arithmetic expression, 2, 3
    definition, 3
    evaluation, 4
    interval, 44
    stabilizing, 28
assignment, 11
automatic differentiation, 2
    backward, 306
    reverse, 305, 306

base, 15
best linear unbiased estimator
        (BLUE), 78
bisection
    midpoint bisection, 242
    secant bisection, 244
    spectral, 250, 254

bracket, 241
Bulirsch sequence, 207

cancellation, 25
Cardano's formulas, 57
characteristic polynomial, 250
Cholesky
    factor, 76
    factorization, 76
        modified, 77
Clenshaw-Curtis formulas, 190
column pivoting, 83, 85
condition, 23, 33
    condition number, 34, 38
    ill-conditioned, 33, 35
    matrix condition number, 99
    number, 99
continuation method, 334, 336
continued fraction, 22
convergence
    factor, 236
    global, 237
    local, 237
    order, 256, 257
    Q-linear, 236
    Q-quadratic, 237
    Q-superlinear, 237
    R-linear, 236
correct rounding, 16
Cramer's rule, 63

data perturbations, 99
defective, 262
deflation, 267
    explicit, 267
    implicit, 267
derivative
    checking correctness, 56
    multivariate, 303
determinant, 67, 88
difference quotient
    central, 149
    forward, 148
differential equation
    Adams-Bashforth method, 215
    Adams-Moulton method, 216
    backwards differentiation formula
        (BDF), 218
    boundary-value problem, 233
    Euler's method, 212
    global error, 223
    local error estimation, 221
    multi-step method, 211, 214
    multi-value method, 211
    numerical solution of, 210
    one-step method, 211
    predictor-corrector method, 216

rigorous solution of, 214
Runge-Kutta method, 211
shooting method, 122, 233
step size control, 219
stiff, 214, 218
Taylor series method, 214
differential number, 7
    constant, 8
differentiation
    analytical, 4
    automatic, 7, 10
        forward, 10
    numerical, 148
        discretization error, 151
        higher derivatives, 152
        optimal step size, 150
        rounding error analysis, 150
discretization error
    in numerical differentiation,
        151
divided difference
    confluent, 136
    first-order, 132
    general, 134
    second-order, 133
double precision format, 15

eigenvalue problem
    definite, 251
    general linear, 250
    nonlinear, 250
    Opitz methods, 262
    QR-algorithm, 250
    QZ-algorithm, 250
    rigorous error bounds, 272
    spectral bisection, 250
equilibration, 84
error
    damping, 35
    magnification, 35
error analysis
    for arithmetic expressions, 26
    for iterative refinement, 110
    for linear systems, 99, 112
    for multivariate zeros, 322
    reducing integration errors, 224
    for triangular factorizations, 82, 90
    for univariate zeros, 265
error bounds
    a posteriori, 324
    for complex zeros, 277
    for nonlinear systems, 323
    for polynomial zeros, 280
    for simple eigenvalues, 272
Euler-MacLaurin Summation
        Formula, 201
exponent range, 15

extrapolation, 145
  Neville formulas, 146

factorization
  Cholesky, 76
    modified, 77
  LDL^H, 75
  LR, LU, 67
  QR, 80
  modified LDL^T, 75
  orthogonal, 80
  SVD, 81
  triangular, 67
    incomplete, 330
fixed point, 308
floating point
  IEEE standard, 15
  number, 14
function
  elementary, 3
  piecewise cubic, 155
  piecewise linear, 153
    in terms of radial basis
      functions, 170
  piecewise polynomial, *see* spline,
    155
  radial basis function, 170
  spline, *see* spline, 155

Gaussian elimination, 63
  rounding error analysis, 90
global optimization, 38
gradual underflow, 18
grid, 153

H-matrix, 69, 97
Halley's Method, 291
Horner scheme, 10
  complete, 54
  for Newton form, 135

IEEE floating point standard, 15
iff, if and only if, vii
ill-conditioned, *see* condition, 33
inclusion isotonicity, 44
integral, 302
integration
  adaptive, 203
  error estimation, 207
  multivariate, 179
  numerical, 179
  quadrature formula, *see* quadrature
    formula, 179
  Romberg's method, 206
  stepsize control, 208
interpolation
  by cubic splines, 153, 156

by polynomials, 131
  convergence, 141
  divergence, 143
  interpolation error, 138
  Lagrange formulas, 131
  Newton formulas, 134
Hermite, 139
in Chebyshev Points, 144
in expanded Chebyshev points, 145
linear, 132
piecewise linear, 153
quadratic, 133
interval, 39
  arithmetic, 38
  arithmetic expression, 44
  eigenvalue enclosure, 272
  evaluation
    mean value form, 47
  hull, $\Box S$, 40
  Krawczyk's method, 116
  Krawczyk iteration, 327
  Krawczyk operator, 327
  linear system, 114
    Gaussian elimination, 118
    solution set, $\Sigma(\mathbf{A}, \mathbf{b})$, 114
  midpoint, mid $\mathbf{x}$, 39
  Newton's method
    multivariate, 326
  Newton method
    univariate, 268
  point interval, 39
  radius, rad $\mathbf{x}$, 39
  set of intervals, 39
INTLAB, 10
INTLAB
  verifylss, 128
iterative method, 20
iterative refinement, 106
  limiting accuracy, 110

Jacobian matrix, 303

Kepler's barrel rule, 189
Krawczyk iteration, 327
Krawczyk operator, 327

Landau symbols, 34
Lapack, 62
Larkin method, 259
least squares
  method, 78
  solution, 78
limiting accuracy
  for iterative refinement, 110
  for multivariate zeros, 322
  for numerical differentiation, 151
  for univariate zeros, 265

linear system
  error bounds, 112
  interval equations, 114
    Gaussian elimination, 118
    Krawczyk's method, 116
  iterative refinement, 106
  overdetermined, 61, 78
  quality factor, 105
  regularization, 81
  rounding errors, 82
  sparse, 64
  triangular, 65
Lipschitz continuous, 304

M-matrix, 98, 124
machine precision, 17
mantissa length, 15
MATLAB, 1
  1:n, 3
  ==, 65
  A\b, 63, 67, 81
  A', 62
  A(:,k), 63
  A(i,:), 62
  A.', 62
  A.*B, 63
  A./B, 63
  A/B, 63
  chol, 76, 127
  cond, 125
  conj, 299
  disp, 32
  eig, 229
  eps, 17
  eye, 63
  false, 242
  feval, 247
  fprintf, 54
  full, 121
  get, 177
  help, 1
  input, 32
  inv, 63
  lu, 122
  max, 40
  mex, 176
  NaN, 15
  num2str, 11
  ones, 63
  plot, 12
  poly, 266
  print, 12
  qr, 80
  randn, 232
  roots, 266, 268
  set, 177
  sign, 104
  size, 11, 63

  sparse, 121
  sprintf, 54
  spy, 121
  subplot, 177
  text, 177
  title, 177
  true, 242
  xlabel, 177
  ylabel, 177
  zeros, 63
  %, 12
  &, 32
  ~, 32
  |, 32
  conjugate transpose, A', 62
  figures, 176
  online help, 1
  passing function names, 247
  precision, 15, 17
  pseudo-, 1
  public domain variants
    OCTAVE, 2
    SCILAB, 2
  sparse matrices, 75
  subarray, 66
  transpose, A.', 62
  vector indexing, 11
matrix
  $A_{:k}=k$th column of A, 62
  $A_{i:}=i$th row of A, 62
  absolute value, 62, 98
  all-one matrix, J, 62
  banded, 74
  condition number, 99
  conjugate transposed, A^H, 62
  diagonal, 65
  diagonally dominant, 97
  factorization, *see* factorization, 67
  H-matrix, 69, 97, 124
  Hermitian, 62
  Hilbert, 126
  inequalities, 62
  inverse, 81
    conjugate transposed, A^{-H}, 62
    transposed, A^{-T}, 62
  Jacobian, 303
  M-matrix, 98
  monomial, 121
  norm, *see* norm, 94
  notation, 62
  orthogonal, 80
  parameter matrix, 250
  pencil, 250
  permutation, 85
    symmetric, 86
  positive definite, 69
  sparse, 75
  symmetric, 62

transposed, A^T, 62
triangular, 65
tridiagonal, 74
unitary, 80, 126
unit matrix, I, 62
mean value form, 30, 48
mesh size, 153
Milne rule, 189
Muller's Method, 264

NETLIB, 62
Newton's method, 281
  discretized, 314
  global monotone convergence, 336
  multivariate, 311
    affine invariant, 329
    convergence, 318
    damped, 315
    error analysis, 322
  multivariate
    interval, 326
    local convergence, 314
  univariate
    damped, 287
    global behavior, 284
    modifed, 289
    vs. secant method, 282
Newton path, 329
Newton step, 287
nodes, 153, 180
nonlinear system, 301
  continuation method, 334
  error bounds, 323
  finding all zeros, 328
  quasi-Newton method, 333
norm, 94
  ∞-norm, 94, 95
  1-norm, 94, 95
  2-norm, 94, 95
  column sum, 95
  Euclidean, 94
  matrix, 95
  maximum, 94
  monotone, 98
  row sum, 95
  spectral, 95
  sum, 94
normal equations, 78
number
  integer, 14
  real, 14
    fixed point, 14
    floating point, 14
numerical differentiation
  limiting accuracy, 151
numerical integration
  rigorous, 185
numerical stability, 23

OCTAVE, 2
operation, 3
Opitz method, 259
optimization problems, 301
orthogonal polynomial
  3-term recurrence relation,
    195
orthogonal polynomials, 191
outward rounding, 42
overflow, 18

parallelization, 73
parameter matrix, 250
partial pivoting, 83
PASCAL-XSC, 42
permutation
  matrix, 85
  symmetric, 86
perturbation theory, 94
pivot elements, 83
pivoting
  column, 85
  complete, 91
pivot search, 83, 88
point interval, 39
pole
  Opitz methods, 259
polynomial
  Cardano's formulas, 57
  characteristic, 250
  Chebyshev, 145
  interpolation, 131
  Lagrange, 131
  Legendre, 194
  Newton form, 135
  orthogonal, 191
    3-term recurrence relation,
      195
positive definite, 69
positive semidefinite, 69
precision, 23
  machine precision, 17
preconditioner, 113, 330

quadratic convergence, 21
quadrature formula
  Clenshaw-Curtis, 190
  closed, 229
  Gaussian, 187, 192
  interpolatory, 188
  Milne rule, 189, 205, 226
  Newton-Cotes, 189
  normalized, 181
  order, 182
  rectangle rule, 204
  Simpson rule, 189, 204, 226
  trapezoidal rule, 189, 196, 226
    equidistant, 197

quality factor
  linear system, 105
  mean value form, 48
quasi-Newton method, 333

range inclusion, 44
rectangle rule, 204
regula falsi, 235
regularization, 81
result adaptation, 19
Romberg triangle, 206
root secant method, 241
rounding, 16
  by chopping, 16
  correct, 16
  optimal, 16, 42
  outward, 42
row equilibration, 84

scaling, 83
  implicit, 88
SCILAB, 2
search directions, 317
secant method, 234
  convergence order, 259
  convergence speed, 237
  root secant, 241
  secant bisection, 244
  vs. Newton's method, 282
sign change, 241
Simpson rule, 189
single precision, 15
singular-value decomposition, 81
slope
  multivariate, 303
spectral bisection, 254
spline, 155
Spline
  Approximation, 165
spline
  approximation error, 162
  B-spline, 155, 176
    in terms of radial basis functions, 172
  basis spline, 155
  complete, 161
  cubic, 155
  free node condition, 158
  interpolation, 153, 156
  optimality of complete splines, 163
  parametric, 169
  periodic, 160
stability, 37
  numerical, 23
  stabilizing expressions, 28
standard deviation, 2, 26, 30
subdistributive law, 59

Theorem
  of Aird and Lynch, 113
  Euler-MacLaurin summation formula,
    201
  fixed point theorem
    by Banach, 309
    by Brouwer, 311
    by Leray and Schauder, 310
  Gauss-Markov, 78
  inertia theorem of Sylvester, 252
  of Oettli and Prager, 104
  of Rouché, 277
  of van der Sluis, 102
trapezoidal rule, 189, 197
triangular
  factorization, 67
    incomplete, 330
  matrix, 65

underflow, 18

vector
  absolute value, 62, 98
  all-one vector, e, 62
  inequalities, 62
  norm, *see* norm, 94
  unit vectors, $e^{(k)}$, 62
vectorization, 73

weight function, 180
weights, 180

zero, 233
  bracket, 241
  cluster, 239
  complex, 273
    rigorous error bounds, 277
    spiral search, 275
  deflation, 267
  finding all zeros, 268, 328
  hyperbolic interpolation, 261
  interval Newton method, 268
  limiting accuracy, 265
  method of Larkin, 259
  method of Opitz, 259
  Muller's Method, 264
  multiple, 239
  multivariate, 301
    limiting accuracy, 322
  polynomial, 280
    Cardano's formulas, 57
  sign change, 241
  simple, 239
  univariate
    Halley's method, 291
    Newton's method, 281